OXFORD LOGIC GUIDES: 14

General Editors

ANGUS MACINTYRE
JOHN SHEPHERDSON
DANA SCOTT

OXFORD LOGIC GUIDES

1. Jane Bridge: *Beginning model theory: the completeness theorem and some consequences*
2. Michael Dummett: *Elements of intuitionism*
3. A. S. Troelstra: *Choice sequences: a chapter of intuitionistic mathematics*
4. J. L. Bell: *Boolean-valued models and independence proofs in set theory* (1st edition)
5. Krister Segerberg: *Classical propositional operators: an exercise in the foundation of logic*
6. G. C. Smith: *The Boole–De Morgan correspondence 1842–1864*
7. Alec Fisher: *Formal number theory and computability: a work book*
8. Anand Pillay: *An introduction to stability theory*
9. H. E. Rose: *Subrecursion: functions and hierarchies*
10. Michael Hallett: *Cantorian set theory and limitation of size*
11. R. Mansfield and G. Weitkamp: *Recursive aspects of descriptive set theory*
12. J. L. Bell: *Boolean-valued models and independence proofs in set theory* (2nd edition)
13. Melvin Fitting: *Computability theory: semantics and logic programming*
14. J. L. Bell: *Toposes and local set theories: an introduction*

Toposes and Local Set Theories
An Introduction

J. L. BELL

Reader in Mathematical Logic
London School of Economics

CLARENDON PRESS · OXFORD
1988

Oxford University Press, Walton Street, Oxford OX2 6DP
Oxford New York Toronto
Delhi Bombay Calcutta Madras Karachi
Petaling Jaya Singapore Hong Kong Tokyo
Nairobi Dar es Salaam Cape Town
Melbourne Auckland
and associated companies in
Berlin Ibadan

Published in the United States
by Oxford University Press, New York

British Library Cataloguing in Publication Data
Bell, J. L. (John Lane), 1945–
Toposes and local set theories.
1. Toposes
I. Title
512′.55
ISBN 0-19-853274-1

Library of Congress Cataloging-in-Publication Data
Bell, J. L. (John Lane)
Toposes and local set theories: an introduction/J. L. Bell.
— (Oxford logic guides; 14)
Bibliography:
Includes indexes.
1. Toposes. 2. Set theory. 3. Logic, Symbolic and mathematical.
I. Title. II. Title: Local set theories. III. Series.
QA169.B45 1988 511.3′2—dc19 88-2973
ISBN 0-19-853274-1

Typeset by The Universities Press (Belfast) Ltd
Printed in Great Britain
at the University Printing House, Oxford
by David Stanford
Printer to the University

To

—J²M²

PREFACE

In recent years category theory has come to play a significant role in the foundations of mathematics. Perhaps the most striking development in this regard is the invention by Lawvere and Tierney of the concept of (elementary) *topos*. This concept unites, in a simple, elegant way, a number of important, but seemingly diverse, notions from algebraic geometry, set theory, and intuitionistic logic and has led to the forging of surprising new links between classical and constructive mathematics.

This book (which with unusual self-restraint I have resisted entitling 'Tractatus Logico-Toposophicus') is a virtually self-contained introduction to what may be termed the logical approach to topos theory, that is, the presentation of toposes as the models of theories—the so-called *local set theories*—formulated within a typed intuitionistic logic. This approach, although involving a considerable amount of logical detail, has the merit of avoiding the difficult category-theoretic arguments originally employed to establish the basic properties of toposes, replacing them instead with comparatively simple 'set-theoretic' reasoning.

The category theory necessary for understanding the text is developed in the first two chapters. In subsequent chapters we introduce local set theories, sheaves, locale-valued sets, and natural and real numbers in local set theories. Chapter 8 contains some thoughts on the foundational and philosophical significance of topos theory. Finally, in the Appendix I sketch one of the leading ideas of the subject, that of *classifying topos*, which is not readily accommodated within the context of local set theories, and so requires harder category theory for its exposition.

In saying that the book is *virtually* self-contained, I mean that, although I have attempted to include most of the material needed for a grasp of the text, some demands on the reader's prior mathematical knowledge will inevitably be made. For instance, an acquaintance with (intuitionistic) logic will considerably aid the reader's progress through Chapter 3, and in the last section of Chapter 6 (which may be omitted without loss of continuity) I assume a knowledge of Boolean-valued models of set theory. Nevertheless, the reader will, I hope, find that only modest prior knowledge will suffice.

A word on the set-theoretical foundations for the material treated in this book. The logical approach to topos theory has the merit of requiring only modest metatheoretical support: indeed, the central soundness, completeness, and equivalence theorems of Chapter 3 could in principle be proved in some form of recursive arithmetic. However, category theory itself does require some set-theoretical underpinning: for our purposes Gödel–Bernays set theory with its familiar distinction between sets and (proper) classes will prove adequate.

I would like to tender my thanks to several colleagues and fellow toposophers who have, in their various ways, contributed to bringing this work into existence. Moshe Machover introduced me to topos theory through his lectures on John Zangwill's work. (Zangwill's unpublished Master's thesis plays an important role in the present book.) My interest in the subject was greatly stimulated by attendance at the meetings of the Peripatetic Seminar on Sheaves and Logic founded by Dana Scott in the 1970s (and, at the time of writing, still going strong!). Lectures by, and conversations and correspondence with, members of this variable, but always dedicated, set of mathematicians have been of particular importance to me: I would especially acknowledge Michael Brockway, Peter Johnstone, Gordon Monro, Chris Mulvey, Gavin Wraith, and of course Scott himself. I have also benefited greatly from my association with the members of the 'Bristol School': Jonathan Chapman, John Mayberry, Frederick Rowbottom, and John Zangwill. Discussions with my friend Alberto Peruzzi have been of real help in clarifying my ideas; I am also grateful to him for reading the typescript and making valuable suggestions for improving it. My wife Mimi patiently and expertly typed much of my manuscript; my love and thanks to her. In addition I would like to thank Sam Fendrich and Marcus Giaquinto for their help in spotting errors. And finally, it is a pleasure to record my gratitude to the staff of the Oxford University Press for once again supporting a project of mine.

London J. L. B.
August 1987

ACKNOWLEDGEMENTS

The author and publisher would like to thank the following for permission to reproduce copyright material in Chapter 8.

D. Reidel Publishing Company: 'From Absolute to Local Mathematics' (by J. L. Bell), *Synthese,* Vol. 69 no. 3, December 1986, pp. 409–26. Copyright © 1986 by D. Reidel Publishing Company.

British Society for the Philosophy of Science: 'Category Theory and the Foundations of Mathematics' (by J. L. Bell), *British Journal for the Philosophy of Science* 32 (1986), pp. 349–58.

CONTENTS

1

Elements of category theory

In this first chapter we introduce and develop those concepts from category theory which we shall be using throughout the book.

Categories

A category $\mathbf{C}$ consists of a class $\mathrm{Ob}(\mathbf{C})$ of elements $X, Y, \ldots, A, B, \ldots$ called *objects* of $\mathbf{C}$ (or $\mathbf{C}$-*objects*) and a class $\mathrm{Arr}(\mathbf{C})$ of elements $f, g, \ldots$ called *arrows* of $\mathbf{C}$ (or $\mathbf{C}$-*arrows*) subject to the following conditions.

(i) Each arrow f of $\mathbf{C}$ is assigned a pair of objects $\mathrm{dom}(f)$, $\mathrm{cod}(f)$ called the *domain* and *codomain* of f: if $X = \mathrm{dom}(f)$ and $Y = \mathrm{cod}(f)$ we say that f is an arrow *from X to Y* and write

$$X \xrightarrow{f} Y \qquad \text{or} \qquad f : X \to Y.$$

(ii) Each pair of arrows $f : X \to Y$ and $g : Y \to Z$ of $\mathbf{C}$ for which $\mathrm{cod}(f) = \mathrm{dom}(g)$ (such arrows are called *composable*) is assigned an arrow $g \circ f : X \to Z$ called the *composition* (or *composite*) of f and g: this composition will also be written gf or

$$X \xrightarrow{f} Y \xrightarrow{g} Z.$$

(iii) Each object X of $\mathbf{C}$ is assigned an arrow $1_X : X \to X$ called the *identity arrow* on X. When confusion is unlikely, we write just 1 for 1_X.

(iv) *Associativity law*. For any arrows $X \xrightarrow{f} Y$, $Y \xrightarrow{g} Z$, $Z \xrightarrow{h} W$ of $\mathbf{C}$, the compositions $X \xrightarrow{h \circ (g \circ f)} W$ and $X \xrightarrow{(h \circ g) \circ f} W$ are equal: the notation $h \circ g \circ f$ or

$$X \xrightarrow{f} Y \xrightarrow{g} Z \xrightarrow{h} W$$

may then be unambiguously employed to denote the resulting arrow.

(v) *Identity law*. For any object Y and any arrows $X \xrightarrow{f} Y$, $Y \xrightarrow{g} Z$ of $\mathbf{C}$, we have

$$1_Y \circ f = f, \qquad g \circ 1_Y = g.$$

(vi) '*Smallness*' *condition*. For any objects X, Y of $\mathbf{C}$, the collection $\mathrm{Hom}_{\mathbf{C}}(X, Y)$ of arrows from X to Y is a *set*.

A category is *small* if its class of objects, and hence also its class of arrows, is a *set*.

The set $\mathrm{Hom}_{\mathbf{C}}(X,Y) = \{f \in \mathrm{Arr}(\mathbf{C}) : \mathrm{dom}(f) = X \text{ and } \mathrm{cod}(f) = Y\}$, which is also written $\mathbf{C}(X, Y)$—or just $\mathrm{Hom}(X,Y)$ if $\mathbf{C}$ is clear from the context—is called the *hom-set* of X,Y. Notice that $\mathrm{Arr}(\mathbf{C})$ and the operations dom and cod are completely determined by specifying the sets $\mathrm{Hom}_{\mathbf{C}}(X,Y)$: this should be borne in mind when considering the examples below.

Examples

(i) *The category* **Set** *of sets* whose class of objects is the class V of all sets and, for $X,Y \in V$, $\mathbf{Set}(X,Y)$ is the set of all functions† from X to Y.

(ii) *The category* **Grp** *of groups* whose class of objects is the class of all groups and, for groups G,H, $\mathbf{Grp}(G,H)$ is the set of all group homomorphisms from G to H. Again, composition and identity arrows are as usual for functions.

(iii) *The category* **Esp** *of topological spaces* whose class of objects is the class of all topological spaces and, for topological spaces X,Y, $\mathbf{Esp}(X,Y)$ is the set of all continuous functions from X to Y. Again, composition and identity arrows are as usual for functions.

(iv) In like manner one obtains the following categories:

Finset	finite sets and functions
Ab	abelian groups and group homomorphisms
(C)Rng	(commutative) rings with identity and ring homomorphisms
Field	fields with $0 \neq 1$ and field homomorphisms
Pos	partially ordered sets and order-preserving functions
Lat	lattices and lattice homomorphisms
Bool	Boolean algebras and Boolean homomorphisms
Haus	Hausdorff spaces and continuous functions
Comp Haus	compact Hausdorff spaces and continuous functions
Top Grp	topological groups and continuous homomorphisms
$\mathbf{Vect}_K$	vector spaces over a field K and linear maps

All the examples of categories we have proposed so far have the property that their arrows are genuine *functions*. The following examples are more 'abstract' in that they no longer have this property.

(v) *The category of sets and relations.* Here the class of objects is the class of all sets, $\mathrm{Hom}(X,Y)$ is the set of all relations $R \subseteq X \times Y$ and composition is the usual composition of relations.

† Here and in some of the examples below we regard a function as *uniquely determining both its domain and codomain*. That is, we must think of a function $f: X \to Y$ as being a triple (f, X, Y).

(vi) *The category* $\mathbf{Mat}_R$ *of R-matrices for a given commutative ring R (with* 1). Here the objects are the positive integers, each hom-set $\mathrm{Hom}(m,n)$ is the set of all $n \times m$ matrices with entries in R, and composition is matrix multiplication.

(vii) A *preordered class* is a class P equipped with a reflexive transitive relation $\leqslant$ called a *preordering* on P. If in addition $\leqslant$ is antisymmetric it is called a *partial ordering* on P. Each preordered class $(P, \leqslant)$ gives rise to a category $\mathbf{P}$ whose objects are the elements of P and such that $\mathrm{Hom}(p,q)$ has exactly one element if $p \leqslant q$ and is empty otherwise. Conversely, any category $\mathbf{C}$ with the property that, for any objects X, Y of $\mathbf{C}$, $\mathrm{Hom}_{\mathbf{C}}(X,Y)$ has at most one element can be obtained in this way. Thus preordered sets—and *a fortiori*, partially ordered sets—may be regarded as categories.

(viii) A *monoid* is a set M equipped with a multiplication operation $\cdot : M \times M \to M$ which is associative and for which there is a 2-sided identity, i.e. an element $e \in M$ such that $e \cdot x = x \cdot e = x$ for all $x \in M$. (Thus a group is a special kind of monoid.) Any monoid M may be regarded as a category $\mathbf{M}$ which has exactly one object, whose arrows are the elements of M and in which composition of arrows is just the multiplication in M. Conversely, any category having exactly one object may be regarded as a monoid.

(ix) If a category has only a few arrows (and hence objects as well), it is often helpful to depict it in the form of a diagram that represents its objects as dots and its *non-identity* arrows as honest-to-goodness arrows. For example

$\mathbf{0} =$ the *empty* category)

$\mathbf{1} = \quad \bullet \quad$ the *degenerate* category)

$\mathbf{2} = \quad \bullet \to \bullet$

$\mathbf{3} =$ 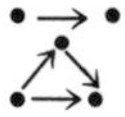

$\rightrightarrows \; = \quad \bullet \rightrightarrows \bullet$

This leads us to the notion of a *diagram* in a category. A *diagram* in a category $\mathbf{C}$ is a set (possibly empty) of objects of $\mathbf{C}$ together with a set (again possibly empty) of arrows between these objects. The objects of a diagram are often called its *vertices*. A diagram $\mathbf{D}$ can be depicted as an array of arrows and letters (or dots):

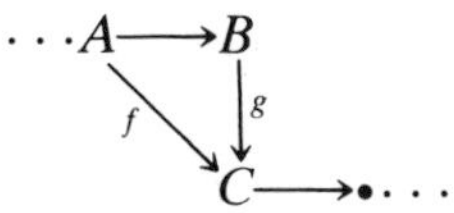

A *path* in a diagram **D** is a finite sequence $(f_1, \ldots, f_n)$ of arrows of **D** such that $\operatorname{cod} f_i = \operatorname{dom} f_{i+1}$ for $i = 1, \ldots, n-1$. The natural number n is called the *length* of the path. **D** is said to *commute* or to be *commutative* if for any path in **D** of length $\geqslant 2$, the arrow obtained by composing along the arrows of the path depends *only* on the endpoints of the path. More precisely, **D** commutes if, for any paths $(f_1, \ldots, f_n)$, $(g_1, \ldots, g_m)$ such that either $m \geqslant 2$ or $n \geqslant 2$ (or both), $\operatorname{dom}(f_1) = \operatorname{dom}(g_1)$ and $\operatorname{cod}(f_n) = \operatorname{cod}(g_m)$, we have $f_n \circ f_{n-1} \circ \cdots \circ f_1 = g_m \circ g_{m-1} \circ \cdots \circ g_1$. For example, consider the diagrams

(a) 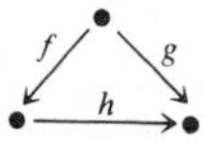

(b) 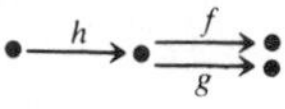

(a) commutes iff $g = h \circ f$, and (b) commutes iff $f \circ h = g \circ h$. A commutative diagram of form (a) is called a *commutative triangle*; we say that $\bullet \xrightarrow{f} \bullet \xrightarrow{h} \bullet$ is a *factorization* of g.

Subcategories. A *subcategory* of a category **C** is a diagram **A** in **C** such that (1) whenever **A** contains an object X of **C**, it also contains the corresponding identity arrow 1_X; (2) whenever **A** contains an arrow f of **C**, it also contains the objects $\operatorname{dom}(f)$ and $\operatorname{cod}(f)$; and (3) whenever **A** contains a pair of composable arrows f and g, it also contains their composition $g \circ f$.

Clearly a subcategory of a category is itself a category.

A subcategory **A** of a category **C** is *full* if, for all $X, Y \in \mathrm{Ob}(\mathbf{A})$, we have $\mathbf{A}(X, Y) = \mathbf{C}(X, Y)$. Note that, since a full subcategory of a given category **C** is uniquely determined by its collection of objects, any collection S of objects of **C** determines a unique full subcategory **S** whose objects are the members of S and, for $X, Y \in S$, $\mathbf{S}(X, Y) = \mathbf{C}(X, Y)$.

Examples

(i) Each category is a full subcategory of itself.

(ii) **Finset** is a full subcategory of **Set**.

(iii) The category of sets and injective (resp. surjective, bijective) mappings is a subcategory of **Set** which is not full.

(iv) The category of sets and relations is not a subcategory of **Set**.

(v) **Bool** is a subcategory of **Lat** and **Lat** of **Pos**. Neither is full.

Slice categories. Let **C** be a category and A an object of **C**. We define the category $\mathbf{C}/A$ of **C**-*objects over* A, sometimes called a *slice* category, to have as objects all **C**-arrows $B \xrightarrow{f} A$ with codomain A, while an *arrow* in $\mathbf{C}/A$ from $B \xrightarrow{f} A$ to $C \xrightarrow{g} A$ is a **C**-arrow $B \xrightarrow{k} C$ such that the triangle

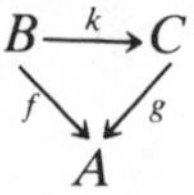

commutes†. Clearly, the identity arrow in $\mathbf{C}/A$ on an object $B \xrightarrow{f} A$ is just 1_B.

Opposite categories. For any category $\mathbf{C}$, the *opposite* or *mirror category* $\mathbf{C}^{\text{op}}$ is defined to have the same objects and arrows as $\mathbf{C}$, but the domain and codomain functions are interchanged and composition in $\mathbf{C}^{\text{op}}$ is 'opposite' to that in $\mathbf{C}$. That is, for any objects X and Y of $\mathbf{C}$ (or $\mathbf{C}^{\text{op}}$),

$$\text{Hom}_{\mathbf{C}^{\text{op}}}(X, Y) = \text{Hom}_{\mathbf{C}}(Y, X)$$

and, for any arrows $Z \xrightarrow{g} Y \xrightarrow{f} X$ in $\mathbf{C}^{\text{op}}$, i.e. arrows $X \xrightarrow{f} Y \xrightarrow{g} Z$ in $\mathbf{C}$,

$$f \circ g \text{ in } \mathbf{C}^{\text{op}} = g \circ f \text{ in } \mathbf{C}.$$

Notice that *identity* arrows in $\mathbf{C}$ and $\mathbf{C}^{\text{op}}$ are the same. Clearly, also, $(\mathbf{C}^{\text{op}})^{\text{op}} = \mathbf{C}$. The category $\mathbf{C}^{\text{op}}$ may be thought of as being obtained from $\mathbf{C}$ by 'reversing all arrows', or, if one prefers, as the 'mirror image' of $\mathbf{C}$.

Duality principle. Let S be a statement about the objects and arrows of a category $\mathbf{C}$. The *dual* S^{op} of S is the corresponding statement about $\mathbf{C}^{\text{op}}$, phrased as a statement about $\mathbf{C}$. That is, S^{op} is the statement obtained from S by 'reversing all arrows'. Since S holds in $\mathbf{C}^{\text{op}}$ iff S^{op} holds in $\mathbf{C}$, and $(\mathbf{C}^{\text{op}})^{\text{op}} = \mathbf{C}$, we get:

Duality principle for categories. If S is a statement true for all categories, then S^{op} is also true for all categories.

We also have the notion of duality for concepts or constructs in categories. If W is any concept or construct defined for all categories, the dual W^{op} or co-W of W is the concept or construct defined for any category $\mathbf{C}$ by formulating W in $\mathbf{C}^{\text{op}}$ and interpreting the result in $\mathbf{C}$. A concept W is *self-dual* if $W = W^{\text{op}}$.

For example, an *initial* object in a category is an object A such that, for any object X, there is exactly one arrow $A \rightarrow X$. The dual of this concept is that of a *terminal* object, i.e. an object A such that, for any object X, there is exactly one arrow $X \rightarrow A$.

Some basic category-theoretic notions

In category theory many concepts ordinarily formulated in terms of *elements* are instead formulated in terms of *arrows*. For example, instead of saying that a set X is empty we can say that there is exactly one arrow from X to any set Y. Many of the basic category-theoretic concepts embody this idea.

† Strictly speaking, we should take the triple (k, f, g) as the $\mathbf{C}/A$-arrow. However, in this and some similar contexts we shall ignore such niceties.

Throughout this discussion we let **C** be a fixed category. $A, B, C, \ldots$ will denote objects of **C** and $f, g, h, \ldots$ arrows of **C**.

An arrow $f: A \to B$ is *invertible* or an *isomorphism* if there is an arrow f' such that $f' \circ f = 1_A$ and $f \circ f' = 1_B$. If such an f' exists, it is unique and is written as $f' = f^{-1}$. Two objects A and B are *isomorphic* in **C**, written $A \cong B$, if there is an invertible arrow $f: A \to B$. Clearly $\cong$ is an equivalence relation. Notice also that the concepts 'invertibility', 'isomorphism', and 'isomorphic' are self-dual. We shall often write $f: A \cong B$ or $A \xrightarrow{f} B$ to indicate that f is an isomorphism.

In **Set**, the isomorphisms are just the bijections, and in **Grp**, the bijective homomorphisms (not, however, in **Esp** where they are the homeomorphisms.)

An arrow $f: A \to B$ is *monic* if for any pair of arrows $g_1, g_2: C \to A$ the equality $f \circ g_1 = f \circ g_2$ implies $g_1 = g_2$, i.e. if f is *left cancellable.* We often write

$$f: A \rightarrowtail B \qquad \text{or} \qquad A \overset{f}{\rightarrowtail} B$$

to indicate that f is a monic arrow. In **Set**, **Esp**, and **Grp** the monic arrows are precisely the one–one functions.

An important special kind of monic arrow in **Set** is the *insertion map* $X \hookrightarrow Y$ of a subset X of a set Y into Y: this is just the identity map on X regarded as a function with codomain Y.

An arrow $f: A \to B$ is a *split monic* if it has a *left inverse,* i.e. if there is $g: B \to A$ such that $g \circ f = 1_A$. Clearly a split monic is monic.

An arrow $f: A \to B$ is *epic* if for any pair of arrows $g_1, g_2: B \to C$ the equality $g_1 \circ f = g_2 \circ f$ implies $g_1 = g_2$, i.e. if f is *right cancellable.* In **Set** and **Esp** the epic arrows are precisely the onto functions. (This is also true for **Grp** but it is nontrivial to prove: for a proof see MacLane (1971), Ch. I, §5, ex. 5). On the other hand, in **Haus** an arrow $X \to Y$ can be epic without being onto in the set-theoretical sense. For example, if X is a dense subspace of a Hausdorff space Y, then the insertion map $X \hookrightarrow Y$ is epic in **Haus** but not onto. This also shows that (in **Haus**) an arrow can be simultaneously monic and epic without being an isomorphism.

We often write

$$f: A \twoheadrightarrow B \qquad \text{or} \qquad A \overset{f}{\twoheadrightarrow} B$$

to indicate that f is an epic arrow.

An arrow $f: A \to B$ is a *split epic* if it has a *right inverse,* i.e. if there is $g: B \to A$ such that $f \circ g = 1_B$. Clearly a split epic is epic.

It is easily shown that the following conditions on an arrow f are equivalent: (i) f is an isomorphism; (ii) f is epic and split monic; (iii) f is monic and split epic.

An object A is said to be *initial* (*terminal*) if for any object X there is a *unique* arrow $A \to X$ (resp. $X \to A$). Clearly the concepts of initial and terminal object are mutually dual. Moreover, an initial object is 'unique up to isomorphism' in the sense that *all initial objects in a category are isomorphic*. To see this, suppose that A and A' are both initial objects. Then there are arrows $A \xrightarrow{f} A'$ and $A' \xrightarrow{f'} A$. Since A and A' are initial, $f \circ f' = 1_{A'}$, and $f' \circ f = 1_A$, so f is an isomorphism and $A \cong A'$. Duality now allows us to infer that *all terminal objects in a category are isomorphic.*

The categories **Set** and **Esp** each have a *unique* initial object—the empty set $\varnothing$ or 0 and the empty space respectively. The ring Z of integers is an initial object in **Rng**, while **Field** has no initial object. An initial object in (the category associated with) a partially ordered set is a *least* element.

The categories **Set**, **Ab**, **Esp**, **Grp**, and **Rng** all have terminal objects: the 'one-element' objects. (Note that in the case of **Grp** and **Ab** a terminal object is also an initial object.) **Field** has no terminal objects. A terminal object in (the category associated with) a partially ordered set is a *largest* element.

We usually employ the symbols 0 and 1 to denote an initial and terminal object, respectively, in a given category. The unique arrow from 0 to A or from A to 1 will be denoted simply by $0 \to A$ or $A \to 1$. In **Set**, arrows $1 \to A$ obviously correspond to *elements* of A. We extend this idea to an arbitrary category **C** with a terminal object by calling any **C**-arrow $1 \to A$ a **C**-*element* of A.

We note that, for any **C**-object A, the slice category **C**$/A$ has a terminal object, namely the identity 1_A on A. This serves to justify the use of the same symbol '1' for identity arrows and terminal objects.

Products and coproducts. Let A_1 and A_2 be objects of **C**. A *product* of A_1 and A_2 in **C** is an object P together with arrows $A_1 \xleftarrow{\pi_1} P \xrightarrow{\pi_2} A_2$ called (canonical) *projections* such that, for each object B and each pair of arrows $A_1 \xleftarrow{f_1} B \xrightarrow{f_2} A_2$ there is a *unique* arrow $g: B \to P$ such that the diagram

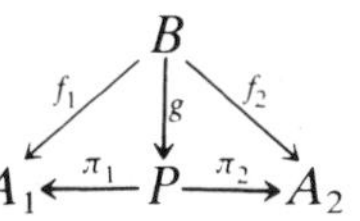

commutes. We show below that this condition specifies P 'uniquely up to isomorphism' in the sense that if $A_1 \xleftarrow{\pi_1'} P' \xrightarrow{\pi_2'} A_2$ is any other diagram satisfying the prescribed conditions, then there is a unique isomorphism $i:P\to P'$ such that the diagram

$$\begin{array}{ccccc} & & P & & \\ & \swarrow^{\pi_1} & \downarrow{\scriptstyle i} & \searrow^{\pi_2} & \\ A_1 & \xleftarrow[\pi_1']{} & P' & \xrightarrow[\pi_2']{} & A_2 \end{array}$$

commutes. We write $A_1\times A_2$ for 'the' product of A_1 and A_2 (when it exists) and $\langle f_1,f_2\rangle$ for the arrow g above (noting that, in **Set**, $\langle f_1,f_2\rangle$ is given by

$$\langle f_1,f_2\rangle(a)=(f_1(a),f_2(a))).$$

It is easily seen that, if $g_1:B\to A_1$ and $g_2:B\to A_2$, then $\langle f_1,f_2\rangle=\langle g_1,g_2\rangle$ iff $f_1=g_1$ and $f_2=g_2$. Also, notice that for any $C\xrightarrow{h}B$, we have

$$\langle f_1,f_2\rangle\circ h=\langle f_1\circ h,f_2\circ h\rangle.$$

Here is a neat way to show that products are unique up to isomorphism. Given objects A_1, A_2 of $\mathbf{C}$ let $\mathbf{C}\,|\,A_1,A_2$ be the category whose objects are pairs of $\mathbf{C}$-arrows $A_1\xleftarrow{f_1}B\xrightarrow{f_2}A_2$ and for which an arrow between two $\mathbf{C}\,|\,A_1,A_2$—objects

$$A_1\xleftarrow{f_1}B\xrightarrow{f_2}A_2,\qquad A_1\xleftarrow{f_1}B'\xrightarrow{f_2}A_2$$

is a $\mathbf{C}$-arrow $g:B\to B'$ such that the diagram

$$\begin{array}{ccccc} & & B & & \\ & \swarrow^{f_1} & \downarrow{\scriptstyle g} & \searrow^{f_2} & \\ A_1 & \xleftarrow{f_1'} & B' & \xrightarrow{f_2'} & A_2 \end{array}$$

commutes. Composition is defined as for $\mathbf{C}$. Then evidently $A_1\xleftarrow{\pi_1}P\xrightarrow{\pi_2}A_2$ is a product of A_1 and A_2 in $\mathbf{C}$ iff it is a terminal object in $\mathbf{C}\,|\,A_1,A_2$. Since a terminal object in $\mathbf{C}\,|\,A_1,A_2$ is unique up to isomorphism, so is the product of A_1 and A_2 in $\mathbf{C}$†.

For any object A, the arrow $\delta_A=\langle 1_A,1_A\rangle:A\to A\times A$ is called the *diagonal* arrow on A.

Note that $1\times A\cong A\times 1\cong A$ for any object A.

We say that $\mathbf{C}$ has *binary products* if $A\times B$ exists in $\mathbf{C}$ for every pair (A,B) of objects of $\mathbf{C}$.

Dually, a *coproduct* of A_1 and A_2 is an object Q together with arrows $A_1\xrightarrow{\sigma_1}Q\xleftarrow{\sigma_2}A_2$ called (canonical) *injections* such that, for each object B and each pair of arrows $A_1\xrightarrow{f_1}B\xleftarrow{f_2}A_2$ there is a unique

† The uniqueness up to isomorphism of virtually all category-theoretic constructs can be established by a method similar to the one just described. That is, to show that a given construct is unique up to isomorphism, one identifies a suitable category in which the construct corresponds to a terminal object.

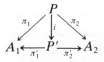

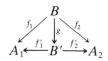

arrow $h: Q \to B$ such that the diagram

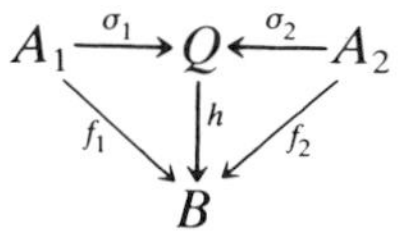

commutes. Again, these conditions determine Q 'uniquely up to isomorphism'. We write $A_1 + A_2$ for 'the' coproduct of A_1 and A_2, when it exists, and $\binom{f_1}{f_2}$ for the arrow h above.

We say that **C** has *binary coproducts* if $A + B$ exists in **C** for every pair (A, B) of objects of **C**.

Note that $A + 0 \cong 0 + A \cong A$ for any object A.

Examples

In **Set**, $\times$ is Cartesian product and $+$ is disjoint union; in **Grp**, $\times$ is Cartesian product while $+$ is free product; in **Esp**, $\times$ is topological product and $+$ is disjoint topological sum. In a preordered class, $\times$ is infimum and $+$ is supremum. A partially ordered set which, considered as a category, has binary products and coproducts is called a *lattice*. If p,q are arbitrary elements of a lattice, we write $p \wedge q$, $p \vee q$ for the infimum (product) and supremum (coproduct), respectively, of p and q.

Given two arrows $A_1 \xrightarrow{f_1} B_1$, $A_2 \xrightarrow{f_2} B_2$, we define their *product* $f_1 \times f_2$ to be the arrow $\langle f_1 \circ \pi_1, f_2 \circ \pi_2 \rangle : A_1 \times A_2 \to B_1 \times B_2$. Thus $f_1 \times f_2$ is the unique arrow making the diagram

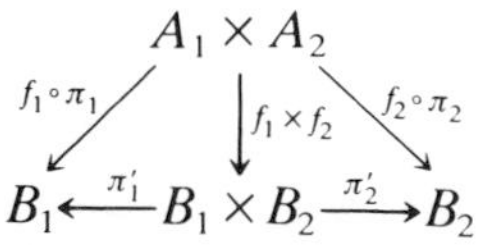

commute.

In **Set**, $f_1 \times f_2$ is given by $(f_1 \times f_2)(a_1, a_2) = (f_1(a_1), f_2(a_2))$.

We now show that the product operation (and hence, by duality, the coproduct) operation in a category is commutative.

1.1 PROPOSITION. In any category we have

$$A_1 \times A_2 \cong A_2 \times A_1.$$

Proof. There are unique arrows i, i' such that the diagrams

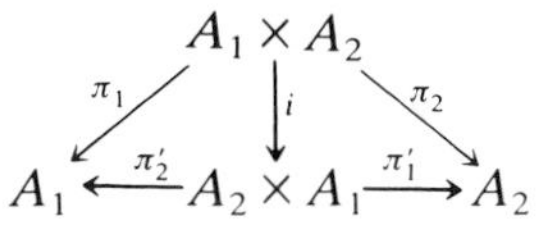

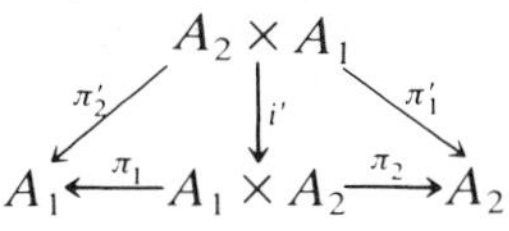

commute. Putting these together in both orders give commutative diagrams

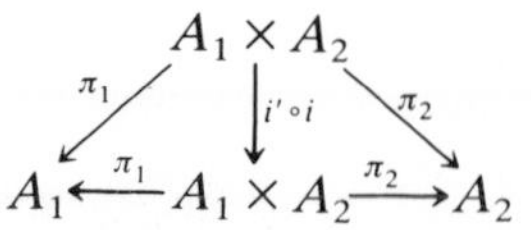

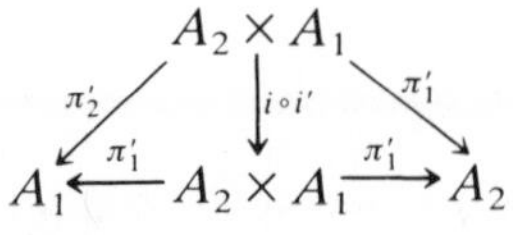

But replacing $i' \circ i$ and $i \circ i'$ by $1_{A_1 \times A_2}$ and $1_{A_2 \times A_1}$ respectively preserves the commutativity of these diagrams, so that $i' \circ i = 1_{A_1 \times A_2}$, $i \circ i' = 1_{A_2 \times A_1}$. Thus i and i' are isomorphisms as required. □

In a similar way, we can prove the associative law for $\times : A_1 \times (A_2 \times A_3) \cong (A_1 \times A_2) \times A_3$. This enables us to write $A_1 \times A_2 \times A_3$ unambiguously, and, by extension, $A_1 \times A_2 \times \cdots \times A_n$.

The isomorphisms between $A_1 \times \cdots \times A_n$ and the various rearrangements of the factors, e.g. $A_1 \times (A_2 \times \cdots \times A_n)$ or $(A_2 \times \cdots \times A_n) \times A_1$, guaranteed by the commutative and associative laws, are called *canonical* isomorphisms.

Just as for sets, the notions of product and coproduct can be extended to arbitrarily many factors. Suppose we are given an indexed set $\{A_i : i \in I\}$ of objects of a category **C**. A *product* of the set is an object, usually denoted by $\prod_{i \in I} A_i$ or $\prod A_i$ together with a set of arrows $\pi_i : \prod_{i \in I} A_i \to A_i$ $(i \in I)$ called (canonical) *projections* such that, for any object B and any set of arrows $f_i : B \to A_i$ $(i \in I)$ there is a *unique* arrow $h : B \to \prod A_i$ such that, for each $i \in I$ the diagram

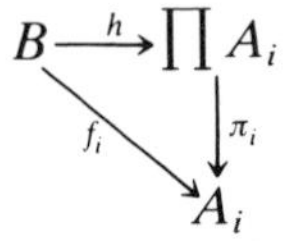

commutes. The product of a set of objects, if it exists, is unique up to isomorphism—the precise sense in which this phrase is being used here will be revealed in the next section when we consider limits. We note that a product of the empty set of objects is just a terminal object, and that the product of the family consisting of a single object A is A itself, with projection $1_A : A \to A$.

A category **C** is said to *have* (*finite*) *products* if the product of any (finite) set of objects exists in **C**. It is easy to see that a category has finite products iff it has binary products and a terminal object.

Given $B \xrightarrow{f_1} A_1, \ldots, B \xrightarrow{f_n} A_n$ we write $\langle f_1, \ldots, f_n \rangle$ for the unique $B \xrightarrow{h} A_1 \times \cdots \times A_n$ such that

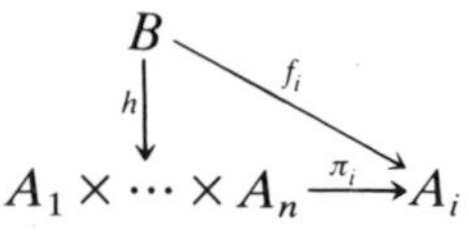

commutes for all i $(i = 1, \ldots, n)$. It is readily seen that, if $C \xrightarrow{g} B$, then

$$\langle f_1, \ldots, f_n \rangle \circ g = \langle f_1 \circ g, \ldots, f_n \circ g \rangle.$$

Dually, a *coproduct* of an indexed set $\{A_i : i \in I\}$ of objects is an object, usually denoted by $\coprod_{i \in I} A_i$ or $\coprod A_i$, together with a set of arrows, called (canonical) *injections*, $\sigma_i : A_i \to \coprod A_i$ $(i \in I)$ such that, for each object B and arrows $f_i : A_i \to B$ $(i \in I)$, there is a unique arrow $h : \coprod A_i \to B$, usually denoted by $(f_i)_{i \in I}$, such that for each $i \in I$ the diagram

$$\begin{array}{ccc} A_i & \xrightarrow{\sigma_i} & \coprod A_i \\ & {\scriptstyle f_i}\searrow & \downarrow{\scriptstyle h} \\ & & B \end{array}$$

commutes. We note that a coproduct of the empty family is just an initial object. If each A_i is the same object A, then $\coprod_{i \in I} A_i$ is written $\coprod_I A$ and is called the *I-fold copower* of A.

A category **C** is said to *have (finite) coproducts* if the coproduct of any (finite) set of objects exists in **C**. It is easy to see that a category has finite coproducts iff it has binary coproducts and an initial object.

Products and coproducts of arbitrary sets of objects in **Set**, **Esp**, and **Grp** are obtained as obvious extensions of the finite case. In a preordered class, product is infimum and coproduct supremum. A partially ordered set which, considered as a category, has products and coproducts is called a *complete lattice*.

Limits and colimits

Products and coproducts are special cases of the more general notions of *limit* and *colimit* of a diagram in a category.

Let **D** be a diagram with vertices $\{D_i : i \in I\}$ in a category **C**. A *cone over* **D** is a family $\{A \xrightarrow{f_i} D_i : i \in I\}$ of arrows from a fixed object A to the objects in **D** such that, for any arrow $D_i \xrightarrow{d} D_j$ in **D**, the diagram

$$\begin{array}{ccc} & A & \\ {\scriptstyle f_i}\swarrow & & \searrow{\scriptstyle f_j} \\ D_i & \xrightarrow{d} & D_j \end{array}$$

commutes. The object A is called the *vertex* of the cone.

An *arrow* from a cone over **D** $\{A \xrightarrow{f_i} D_i : i \in I\}$ to a cone over **D** $\{A' \xrightarrow{f_i} D_i : i \in I\}$ is a **C**-arrow $A \xrightarrow{g} A'$ such that the diagram

$$\begin{array}{ccc} A & \xrightarrow{g} & A' \\ & {\scriptstyle f_i}\searrow & \downarrow{\scriptstyle f_i'} \\ & & D_i \end{array}$$

commutes for each $i \in I$. If such an arrow $g: A \to A'$ exists, we say that the cone $\{A \xrightarrow{f_i} D_i : i \in I\}$ *factors through* the cone $\{A' \xrightarrow{f_i} D_i : i \in I\}$. The cones over **D** then form a category in the resulting way. We define a *limit for the diagram* **D** to be a *terminal* object in this category. That is, a limit for **D** is a cone $\{A \xrightarrow{f_i} D_i : i \in I\}$ such that for *all* cones $\{A' \xrightarrow{f_i} D_i : i \in I\}$ there is a *unique* arrow $A' \xrightarrow{g} A$ for which the diagram

$$\begin{array}{ccc} A' & \xrightarrow{g} & A \\ & {\scriptstyle f'_i}\searrow & \downarrow{\scriptstyle f_i} \\ & & D_i \end{array}$$

commutes for each $i \in I$. (In other words, each cone over **D** *factors uniquely* through a limit for **D**: this is called the *universal property of limits*.) *Since terminal objects are unique up to isomorphism, so are limits.* We write $\lim \mathbf{D}$ for 'the' limit of **D**, when it exists. Instead of the term 'limit', we occasionally use the phrase 'limit diagram'.

Dually, we define a *cone under* **D** to be a cone over **D** considered in $\mathbf{C}^{\mathrm{op}}$, and a *colimit* for **D** to be a limit for **D** considered in $\mathbf{C}^{\mathrm{op}}$. Thus a cone under **D** is a family $\{D_i \xrightarrow{f_i} A : i \in I\}$ such that for any arrow $D_i \xrightarrow{d} D_j$ in **D**, the diagram

$$\begin{array}{ccc} D_i & \xrightarrow{d} & D_j \\ & {\scriptstyle f_i}\searrow \;\; \swarrow{\scriptstyle f_j} & \\ & A & \end{array}$$

commutes. (The object A is again called the *vertex* of the cone.) And a *colimit* for **D** is a cone under **D**, $\{D_i \xrightarrow{f_i} A : i \in I\}$, such that for any cone under **D** $\{D_i \xrightarrow{f'_i} A' : i \in I\}$ there is a *unique* arrow $A \xrightarrow{g} A'$ for which the diagram

$$\begin{array}{ccc} D_i & \xrightarrow{f_i} & A \\ & {\scriptstyle f'_i}\searrow & \downarrow{\scriptstyle g} \\ & & A' \end{array}$$

commutes for each $i \in I$. We write $\operatorname{colim} \mathbf{D}$ for 'the' colimit of **D** if it exists. Instead of the term 'colimit', we occasionally use the phrase 'colimit diagram'.

For simplicity, we frequently identify a limit or colimit with its vertex.

Examples

(i) For **C**-objects A and B let **D** be the arrowless diagram

$$A \qquad\qquad B$$

A cone over **D** is then an object C together with two arrows f and g of

the form

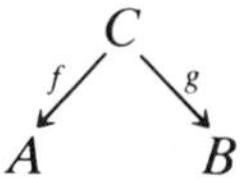

And lim **D** is $A \times B$. Dually, colim $\mathbf{D} = A + B$.

(ii) Given a family $\{A_i : i \in I\}$ of **C**-objects let **D** be the arrowless diagram $\{A_i : i \in I\}$. A cone over **D** is then an object C together with arrows $C \xrightarrow{f_i} A_i$. And lim **D** is $\prod_{i \in I} A_i$. Dually, colim $\mathbf{D} = \coprod_{i \in I} A_i$.

(iii) Let **D** be the *empty* diagram. A cone over **D** is then identifiable with a **C**-object, and lim **D** is a terminal object in **C**. Dually colim **D** is an initial object in **C**.

(iv) *Equalizers and coequalizers.* Let **D** be the diagram

$$A \overset{f}{\underset{g}{\rightrightarrows}} B$$

A cone over **D** is a pair of arrows $C \xrightarrow{h} A$, $C \xrightarrow{j} B$ such that the diagrams

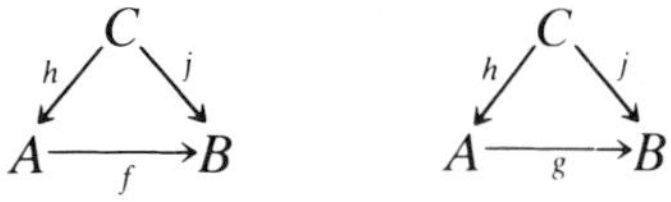

commute. Then $j = f \circ h = g \circ h$, so in this case a cone over **D** is essentially just an arrow $h : C \to A$ such that

$$C \xrightarrow{h} A \overset{f}{\underset{g}{\rightrightarrows}} B$$

commutes, i.e. $f \circ h = g \circ h$. (Under these conditions we say that h *equalizes* f and g.) A limit for **D** is an arrow $C \xrightarrow{e} A$ such that $f \circ e = g \circ e$ and for any arrow $C' \xrightarrow{e'} A$ such that $f \circ e' = g \circ e'$ there is a unique $u : C' \to C$ such that the diagram

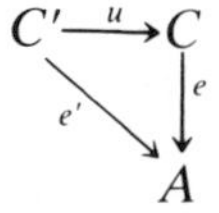

commutes. Such an arrow $C \xrightarrow{e} A$ is called an *equalizer* of f and g.

A category **C** is said to *have equalizers* if the equalizer of any pair of arrows exists in **C**.

In **Set**, the equalizer of a pair of maps $f, g : A \to B$ is 'the largest subset on which f and g coincide', i.e.

$$\{x \in A : f(x) = g(x)\} \hookrightarrow A$$

where $\hookrightarrow$ is the insertion map. Similarly for **Grp** and **Esp**.

Dually, a *colimit* for the diagram $A \overset{f}{\underset{g}{\rightrightarrows}} B$ is called a *coequalizer* of f and g. Thus a coequalizer of f and g is an arrow $h : B \to C$ such that

$h \circ f = h \circ g$ and whenever $h' : B \to C'$ satisfies $h' \circ f = h' \circ g$ there is a unique arrow $u : C \to C'$ such that the diagram

$$\begin{array}{ccc} B & \xrightarrow{h} & C \\ & \searrow^{h'} & \downarrow{\scriptstyle u} \\ & & C' \end{array}$$

commutes.

A category **C** is said to *have coequalizers* if the coequalizer of any pair of arrows exists in **C**.

In **Set**, the coequalizer of a pair of maps $f,g : A \to B$ is the quotient of B by the least equivalence relation which identifies $f(a)$ and $g(a)$ for each $a \in A$. Similar constructions work in **Grp** and **Esp**.

1.2 PROPOSITION. Any equalizer is monic and so, dually, any coequalizer is epic.

Proof. Let $C \xrightarrow{e} A \overset{f}{\underset{g}{\rightrightarrows}} B$ be an equalizer diagram and suppose $D \overset{h}{\underset{k}{\rightrightarrows}} C \xrightarrow{e} A$ commutes. Then $f \circ (e \circ h) = g \circ (e \circ h)$ and so there is a *unique* $u : D \to C$ such that $e \circ h = e \circ u$. But also $e \circ k = e \circ u$ and so, by uniqueness of u, $h = u = k$. □

(v) *Pullbacks.* A *pullback* of a pair $A \xrightarrow{f} C \xleftarrow{g} B$ of arrows with a common codomain C is a limit for the diagram

$$\begin{array}{ccc} & & B \\ & & \downarrow{\scriptstyle g} \\ A & \xrightarrow[f]{} & C \end{array}$$

A cone for this diagram consists of three arrows f', h, g' such that

$$\begin{array}{ccc} D & \xrightarrow{f'} & B \\ {\scriptstyle g'}\downarrow & \searrow^{h} & \downarrow{\scriptstyle g} \\ A & \xrightarrow{f} & C \end{array}$$

commutes. Then $h = g \circ f' = f \circ g'$, so a cone in this case reduces to a pair $A \xleftarrow{g'} D \xrightarrow{f'} B$ such that the square

$$\begin{array}{ccc} D & \xrightarrow{f'} & B \\ {\scriptstyle g'}\downarrow & & \downarrow{\scriptstyle g} \\ A & \xrightarrow{f} & C \end{array}$$

commutes. So a pullback of the pair $A \xrightarrow{f} C \xleftarrow{g} B$ is a pair of arrows $A \xleftarrow{g'} D \xrightarrow{f'} B$ such that (i) $f \circ g' = g \circ f'$ and (ii) whenever $A \xleftarrow{h} E \xrightarrow{j} B$

are such that $f \circ h = g \circ j$, then there is a unique arrow $k : E \to D$ such that $h = g' \circ k$ and $j = f' \circ k$.

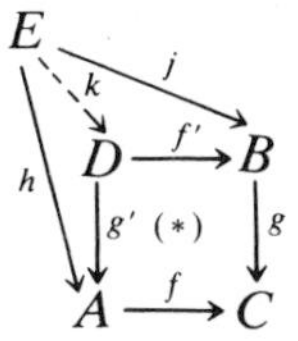

The inner square (*) here is called a *pullback square* or *diagram* and we say that

f' arises by pulling back f along g;
g' arises by pulling back g along f;
D arises by pulling back A along g, or B along f.

In **Set**, the pullback is given by

$$D = \{(x,y) \in A \times B : f(x) = g(y)\}$$

while k is the function defined by

$$k(z) = (h(z), j(z)) \quad \text{for } z \in E.$$

Pullbacks are similarly constructed in **Grp**, **Rng**, **Esp**.
In **Set**, there are several interesting cases of pullbacks.

• If $C = 1$, then $D \cong A \times B$.
• If $A \subseteq C$ and f is the insertion map, then $D \cong g^{-1}[A]$. More generally, if f is any injective map, then $D \cong g^{-1}[f[A]]$.
• If $A, B \subseteq C$ and f and g are insertions, then $D \cong A \cap B$.

Here are some basic facts concerning pullbacks.

1.3 PROPOSITION. *Pullbacks of monic arrows are monic*. That is, if the arrow $A \xrightarrow{f} C$ in a pullback square

$$\begin{array}{ccc} D & \xrightarrow{f'} & B \\ {\scriptstyle g'}\downarrow & & \downarrow{\scriptstyle g} \\ A & \rightarrowtail{\scriptstyle f} & C \end{array} \qquad (*)$$

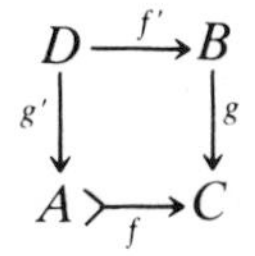

is monic, so is $D \xrightarrow{f'} B$.

Proof. Suppose f is monic, and $p, q : D' \to D$ satisfy $f'p = f'q$. Then $gf'p = gf'q$, so that

$$fg'p = gf'p = gf'q = fg'q.$$

Since f is monic, we may cancel it on the left, obtaining $g'p = g'q$. Since (*) is a pullback, there must be unique $h: D' \to D$ such that $f'h = f'p$ and $g'h = g'q$. But by the above, both p and q satisfy these equations when substituted for h. It follows that $p = q$ and so f' is monic. □

The proof of the next proposition is left as a straightforward, if a trifle tedious, exercise for the reader.

1.4 PROPOSITION. Consider a commutative diagram of the form

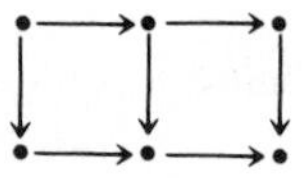

(i) If both squares are pullbacks, then the outer rectangle (with top and bottom edges the evident compositions) is a pullback.

(ii) If the outside rectangle and the right-hand square are pullbacks, so is the left-hand square. □

Existence of limits and colimits

A category **C** is said to be (finitely) *complete* or *cocomplete* if the limit or colimit of any (finite) diagram in **C** exists in **C**. (Notice that completeness and cocompleteness are mutually dual notions.) There is a simple criterion for this to be the case, viz.,

1.5 THEOREM. **C** is (finitely) complete iff **C** has (finite) products and equalizers. Dually, **C** is (finitely) cocomplete iff **C** has (finite) coproducts and coequalizers.

Proof. Since products and equalizers are limits, one way round is obvious.

Conversely, suppose that **C** has (finite) products and equalizers. Let **D** be a (finite) diagram in **C** with vertices $\{D_i : i \in I\}$. Let $D = \prod_{i \in I} D_i$ and for each $i \in I$ let $\pi_i : D \to D_i$ be the projection arrow. For each arrow $d: D_i \to D_j$ of **D**, write $i = i(d)$ and $j = j(d)$. Let $D' = \prod_{d \in D} D_{j(d)}$, and for each $d \in \mathbf{D}$ let $\bar{\pi}_{j(d)} : D' \to D_{j(d)}$ be the projection arrow.

Define arrows $D \underset{g}{\overset{f}{\rightrightarrows}} D'$ as follows. First, f is the unique arrow making the diagrams

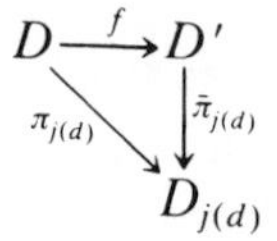

(1)

commute for all $d \in D$. Secondly, g is the unique arrow making the diagrams

$$\begin{array}{ccc} D & \xrightarrow{g} & D' \\ & \searrow^{d\circ\pi_{i(d)}} & \downarrow{\scriptstyle \bar{\pi}_{j(d)}} \\ & & D_{j(d)} \end{array} \qquad (2)$$

commute for all $d \in D$.

Now let $A \xrightarrow{h} D$ be the equalizer of $D \underset{g}{\overset{f}{\rightrightarrows}} D'$. We claim that the family

$$F = \{A \xrightarrow{\pi_i \circ h} D_i : i \in I\}$$

is a limit for **D** in **C**.

First, F is a cone over **D**. For if d is any arrow in **D**, we have

$$\begin{aligned} d \circ \pi_{i(d)} \circ h &= \bar{\pi}_{j(d)} \circ g \circ h && \text{(by (2))} \\ &= \bar{\pi}_{j(d)} \circ f \circ h && (h \text{ equalizes } f \ \& \ g) \\ &= \pi_{j(d)} \circ h && \text{(by (1))}. \end{aligned}$$

Therefore the diagram

$$\begin{array}{ccccc} & & A & & \\ & \swarrow^{\pi_{i(d)}\circ h} & & \searrow^{\pi_{j(d)}\circ h} & \\ D_{i(d)} & & \xrightarrow{d} & & D_{j(d)} \end{array}$$

commutes.

Next, let $F' = \{A' \xrightarrow{k_i} D_i : i \in I\}$ be any cone over **D**. Let k be the unique arrow making the diagrams

$$\begin{array}{ccc} A' & \xrightarrow{k} & D \\ & \searrow^{k_i} & \downarrow{\scriptstyle \pi_i} \\ & & D_i \end{array} \qquad (3)$$

commute for all $i \in I$. Then $f \circ k = g \circ k$. To establish this it suffices to show that $\bar{\pi}_{j(d)} \circ f \circ k = \bar{\pi}_{j(d)} \circ g \circ k$ for any $d \in D$, and this is so since

$$\begin{aligned} \bar{\pi}_{j(d)} \circ f \circ k &= \pi_{j(d)} \circ k && \text{(by (1))} \\ &= k_{j(d)} && \text{(by (3))} \\ &= d \circ k_{i(d)} && \text{(since } F' \text{ is a cone)} \\ &= d \circ \pi_{i(d)} \circ k && \text{(by (3))} \\ &= \bar{\pi}_{j(d)} \circ g \circ k && \text{(by (2))}. \end{aligned}$$

Since h is the equalizer of f and g, there is a unique $l : A' \to A$ making the diagram

$$\begin{array}{ccc} A' & \xrightarrow{l} & A \\ & \searrow^{k} & \downarrow{\scriptstyle h} \\ & & D_i \end{array} \qquad (4)$$

commute. It follows immediately from this and (3) that the diagrams

$$\begin{array}{ccc} A' & \xrightarrow{l} & A \\ & {\scriptstyle k_i}\searrow & \downarrow{\scriptstyle \pi_i \circ h} \\ & & D_i \end{array} \tag{5}$$

commute for all $i \in I$. And since the commutativity of diagrams (5) for all $i \in I$ is equivalent to the commutativity of (4), the arrow $A' \xrightarrow{l} A$ making diagrams (5) commute is unique.

Accordingly, any cone F' over **D** factors uniquely through F; the latter is therefore a limit for **D**. □

It follows from this result that, since **Set**, **Grp**, and **Esp** all have products, coproducts, equalizers, and coequalizers, they are complete and cocomplete.

1.6 EXERCISE. Show that, if **C** has a terminal object and pullbacks of pairs of arrows with common codomain, then **C** is finitely complete.

Functors

Let **C** and **D** be categories. A *functor* $F:\mathbf{C} \to \mathbf{D}$ or $\mathbf{C} \xrightarrow{F} \mathbf{D}$ is a function F which assigns to each **C**-object X a **D**-object $F(X)$ and to each **C**-arrow $f:X \to Y$ a **D**-arrow $F(f):F(X) \to F(Y)$ in such a way that

- if $X \xrightarrow{f} Y \xrightarrow{g} Z$ in **C**, then $F(g \circ f) = F(g) \circ F(f)$ in **D**.
- $F(1_X) = 1_{F(X)}$ for all $X \in \mathrm{Ob}(\mathbf{C})$.

We sometimes write FX for $F(X)$ and Ff for $F(f)$. The function $X \mapsto FX$ (resp. $f \mapsto Ff$) is called the *object* (resp. *arrow*) function of F.

A functor $\mathbf{C} \to \mathbf{Set}$ is called a *set-valued* functor on **C**.

Examples

(i) The obvious *identity functor* $1_{\mathbf{C}}:\mathbf{C} \to \mathbf{C}$, and the obvious *insertion functor* of any subcategory **A** of **C** into **C**.

(ii) The *power set functor* $P:\mathbf{Set} \to \mathbf{Set}$. Here P assigns to each X its power set PX and to each map $f:X \to Y$ that map $Pf:PX \to PY$ which sends each $S \subseteq X$ to its image $f[S] \subseteq Y$.

(iii) For any commutative ring R the set of all nonsingular $n \times n$ matrices is the familiar *general linear group* $\mathrm{GL}_n(R)$; each homomorphism $f:R \to R'$ induces in the obvious way a homomorphism $\mathrm{GL}_n(f):\mathrm{GL}_n(R) \to \mathrm{GL}_n(R')$. Thus we get the *general linear group functor* $\mathrm{GL}_n:\mathbf{CRng} \to \mathbf{Grp}$.

(iv) The function $F:\mathbf{Set} \to \mathbf{Grp}$ that assigns to each set A the free group generated by A and to each map f the induced homomorphism which coincides with f on the free generators is a functor, called the *free group functor.* One similarly defines the *free abelian group functor,* the *free ring functor,* the *free K-vector space functor,* etc.

(v) The 'forgetful' functor $U:\mathbf{Grp}\to\mathbf{Set}$ that assigns to each group G its underlying set and to each group homomorphism the same function regarded as a map between sets. (That is, U 'forgets' the group structure.) Similarly, we have forgetful functors $\mathbf{Esp}\to\mathbf{Set}$, $\mathbf{Rng}\to\mathbf{Set}$ and also, for example, $\mathbf{(C)Rng}\to\mathbf{Ab}$ which 'forgets' multiplication.

(vi) For each object A of a category $\mathbf{C}$ we have the (set-valued) *hom-functor* $H_A:\mathbf{C}\to\mathbf{Set}$ defined by $H_A(X)=\mathbf{C}(A,X)$ and for $X\xrightarrow{f}Y$, $H_A(f):\mathbf{C}(A,X)\to\mathbf{C}(A,Y)$ is the map $g\mapsto f\circ g$.

(vii) *Pullback functors.* Let $\mathbf{C}$ be a category in which the pullback of any pair of arrows exists. Then any $\mathbf{C}$-arrow $A\xrightarrow{f}B$ induces a *pullback functor* $f^*:\mathbf{C}/B\to\mathbf{C}/A$ between slice categories which assigns, to each object $Y\xrightarrow{g}B$ in $\mathbf{C}/B$, the pullback $X\to A$ of g along f.

(viii) The following examples of functors serve to reveal the true generality of the notion.

- A functor between two (pre) ordered sets is just an order-preserving map.
- A functor between two groups (regarded as single object categories) is a group homomorphism.
- If G is a group, a functor $\mathbf{G}\to\mathbf{Set}$ is a permutation representation of G, while a functor $\mathbf{G}\to\mathbf{Matr}_R$ is a matrix representation of G.

These examples are all what are known as *covariant* functors; they preserve the 'direction' of arrows. A *contravariant* functor, on the other hand, is one that 'reverses' arrows. More precisely, a functor $F:\mathbf{C}\to\mathbf{D}^{\mathrm{op}}$ (or equivalently $F:\mathbf{C}^{\mathrm{op}}\to\mathbf{D}$) is called a contravariant functor from $\mathbf{C}$ to $\mathbf{D}$.

Examples

(i) The contravariant power set functor $\bar{P}:\mathbf{Set}\to\mathbf{Set}$. Here $\bar{P}$ assigns to each set X its power set $P(X)$ and to each map $f:X\to Y$ the 'inverse image' map $\bar{P}(f):P(Y)\to P(X)$ which sends each $S\subseteq Y$ to $f^{-1}[S]$.

(ii) The functor $\mathbf{Esp}\to\mathbf{CRng}^{\mathrm{op}}$ which assigns to each topological space X its ring of continuous real-valued functions is a contravariant functor from **Esp** to **CRng**.

(iii) The functor $\mathbf{Bool}\to\mathbf{Esp}^{\mathrm{op}}$ which assigns to each Boolean algebra its Stone space, and the functor $\mathbf{Esp}\to\mathbf{Bool}^{\mathrm{op}}$ which assigns to each topological space its algebra of clopen subsets, are contravariant functors from **Bool** to **Esp** and from **Esp** to **Bool**, respectively.

(iv) If $\mathbf{Q}$ is the category of finite-dimensional vector spaces over a field K, then the functor $\hat{}:\mathbf{Q}\to\mathbf{Q}^{\mathrm{op}}$ which assigns to each vector space X the space $\hat{X}$ of linear functionals to K and to each linear map $f:X\to Y$ the linear map $\hat{f}:\hat{Y}\to\hat{X}$ defined by $\hat{f}(g)=g\circ f$, is a contravariant functor from $\mathbf{Q}$ to $\mathbf{Q}$.

(v) For each object $\mathbf{A}$ of a category $\mathbf{C}$ we have the *contravariant hom-functor* $H^A:\mathbf{C}^{\mathrm{op}}\to\mathbf{Set}$ defined by $H^A(X)=\mathbf{C}(X,A)$ and for $X\xrightarrow{f}Y$, $H^A(f):\mathbf{C}(Y,A)\to\mathbf{C}(X,A)$ is the map $g\mapsto g\circ f$.

Note that functors may be *composed* in the natural way: if $\mathbf{C}\xrightarrow{F}\mathbf{D}$ and $\mathbf{D}\xrightarrow{G}\mathbf{E}$ are functors, so is $\mathbf{C}\xrightarrow{G\circ F}\mathbf{E}$. We sometimes write GF for $G\circ F$. Note also that if $F:\mathbf{C}\to\mathbf{D}$, then $F^{\mathrm{op}}:\mathbf{C}^{\mathrm{op}}\to\mathbf{D}^{\mathrm{op}}$ may be defined in the evident manner.

Sometimes functors, like functions, have more than one argument. To allow for this possibility, we introduce the notion of the *product* of categories. Given categories $\mathbf{C}$ and $\mathbf{D}$, the *product category* $\mathbf{C}\times\mathbf{D}$ has as objects all pairs of objects (A,A') with $A\in\mathrm{Ob}(\mathbf{C})$ and $A'\in\mathrm{Ob}(\mathbf{D})$; an *arrow* from (A,A') to (B,B') in $\mathbf{C}\times\mathbf{D}$ is just a pair (f,f') of arrows with $f:A\to B$ in $\mathbf{C}$ and $f':A'\to B'$ in $\mathbf{D}$; composition is defined by $(f,f')\circ(g,g')=(f\circ g, f'\circ g')$.

A functor whose domain is the product of two categories is often called a *bifunctor*. As examples, consider

- the *Cartesian product bifunctor* $\cdot\times\cdot:\mathbf{C}\times\mathbf{C}\to\mathbf{C}$, defined by $\cdot\times\cdot(A,B)=A\times B$, $\cdot\times\cdot(f,g)=f\times g$ for any category $\mathbf{C}$ with binary products;
- the *tensor product bifunctor* $\cdot\otimes\cdot:\mathbf{Vect}_K\times\mathbf{Vect}_K\to\mathbf{Vect}_K$ defined by

$$\cdot\otimes\cdot(A,B)=A\otimes B,\qquad \cdot\otimes\cdot(f,g):A\otimes B\to C\otimes D$$

where

$$f\otimes g\Big(\sum_i a_i\otimes b_i\Big)=\sum_i f(a_i)\otimes g(b_i).$$

Each bifunctor $F:\mathbf{C}\times\mathbf{D}\to\mathbf{E}$ gives rise to a family of *associated* functors. For each object A of $\mathbf{C}$, we have the *right-associated functor* $F(A,.):\mathbf{D}\to\mathbf{E}$ defined by $F(A,.)(B)=F(A,B)$ and $F(A,.)(h)=F(1_A,h)$. Similarly, for each object B of $\mathbf{D}$ we have the *left-associated functor* $F(.,B):\mathbf{C}\to\mathbf{E}$ given by $F(.,B)=F(A,B)$ and $F(.,B)(g)=F(g,1_B)$.

The most important sort of bifunctor is the *set-valued hom-functor* on a category $\mathbf{C}$. This is the bifunctor $\mathrm{hom}_{\mathbf{C}}:\mathbf{C}^{\mathrm{op}}\times\mathbf{C}\to\mathbf{Set}$ where $\mathrm{hom}_{\mathbf{C}}(A,B)=\mathbf{C}(A,B)$ and, for $f:A'\to A$, $g:B\to B'$, $\mathrm{hom}_{\mathbf{C}}(f,g):\mathbf{C}(A,B)\to\mathbf{C}(A',B')$ is given by $\mathrm{hom}_{\mathbf{C}}(f,g)(x)=g\circ x\circ f$:

$$\begin{array}{c}\mathrm{hom}_{\mathbf{C}}(A,B)\\ \Big\downarrow{\scriptstyle \mathrm{hom}(f,g)}\\ \mathrm{hom}_{\mathbf{C}}(A',B')\end{array}\qquad\qquad \begin{array}{ccc} A & \xrightarrow{\;x\;} & B\\ {\scriptstyle f}\Big\uparrow & & \Big\downarrow{\scriptstyle g}\\ A' & \xrightarrow{\;g\circ x\circ f\;} & B'\end{array}$$
$$=\mathrm{hom}(f,g)(x)$$

To verify that this is a functor, observe that (writing hom for $\hom_{\mathbf{C}}$)

$$\hom(1_A, 1_B)(x) = 1_B \circ x \circ 1_A = x,$$

so that $\hom(1_A, 1_B)$ is the identity on $\hom(A, B)$. And

$$\begin{aligned}\hom((f,g) \circ (h,k))(x) &= \hom(h \circ f, g \circ k)(x) \\ &= g \circ k \circ x \circ h \circ f = g \circ \hom(h,k)(x) \circ f \\ &= \hom(f,g)(\hom(h,k)(x)) \\ &= \hom(f,g) \circ \hom(h,k)(x).\end{aligned}$$

A functor $F: \mathbf{C} \to \mathbf{D}$ is said to be

- *full* if for each pair A, B of **C**-objects, F carries $\mathbf{C}(A,B)$ onto $\mathbf{D}(F(A), F(B))$;
- *faithful* if for each pair A, B of **C**-objects, F is one-to-one on $\mathbf{C}(A,B)$;
- *dense* if for each **D**-object B there is a **C**-object A such that $F(A) \cong B$;
- an *embedding* if it is faithful and *injective on objects*, i.e. $F(A) = F(B)$ implies $A = B$.

Examples

(i) Every insertion functor and every forgetful functor is faithful, but not in general full, and the former, but not the latter, are embeddings.

(ii) The forgetful functor from **Field** to **Set** or **Bool** to **Set** is neither full nor dense. (There is no field with 6 elements and no Boolean algebra with 3 elements.)

(iii) The forgetful functor from **Grp** to **Set** is dense, but not full.

(iv) The insertion functor from **Haus** to **Esp** is full, but not dense.

Natural transformations and functor categories

A functor may be thought of as an *arrow* between categories. In fact, functors between (small) categories constitute the class of arrows of a certain category, namely, the category **Cat** whose objects are all small categories.

We can now introduce the idea of an *arrow between functors,* usually called a *natural transformation.* Given two functors $F, G: \mathbf{C} \to \mathbf{D}$, a *natural transformation* between F and G is a map η from $\mathrm{Ob}(\mathbf{C})$ to $\mathrm{Arr}(\mathbf{D})$ satisfying the following conditions.

- For each $A \in \mathrm{Ob}(\mathbf{C})$, $\eta(A)$, which is usually written η_A, is a **D**-arrow $\eta_A: F(A) \to G(A)$.

• For each **C**-arrow $A \xrightarrow{f} A'$, the diagram

$$\begin{array}{ccc} A & \qquad & F(A) \xrightarrow{\eta_A} G(A) \\ {\scriptstyle f}\downarrow & & {\scriptstyle F(f)}\downarrow \qquad \downarrow{\scriptstyle G(f)} \\ A' & & F(A') \xrightarrow{\eta_{A'}} G(A') \end{array}$$

commutes.

F and G are called the *domain* and *codomain* of η, and we write $\eta: F \to G$ or $F \xrightarrow{\eta} G$. The maps η_A are called the *components* of η, and we sometimes write $\eta = (\eta_A)_{A \in \mathrm{Ob}(\mathbf{C})}$ or simply $\eta = (\eta_A)$.

If each component η_A of η is an *isomorphism* in **D**, η is called a *natural isomorphism*; under these conditions we write $\eta: F \cong G$ and say that F and G are *naturally isomorphic.*

If F is a (covariant) *set-valued* functor on **C**, we say that F is *representable* if there is a **C**-object A such that $F \cong H_A$. Dually, we call a contravariant set-valued functor G representable if there is a **C**-object A such that $G \cong H^A$. An object A satisfying these conditions is called a *representing object* for F or G.

Examples

(i) The identity natural transformation $1_F: F \to F$ with $(1_F)A = 1_{FA}$.

(ii) *Determinants are natural transformations.* Given a commutative ring R and an $n \times n$ matrix M with entries in R, let R^* be the group of invertible elements of R. Then M is nonsingular when the determinant $\det_R M$ is invertible in R, and $\det_R: \mathrm{GL}_n(R) \to R^*$ is an arrow in **Grp**. Since the determinant is defined by the same formula for all rings R, each arrow $f: R \to R'$ in **CRng** leads to a commutative diagram

$$\begin{array}{ccc} GL_n(R) & \xrightarrow{\det_R} & R^* \\ {\scriptstyle GL_n(f)}\downarrow & & \downarrow{\scriptstyle f \mid R^*} \\ GL_n(R') & \xrightarrow{\det_{R'}} & R'^* \end{array}$$

so $\det_R: \mathrm{GL}_n(.) \to (.)^*$ is a natural transformation.

(iii) *Character groups.* The *character group* $D(G)$ of an abelian group G is the set $\hom(G, R/Z)$ of all homomorphisms $t: G \to R/Z$ with the obvious group structure, where R/Z is the additive group of the reals mod 1. This gives rise to the contravariant functor $D = \hom(.,R/Z): \mathbf{Ab} \to \mathbf{Ab}$. The composite $D \circ D$ is the 'double character group' functor. For each abelian group G there is a homomorphism $\tau_G: G \to D(DG)$ obtained by evaluation: to each $g \in G$ assign the map $\tau_G(g): DG \to R/Z$ given by $\tau_G(g)(t) = t(g)$ for $t \in DG$. If I is the identity functor in **Ab**, $\tau = (\tau_G)$ defines a natural transformation $I \to D \circ D$: this statement means that the definition of τ does not depend on artificial choices of bases, generators etc. If G is finite, then τ_G is an isomorphism, so if we restrict to the category of *finite* abelian groups, τ is a natural isomorphism.

(iv) Writing $2=\{0,1\}$, there is a natural isomorphism η from the contravariant hom-functor $H^2:\mathbf{Set}\to\mathbf{Set}$ to the contravariant power set functor $\bar{P}$, given by $\eta_A(g)=g^{-1}(0)$. So $\bar{P}$ is representable, and 2 a representing object for it.

If F and G are set-valued functors on $\mathbf{C}$, write $\mathrm{Nat}(F,G)$ for the collection of natural transformations from F to G. We now prove the fundamental

1.7 YONEDA LEMMA. Let $\mathbf{C}$ be a category, F a set-valued functor on $\mathbf{C}$, and A an object of $\mathbf{C}$. Then the map $\theta:\mathrm{Nat}(H_A,F)\to F(A)$ defined by $\theta:\eta\mapsto\eta_A(1_A)$ is a bijection.

Proof. Let $\eta\in\mathrm{Nat}(H_A,F)$ and $A\xrightarrow{f}B$ in $\mathbf{C}$. Then, since η is a natural transformation, the diagram

$$\begin{array}{ccc} H_A(A) & \xrightarrow{\eta_A} & F(A) \\ {\scriptstyle H_A(f)}\downarrow & & \downarrow{\scriptstyle F(f)} \\ H_A(B) & \xrightarrow{\eta_B} & F(B) \end{array}$$

commutes, so

$$\eta_B(f)=\eta_B(H_A(f)(1_A))=F(f)(\eta_A(1_A)).$$

Therefore η is completely determined by $\eta_A(1_A)$, and it follows that θ is injective.

To show that θ is surjective, take any $a\in F(A)$ and define for each object B of $\mathbf{C}$ the map $a_B^*:H_A(B)\to F(B)$ by

$$a_B^*:f\mapsto F(f)(a).$$

Then the a_B^* form the components of a natural transformation $a^*:H_A\to F$. For if $g:B\to C$, then for each $f\in H_A(B)$ we have

$$F(g)(a_B^*(f))=F(g)(F(f)(a))=(F(g\circ f))(a),$$

and

$$a_C^*(H_A(g))(f)=a_C^*(g\circ f)=(F(g\circ f))(a).$$

So the diagram

$$\begin{array}{ccc} H_A(B) & \xrightarrow{a_B^*} & F(B) \\ {\scriptstyle H_A(g)}\downarrow & & \downarrow{\scriptstyle F(g)} \\ H_A(C) & \xrightarrow{a_C^*} & F(C) \end{array}$$

commutes, and a^* is a natural transformation as claimed. Clearly $\theta(a^*)=a$ so θ is onto and hence a bijection. □

It follows immediately from this result that, for any objects A,B, $\mathrm{Nat}(H_A,H_B)\cong\mathbf{C}(B,A)$, that is, every natural transformation from H_A to H_B arises from an arrow from B to A. By duality, we also have

(1.8) $$\mathrm{Nat}(H^A,H^B)\cong\mathbf{C}^{\mathrm{op}}(B,A)\cong\mathbf{C}(A,B).$$

Natural transformations can be *composed* in the obvious way: given two natural transformations $F\xrightarrow{\eta}G$ and $G\xrightarrow{\xi}H$, their *composition* $\xi\circ\eta$ is the natural transformation $(\xi_A\circ\eta_A)$.

Given categories $\mathbf{C}$,$\mathbf{E}$ with C *small*, the *functor category* $\mathbf{E}^{\mathbf{C}}$ has as objects all functors $F:\mathbf{C}\rightarrow\mathbf{E}$ and as arrows all natural transformations between such functors, with composition defined as above.

For any small category $\mathbf{C}$, we consider the functor category $\mathbf{Set}^{\mathbf{C}^{\mathrm{op}}}$ whose objects are all *contravariant* set-valued functors on $\mathbf{C}$ (such functors are called *presheaves* on $\mathbf{C}$). Define a map $Y:\mathbf{C}\rightarrow\mathbf{Set}^{\mathbf{C}^{\mathrm{op}}}$ by $Y(A)=H^A$ for each $\mathbf{C}$-object A and

$$Y(f)_C(g)=f\circ g$$

for $\mathbf{C}$-arrows $A\xrightarrow{f}B$ and $C\xrightarrow{g}A$. It is readily checked that Y is a functor and that it is injective on objects. Moreover, (1.8) asserts that Y is full and faithful. That is, we have proved the

1.9 YONEDA EMBEDDING THEOREM. For any small category $\mathbf{C}$, the functor Y is a full embedding of $\mathbf{C}$ into $\mathbf{Set}^{\mathbf{C}^{\mathrm{op}}}$. □

Y is called the *Yoneda embedding* of $\mathbf{C}$ into $\mathbf{Set}^{\mathbf{C}^{\mathrm{op}}}$.

To close this section we show that every functor in $\mathbf{Set}^{\mathbf{C}^{\mathrm{op}}}$ can be 'approximated' by representable functors, in the sense of

1.10 THEOREM. Let $\mathbf{C}$ be a small category. Then for any object $F\in\mathbf{Set}^{\mathbf{C}^{\mathrm{op}}}$ there is a diagram $\mathbf{D}$ in $\mathbf{Set}^{\mathbf{C}^{\mathrm{op}}}$ with vertices consisting of representable functors (i.e. of the form H^A) such that $F=\mathrm{colim}\,\mathbf{D}$. That is, *any* $F\in\mathbf{Set}^{\mathbf{C}^{\mathrm{op}}}$ *is the colimit of representable functors.*

Proof. Given $F\in\mathbf{Set}^{\mathbf{C}^{\mathrm{op}}}$, we may assume without loss of generality that the sets $\{FA:A\in\mathrm{Ob}(\mathbf{C})\}$ are disjoint. Then for each $x\in FA$, $A\in\mathrm{Ob}(\mathbf{C})$, write H_x for H^A. Let $\mathbf{D}$ be the diagram in $\mathbf{Set}^{\mathbf{C}^{\mathrm{op}}}$ whose set of vertices is

$$\{H_x:x\in FA,\ A\in\mathrm{Ob}(\mathbf{C})\}.$$

Recall that for each $x\in FA$ we defined in the proof of the Yoneda lemma a natural transformation $x^*:H^A\rightarrow F$ by

$$x^*_B:H^A(B)\rightarrow FB:f\mapsto(Ff)x.$$

We take the *arrows* of the diagram $\mathbf{D}$ to be all $H_x\xrightarrow{\eta}H_y$ in $\mathbf{Set}^{\mathbf{C}^{\mathrm{op}}}$ with

$x \in FA$, $y \in FB$ such that the diagram

$$\begin{array}{ccc} H_x & \xrightarrow{\eta} & H_y \\ & {\scriptstyle x^*}\searrow \quad \swarrow{\scriptstyle y^*} & \\ & F & \end{array} \tag{1}$$

commutes.

Clearly

$$K = \{H_x \xrightarrow{x^*} F : x \in FA,\ A \in \mathrm{Ob}(\mathbf{C})\}$$

is a cone under $\mathbf{D}$. We claim that $K = \operatorname{colim} \mathbf{D}$.

For suppose

$$K' = \{H_x \xrightarrow{\xi_x} G : x \in FA,\ A \in \mathrm{Ob}(\mathbf{C})\}$$

is any cone under $\mathbf{D}$. Then the diagram

$$\begin{array}{ccc} H_x & \xrightarrow{\eta} & H_y \\ & {\scriptstyle \xi_x}\searrow \quad \swarrow{\scriptstyle \xi_y} & \\ & G & \end{array} \tag{2}$$

commutes whenever the diagram (1) commutes. We have to show that there is a unique natural transformation $\alpha : F \to G$ such that the diagram

$$\begin{array}{ccc} H_x & \xrightarrow{x^*} & F \\ & {\scriptstyle \xi_x}\searrow & \downarrow{\scriptstyle \alpha} \\ & & G \end{array} \tag{3}$$

commutes for all $x \in FA$, $A \in \mathrm{Ob}(\mathbf{C})$.

In order to establish this we first need the following: for any $B \xrightarrow{f} A$, and any $x \in FA$, putting $y = (Ff)x \in FB$,

$$(\xi_y)_B(1_B) = (\xi_x)_B(f). \tag{4}$$

To prove (4), we note that since the diagram

$$\begin{array}{ccc} H_y & \xrightarrow{Yf} & H_x \\ & {\scriptstyle y^*}\searrow \quad \swarrow{\scriptstyle x^*} & \\ & F & \end{array}$$

commutes, so therefore does the diagram

$$\begin{array}{ccc} H_y & \xrightarrow{Yf} & H_x \\ & {\scriptstyle \xi_y}\searrow \quad \swarrow{\scriptstyle \xi_x} & \\ & G & \end{array}$$

Hence in particular the diagram

$$\begin{array}{ccc} H^B(B) = H_y(B) & \xrightarrow{(Yf)_B} & H_x(B) = H^A(B) \\ & {\scriptstyle (\xi_y)_B}\searrow \quad \swarrow{\scriptstyle (\xi_x)_B} & \\ & GB & \end{array}$$

commutes, whence

$$(\xi_y)_B(1_B) = (\xi_x)_B((Yf)_B(1_B)) = (\xi_x)_B(f)$$

as required.

Now we define, for each $A \in \mathrm{Ob}(\mathbf{C})$, $\alpha_A : FA \to GA$ by

$$\alpha_A(x) = (\xi_x)_A(1_A)$$

for $x \in FA$. Then $\alpha = (\alpha_A)_{A \in \mathrm{Ob}(\mathbf{C})}$ is a natural transformation $F \to G$. To see this, let $f : B \to A$, $x \in FA$, and put $y = (Ff)x$. Then

$$\begin{aligned} (GF)(\alpha_A(x)) &= (GF)((\xi_x)_A(1_A)) \\ &= (\xi_x)_B(H^A(f)(1_A)) \quad && (\xi \text{ is a natural transformation}) \\ &= (\xi_x)_B(f) \\ &= (\xi_y)_B(1_B) && (\text{by } (4)) \\ &= \alpha_B(y) \\ &= \alpha_B((Ff)x). \end{aligned}$$

So the diagram

$$\begin{array}{ccc} FA & \xrightarrow{\alpha_A} & GA \\ {\scriptstyle Ff}\downarrow & & \downarrow{\scriptstyle Gf} \\ FB & \xrightarrow{\alpha_B} & GB \end{array}$$

commutes for each $A \in \mathrm{Ob}(\mathbf{C})$, and α is a natural transformation as claimed.

Also, (3) commutes, since for any $B \xrightarrow{f} A$ and any $x \in FA$, putting $y = (Ff)x$,

$$\alpha_B(x^*_B(f)) = \alpha_B(y) = (\xi_y)_B(1_B) = (\xi_x)_B(f)$$

by (4).

Finally, α is the unique natural transformation $F \to G$ making the diagrams (3) commute, since for $A \in \mathrm{Ob}(\mathbf{C})$, $x \in FA$, $\alpha_A(x)$ is uniquely determined by

$$\alpha_A(x) = \alpha_A(x^*_A(1_A)) = (\xi_x)_A(1_A). \quad \square$$

Completeness of functor categories

We next show that a functor category inherits completeness from its base category.

1.11 THEOREM. If **E** is complete (or, dually, cocomplete) and **C** is small, then $\mathbf{E}^{\mathbf{C}}$ is complete (or cocomplete).

Proof. Let **D** be a diagram in $\mathbf{E}^{\mathbf{C}}$ with vertices $\{D_i : i \in I\}$. For each $A \in \mathrm{Ob}(\mathbf{C})$ let $\mathbf{D}(A)$ be the diagram in **E** with vertices $\{D_i(A) : i \in I\}$ and arrows $\{\eta_A : \eta \text{ an arrow of } \mathbf{D}\}$. Then, since **E** is complete, $\mathbf{D}(A)$ has a limit

$$\{A^* \xrightarrow{f_i^A} D_i(A) : i \in I\}.$$

Now define a functor $F : \mathbf{C} \to \mathbf{E}$ as follows. For each $A \in \mathrm{Ob}(\mathbf{C})$, we put

$$F(A) = A^* = \lim \mathbf{D}(A) \tag{1}$$

If $A \xrightarrow{g} B$ in **C**, it is easy to check that

$$\{A^* \xrightarrow{D_i(g) \circ f_i^A} D_i(B) : i \in I\}$$

is a cone over $\mathbf{D}(B)$, so by the limit property of B^* there is a unique $h : A^* \to B^*$ such that the diagrams

$$\begin{array}{ccc} A^* & \xrightarrow{h} & B^* \\ & {\scriptstyle D_i(g) \circ f_i^A} \searrow & \downarrow {\scriptstyle f_i^B} \\ & & D_i(B) \end{array} \tag{2}$$

commute for all $i \in I$. We put

$$F(g) = h.$$

Note that diagram (2) can now be written

$$\begin{array}{ccc} FA & \xrightarrow{Fg} & FB \\ {\scriptstyle f_i^A} \downarrow & & \downarrow {\scriptstyle f_i^B} \\ D_i(A) & \xrightarrow{D_i(g)} & D_i(B) \end{array} \tag{3}$$

Diagram (3) shows that the f_i^A form the components of a natural transformation $\alpha_i : F \to D_i$ for each $i \in I$. It is now easy to show that

$$K = \{F \xrightarrow{\alpha_i} D_i : i \in I\}.$$

is a limit for **D**. □

We note that equation (1) in the proof of this theorem may be written

$$(\lim \mathbf{D})(A) = \lim(\mathbf{D}(A)).$$

In other words, the limit of a diagram in a functor category may be defined 'pointwise'. (Of course, this also holds for colimits.)

Since **Set** is complete and cocomplete, we have, as an immediate consequence of (1.11),

1.12 COROLLARY. For any small category $\mathbf{C}$, the category $\mathbf{Set}^{\mathbf{C}}$ is complete and cocomplete. □

Hence the Yoneda embedding theorem yields

1.13 COROLLARY. Any small category is fully embeddable in a complete and cocomplete category. □

Equivalence of categories

A bijective functor between categories is called an *isomorphism*; two categories are *isomorphic* if there is an isomorphism between them. A weaker but more useful notion is the following. A full, faithful and dense functor is called an *equivalence*; two categories $\mathbf{C}$,$\mathbf{D}$ are *equivalent*, written $\mathbf{C} \simeq \mathbf{D}$, if there is an equivalence between them. Obviously any isomorphism is an equivalence.

Let us call a category *skeletal* if isomorphic objects are identical, and let us call a maximal full skeletal subcategory of a category $\mathbf{C}$ a *skeleton* of $\mathbf{C}$. Thus a full subcategory $\mathbf{D}$ of $\mathbf{C}$ is a skeleton of $\mathbf{C}$ iff for each $\mathbf{C}$-object A there is a *unique* $\mathbf{D}$-object B such that $A \cong B$. It is easy to prove (using the axiom of choice) that every category has a skeleton. Note also that skeletal categories are equivalent iff they are isomorphic.

Let $\mathbf{D}$ be a skeleton of $\mathbf{C}$ and let $Q:\mathrm{Ob}(\mathbf{C}) \to \mathrm{Ob}(\mathbf{D})$ be the map defined by $QA \cong A$ for each $A \in \mathrm{Ob}(\mathbf{C})$. Q can be turned into a functor in the following way. For each $A \in \mathrm{Ob}(\mathbf{C})$ choose an isomorphism $\theta_A : A \cong QA$; for each $A \xrightarrow{f} B$ in $\mathbf{C}$ define

$$Qf = \theta_B \circ f \circ \theta_A^{-1}.$$

The resulting functor $Q:\mathbf{C} \to \mathbf{D}$ is called a *projection functor*; it is easy to see that it is an equivalence.

We can now prove

1.14 PROPOSITION. Two categories are equivalent iff they have isomorphic skeletons.

Proof. Let $\mathbf{C} \xrightarrow{F} \mathbf{D}$ be an equivalence, and let $\hat{\mathbf{C}}$,$\hat{\mathbf{D}}$ be skeletons of $\mathbf{C}$,$\mathbf{D}$ respectively. Let $\hat{\mathbf{C}} \overset{K}{\hookrightarrow} \mathbf{C}$ be the insertion functor and $\mathbf{D} \xrightarrow{Q} \hat{\mathbf{D}}$ a projection functor. Clearly the composite functor QFK is an equivalence, and hence an isomorphism, between $\hat{\mathbf{C}}$ and $\hat{\mathbf{D}}$.

Conversely, suppose that $\mathbf{C}$ and $\mathbf{D}$ have isomorphic skeletons $\hat{\mathbf{C}}$ and $\hat{\mathbf{D}}$. Let $\hat{\mathbf{C}} \xrightarrow{J} \hat{\mathbf{D}}$ be an isomorphism. If $\mathbf{C} \xrightarrow{R} \hat{\mathbf{C}}$ and $\hat{\mathbf{D}} \overset{E}{\hookrightarrow} \mathbf{D}$ are projection and insertion functors, respectively, then $EJR:\mathbf{C} \to \mathbf{D}$ is obviously an equivalence. □

It follows immediately from (1.14) that $\simeq$ is an equivalence relation on categories.

1.15 PROPOSITION. The following are equivalent for a functor $F:\mathbf{C}\to\mathbf{D}$:

(i) F is an equivalence:

(ii) F has a 'quasi-inverse' $G:\mathbf{D}\to\mathbf{C}$ such that $FG\cong 1_{\mathbf{D}}$ and $GF\cong 1_{\mathbf{C}}$.

Proof. (i)$\Rightarrow$(ii). For each $\mathbf{D}$-object B, we define GB to be some $\mathbf{C}$-object A for which $FA\cong B$; and for each $\mathbf{D}$-object B, we choose an isomorphism $\varepsilon_B:FGB\cong B$. To define the action of G on the arrows of $\mathbf{D}$, let $B\xrightarrow{g}B'$ be a $\mathbf{D}$-arrow; then the square

$$\begin{array}{ccc} FGB & \xrightarrow{\varepsilon_B} & B \\ {\scriptstyle \varepsilon_{B'}^{-1}\circ g\circ\varepsilon_B}\downarrow & & \downarrow{\scriptstyle g} \\ FGB' & \xrightarrow{\varepsilon_{B'}} & B' \end{array}$$

commutes. Since F is full and faithful, there is a unique $\mathbf{C}$-arrow $f:GB\to GB'$ such that $Ff=\varepsilon_{B'}^{-1}\circ g\circ\varepsilon_B$. We put $Gg=f$. It is now easily verified that G is a functor and that $\varepsilon:FG\cong 1_{\mathbf{D}}$.

To get $\eta:1_{\mathbf{C}}\cong GF$, we observe that for each $\mathbf{C}$-object A, $\varepsilon_{F(A)}:FGFA\to FA$ is an isomorphism. So since F is full and faithful, there is a unique $\mathbf{C}$-isomorphism $\eta_A:A\cong GFA$ such that $F(\eta_A)=\varepsilon_{FA}^{-1}$. It is easy to verify that η is a natural isomorphism.

(ii)$\Rightarrow$(i). Suppose that F has a quasi-inverse $G:\mathbf{D}\to\mathbf{C}$ and let $\eta:FG\to 1_{\mathbf{D}}$, $\theta:GF\to 1_{\mathbf{C}}$ be natural isomorphisms. Clearly F is then dense since, for any $\mathbf{D}$-object B, $FGB\cong B$. It remains to show that F is full and faithful.

To establish the fidelity of F, let $A\overset{f}{\underset{f'}{\rightrightarrows}}A'$ in $\mathbf{C}$ and suppose that $Ff=Ff'$. Since θ is a natural transformation, we then have

$$f\circ\theta_A=\theta_{A'}\circ GFf=\theta_{A'}\circ GFf'=f'\circ\theta_A. \qquad (*)$$

Since θ_A is an isomorphism, it may be cancelled to yield $f=f'$, as required. Therefore F is faithful; replacing F by G in the preceding argument shows that G is also faithful.

To prove the fullness of F, let $FA\xrightarrow{g}FA'$ in $\mathbf{D}$ and let $f=\theta_{A'}\circ Gg\circ\theta_A^{-1}:A\to A'$ in $\mathbf{C}$. Then by (*)

$$GFf=\theta_{A'}^{-1}\circ f\circ\theta_A=Gg.$$

Since G is faithful, $Ff=g$ and so F is full. □

Note that a quasi-inverse of an equivalence is also an equivalence.

Examples

(i) Any category is equivalent to any of its skeletons.

(ii) A category **C** is equivalent to the degenerate category **1** iff all objects of **C** are terminal (or initial).

(iii) Any category **C** with a terminal object 1 is equivalent (in fact isomorphic) to the slice category **C**/1.

(iv) For any field F, the category of finite-dimensional vector spaces over F is equivalent to the category of F-matrices.

(v) $\mathbf{Set}^{\mathrm{op}}$ is equivalent to the category of complete atomic Boolean algebras.

(vi) $\mathbf{Bool}^{\mathrm{op}}$ is equivalent to the category of Boolean spaces.

(vii) Let **Comp Ab** be the category of compact abelian groups. The functor $\hom(-, R/Z)\colon \mathbf{Comp\ Ab}^{\mathrm{op}} \to \mathbf{Ab}$ which sends each compact group to its discrete character group is an equivalence (Pontryagin duality).

(viii) If **C*-alg** is the category of C^*-algebras with norm-decreasing homomorphisms, the functor $\hom(-, \mathbb{C})\colon \mathbf{C^*\text{-}Alg}^{\mathrm{op}} \to \mathbf{Comp\ Haus}$ that sends each C^*-algebra A to its carrier space, i.e. $\mathrm{Hom}(A, \mathbb{C})$ considered, as a subspace of $\mathbb{C}^A$, is an equivalence (Gelfand duality).

(ix) An equivalence between two partially ordered sets *qua* categories is an order-isomorphism.

Call a property P of categories *essentially categorical* if it is preserved under equivalence. It follows from 1.14 that P is essentially categorical iff it is preserved under isomorphism of categories and, for any skeleton $\hat{\mathbf{C}}$ of any category **C**,

$$\hat{\mathbf{C}} \text{ has } P \Leftrightarrow \mathbf{C} \text{ has } P.$$

Using this characterization, it is readily established that most of the properties of categories so far introduced (e.g., completeness, cocompleteness, possession of a terminal object, etc.) are essentially categorical.

Adjunctions

Let F be an equivalence with quasi-inverse G between categories **C**,**D** and let ε be a natural isomorphism $\varepsilon\colon FG \cong 1_{\mathbf{D}}$ given by (1.15). Since F is full and faithful, for any objects A of **C** and B of **D**, the map $f \mapsto Ff$ is a bijection between $\mathbf{C}(A, GB)$ and $\mathbf{D}(FA, FGB)$. Moreover, since $FGB \xrightarrow{\varepsilon_B} B$ is an isomorphism, the map $g \mapsto \varepsilon_B \circ g$ is a bijection between $\mathbf{D}(FA, FGB)$ and $\mathbf{D}(FA, B)$. Composing these two maps yields a bijection

$$\phi_{AB}\colon \mathbf{C}(A, GB) \cong \mathbf{D}(FA, B)$$

given by

(1.16) $$\phi_{AB}(h) = \varepsilon_B \circ Fh$$

for $h: A \to GB$.

The bijections ϕ_{AB} are *natural* in A and B in the sense that they form the components of a natural isomorphism ϕ when both sides of the bijection $\mathbf{C}(A, GB) \cong \mathbf{D}(FA, B)$ are regarded as functors of A and B. Naturality in A means that, for each $A \xrightarrow{f} A'$ in $\mathbf{C}$ the diagram

$$\begin{array}{ccc} \mathbf{C}(A', GB) & \xrightarrow{\phi_{A'B}} & \mathbf{D}(FA', B) \\ {\scriptstyle *f}\downarrow & & \downarrow{\scriptstyle *(Ff)} \\ \mathbf{C}(A, GB) & \xrightarrow{\phi_{AB}} & \mathbf{D}(FA, B) \end{array}$$

commutes, where $*f(h) = h \circ f$. That is, for any $A' \xrightarrow{h} GB$ we must have

(1.17) $$\phi_{AB}(h \circ f) = \phi_{A'B}(h) \circ Ff.$$

To verify this, we observe that, by (1.16),

$$\begin{aligned} \phi_{AB}(h \circ f) = \varepsilon_B \circ F(h \circ f) &= \varepsilon_B \circ Fh \circ Ff \\ &= \phi_{A'B}(h) \circ Ff. \end{aligned}$$

Similarly, naturality in B means that for any arrows $B' \xrightarrow{g} B$ in $\mathbf{D}$, $A \xrightarrow{h} GB'$ in $\mathbf{C}$, we have

(1.18) $$\phi_{AB}(Gg \circ h) = g \circ \phi_{AB'}(h).$$

To verify this, we use (1.16) and the naturality of ε to compute

$$\begin{aligned} \phi_{AB}(Gg \circ h) &= \varepsilon_B \circ FGg \circ Fh \\ &= g \circ \varepsilon_{B'} \circ Fh = g \circ \phi_{AB'}(h). \end{aligned}$$

This leads to the concept of an *adjunction* between $\mathbf{C}$ and $\mathbf{D}$. Suppose we are given functors $F: \mathbf{C} \to \mathbf{D}$ and $G: \mathbf{D} \to \mathbf{C}$, and for each pair of objects A of $\mathbf{C}$, B of $\mathbf{D}$ a bijection

(1.19) $$\phi_{AB}: \mathbf{C}(A, GB) \cong \mathbf{D}(FA, B)$$

for which (1.17) and (1.18) hold. Under these conditions we call the triple (F, G, ϕ) an *adjunction* between $\mathbf{C}$ and $\mathbf{D}$. F is called the *left adjoint* of G and G the *right adjoint* of F (with respect to the given adjunction). We shall also write

$$F \dashv G$$

to assert that F and G are the left and right adjoints of an adjunction, and also as a succinct notation for the adjunction itself.

Examples of adjunctions

C	**D**	F————————	$\vert G$
Set	**Grp**	free group functor	forgetful functor
Set	**Vect**	free vector space functor	forgetful functor
Set	**Top**	discrete space functor	forgetful functor
Top	**Comp Haus**	Stone-Čech compact-ification functor	insertion functor

Units and counits of adjunctions

Given an adjunction (F, G, ϕ) between **C** and **D**, let A be any **C**-object and put $B = FA$ in (1.19). Let $\eta_A : A \to GFA$ be such that $\phi_{A,FA}(\eta_A) = 1_{FA}$. Given a **D**-object B and an arrow $FA \xrightarrow{k} B$, consider

$$\hat{k} = Gk \circ \eta_A : A \to GB$$

Putting $B' = FA$ in (1.18) gives

$$\begin{aligned} \phi_{AB}(\hat{k}) &= \phi_{AB}(Gk \circ \eta_A) = k \circ \phi_{A,FA}(\eta_A) \\ &= k \circ 1_{FA} = k. \end{aligned}$$

Thus the map $k \mapsto \hat{k}$ is inverse to the bijection ϕ_{AB}. It follows that, *for every $A \xrightarrow{h} GB$, there is a unique $FA \xrightarrow{k} B$ such that the diagram*

$$\begin{array}{ccc} A & \xrightarrow{\eta_A} & GFA \\ & {\scriptstyle h}\searrow & \downarrow{\scriptstyle Gk} \\ & & GB \end{array}$$

commutes.

Moreover, the η_A form the components of a *natural transformation*

$$\eta : 1_{\mathbf{C}} \to GF.$$

That is, for each $A \xrightarrow{f} A'$ the diagram

$$\begin{array}{ccc} A & \xrightarrow{\eta_A} & GFA \\ {\scriptstyle f}\downarrow & & \downarrow{\scriptstyle GFf} \\ A' & \xrightarrow{\eta_{A'}} & GFA' \end{array}$$

commutes. To see this, take $B = FA'$ in (1.17) and obtain

$$\begin{aligned}\phi_{A,FA'}(\eta_{A'} \circ f) &= \phi_{A',FA'}(\eta_{A'}) \circ Ff \\ &= 1_{FA'} \circ Ff \\ &= Ff.\end{aligned}$$

Taking $B = FA'$, $B' = FA$ in (1.18) gives

$$\begin{aligned}\phi_{A,FA'}(GFf \circ \eta_A) &= Ff \circ \phi_{A,FA}(\eta_A) \\ &= Ff \circ 1_{FA} \\ &= Ff.\end{aligned}$$

Hence $\phi_{A,FA'}(\eta_{A'} \circ f) = \phi_{A,FA'}(GFf \circ \eta_A)$ so that, since $\phi_{A,FA'}$ is a bijection, $\eta_{A'} \circ f = GFf \circ \eta_A$ and we are done.

The natural transformation $\eta : 1_{\mathbf{C}} \to GF$ is called the *unit* of the adjunction (F, G, ϕ).

Dually, if we define, for each **D**-object B,

$$\varepsilon_B = \phi_{GB,B}(1_{GB}) : FGB \to B,$$

then, using (1.17), we have, for any $A \xrightarrow{h} GB$,

$$\varepsilon_B \circ Fh = \phi_{GB,B}(1_{GB}) \circ Fh = \phi_{AB}(1_{GB} \circ h) = \phi_{AB}(h). \tag{*}$$

It follows that, *for any $FA \xrightarrow{k} B$, there is a unique $A \xrightarrow{h} GB$ such that the diagram*

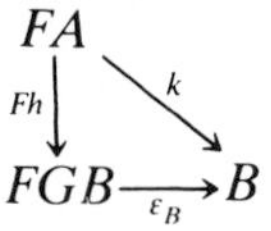

commutes.

Moreover, the ε_B form the components of a natural transformation

$$\varepsilon : FG \to 1_{\mathbf{D}}.$$

To verify this, using (*) and (1.18) we compute, for any $B \xrightarrow{g} B'$,

$$\begin{aligned}\varepsilon_{B'} \circ FGg &= \phi_{GB,B'}(Gg) \\ &= \phi_{GB,B'}(Gg \circ 1_{GB}) \\ &= g \circ \phi_{GB,B}(1_{GB}) \\ &= g \circ \varepsilon_B.\end{aligned}$$

The natural transformation ε is called the *counit* of the adjunction (F, G, ϕ).

Let us summarize. Let (F, G, ϕ) be an adjunction between **C** and **D** with unit η and counit ε. For any pair of objects A of **C**, B of **D**, the map

$$k \mapsto \hat{k} = Gk \circ \eta_A$$

is the bijection ϕ_{AB}^{-1} between $\mathbf{D}(FA,B)$ and $\mathbf{C}(A,GB)$. For $k \in \mathbf{D}(FA,B)$ the arrow $\hat{k} \in \mathbf{C}(A,GB)$ is called the (left) *transpose of k across the adjunction* $F \dashv G$. Dually, the map

$$h \mapsto \tilde{h} = \varepsilon_B \circ Fh$$

is the bijection ϕ_{AB} between $\mathbf{C}(A,GB)$ and $\mathbf{D}(FA,B)$. For $h \in \mathbf{C}(A,GB)$, the arrow $\tilde{h} \in \mathbf{D}(FA,B)$ is called the (right) *transpose of h across the adjunction* $F \dashv G$. We note that, for any $A \xrightarrow{h} GB$, $FA \xrightarrow{k} B$, we have

$$\tilde{\eta}_A = 1_{FA}, \qquad \hat{\varepsilon}_B = 1_{GB}$$

(1.20) $$\hat{1}_{FA} = \eta_A, \qquad \tilde{1}_{GB} = \varepsilon_B$$

$$(\tilde{h})\hat{\ } = h, \qquad (\hat{k})\tilde{\ } = k;$$

(1.21) $$\varepsilon_{FA} \circ F\eta_A = 1_{FA}, \qquad G\varepsilon_B \circ \eta_{GB} = 1_{GB};$$

(1.22) $$(\eta_{A'} \circ f)\tilde{\ } = Ff, \qquad (g \circ \varepsilon_B)\hat{\ } = Gg,$$

for $A \xrightarrow{f} A'$ in $\mathbf{C}$, $B \xrightarrow{g} B'$ in $\mathbf{D}$;

(1.23) $$(h \circ v)\tilde{\ } = \tilde{h} \circ Fv, \qquad (k \circ Fv)\hat{\ } = \hat{k} \circ v,$$

for $C \xrightarrow{v} A$ in $\mathbf{C}$.

Equations (1.20) are clear. To prove (1.21), we observe that, by (1.20),

$$\varepsilon_{FA} \circ F\eta_A = (\eta_A)\tilde{\ } = 1_{FA}$$

and

$$G\varepsilon_B \circ \eta_{GB} = (\varepsilon_B)\hat{\ } = 1_{GB}.$$

To prove (1.22), we apply (1.21) to obtain

$$\begin{aligned}(\eta_{A'} \circ f)\tilde{\ } &= \varepsilon_{FA'} \circ F\eta_{A'} \circ Ff \\ &= 1_{FA'} \circ Ff = Ff\end{aligned}$$

and

$$\begin{aligned}(g \circ \varepsilon_B)\hat{\ } &= Gg \circ G\varepsilon_B \circ \eta_{GB} \\ &= Gg \circ 1_{GB} = Gg.\end{aligned}$$

Finally, to prove (1.23), we calculate

$$(h \circ v)\tilde{\ } = \varepsilon_B \circ Fh \circ Fv = \tilde{h} \circ Fv,$$

which is the first equation in (1.23). It follows that

$$h \circ v = ((h \circ v)\tilde{\ })\hat{\ } = (\tilde{h} \circ Fv)\hat{\ }.$$

Putting $h = \hat{k}$, so that $\tilde{h} = k$, we get the second equation of (1.23).

Equations (1.22) are known as the *triangular identities*, since they assert the commutativity of the triangles

$$\begin{array}{ccc} FA & \xrightarrow{F\eta_A} & FGFA \\ & {}_{1_{FA}}\searrow & \downarrow{\scriptstyle \varepsilon_{FA}} \\ & & FA \end{array} \qquad \begin{array}{ccc} GB & \xrightarrow{\eta_{GB}} & GFGB \\ & {}_{1_{GB}}\searrow & \downarrow{\scriptstyle G\varepsilon_B} \\ & & GB \end{array}$$

1.24 EXERCISE. Show that the left adjoint in an adjunction is an equivalence iff both the unit and counit of the adjunction are natural isomorphisms.

Freedom and cofreedom

Units and counits of adjunctions are instances of more general concepts. Let $G:\mathbf{D}\to\mathbf{C}$ be a functor and let A be a $\mathbf{C}$-object. A pair (B,e) consisting of a $\mathbf{D}$-object B and a $\mathbf{C}$-arrow $A \xrightarrow{e} GB$ is called *G-free over A* if for any $\mathbf{C}$-arrow $A \xrightarrow{f} GC$ there is a unique $\mathbf{D}$-arrow $B \xrightarrow{g} C$ such that the diagram

$$\begin{array}{ccc} A & \xrightarrow{e} & GB \\ & {}_{f}\searrow & \downarrow{\scriptstyle Gg} \\ & & GC \end{array}$$

commutes. (Such a pair is also called a *universal arrow from A to G.*)

Dually, given a functor $F:\mathbf{C}\to\mathbf{D}$ and a $\mathbf{D}$-object B, a pair (A,i) consisting of a $\mathbf{C}$-object A and a $\mathbf{D}$-arrow $FA \xrightarrow{i} B$ is called *F-cofree over B* if for any arrow $FC \xrightarrow{g} B$ there is a unique $C \xrightarrow{f} A$ such that the diagram

$$\begin{array}{ccc} FC & & \\ {\scriptstyle Ff}\downarrow & \searrow^{g} & \\ FA & \xrightarrow{i} & B \end{array}$$

commutes. (Such a pair is also called a *universal arrow from F to B.*)

If $F \dashv G$ and η,ε are the unit and counit of the adjunction, then, for any $\mathbf{C}$-object A, the pair (FA,η_A) is G-free over A and, for any $\mathbf{D}$-object B, the pair (GB, ε_B) is F-cofree over B.

Conversely, adjunctions can be described in terms of these concepts. For suppose we are given functors $F:\mathbf{C}\to\mathbf{D}$, $G:\mathbf{D}\to\mathbf{C}$ and, say, for each $\mathbf{C}$-object A an arrow $A \xrightarrow{\eta_A} GFA$ such that (FA, η_A) is G-free over A and the η_A form the components of a natural transformation $\eta: 1_{\mathbf{C}}\to GF$. If we now define, for each pair of objects A of $\mathbf{C}$, B of $\mathbf{D}$, a map

$$\phi_{AB}:\mathbf{C}(A,GB)\to\mathbf{D}(FA,B)$$

by setting, for each $h \in \mathbf{C}(A, GB)$,

$$\phi_{AB}(h) = \text{unique } k \in \mathbf{D}(FA, B) \text{ such that } h = Gk \circ \eta_A,$$

then clearly ϕ_{AB} is a bijection $\mathbf{C}(A, GB) \cong \mathbf{D}(FA, B)$. Moreover, ϕ_{AB} satisfies (1.17) and (1.18). Let us verify (1.17). Given $A \xrightarrow{f} A'$, $A' \xrightarrow{h'} GB$, we have

$$\phi_{A'B}(h') = \text{unique } FA' \xrightarrow{k'} B \text{ such that } h' = Gk' \circ \eta_{A'} \qquad (*)$$

and so

$$\phi_{A'B}(h') \circ Ff = k' \circ Ff. \qquad (**)$$

Also

$$\begin{aligned} G(k' \circ Ff) \circ \eta_A &= Gk' \circ GFf \circ \eta_A \\ &= Gk' \circ \eta_{A'} \circ f \qquad (\eta \text{ being a natural transformation}) \\ &= h' \circ f \qquad (\text{by } (*)). \end{aligned}$$

But by definition

$$\phi_{AB}(h' \circ f) = \text{unique } FA \xrightarrow{k} B \text{ such that } h' \circ f = Gk \circ \eta_A.$$

Comparing this with the previous equation gives

$$\phi_{AB}(h' \circ f) = k' \circ Ff,$$

which, according to (**), implies

$$\phi_{AB}(h' \circ f) = \phi_{A'B}(h') \circ Ff,$$

i.e. (1.17). Equation (1.18) is verified similarly.

We have thus shown that (F, G, ϕ) is an adjunction between $\mathbf{C}$ and $\mathbf{D}$. Moreover, it is evident that the η_A form the components of the unit of the adjunction.

Dually, suppose we are given functors $F: \mathbf{C} \to \mathbf{D}$, $G: \mathbf{D} \to \mathbf{C}$ and for each $\mathbf{D}$-object B an arrow $FGB \xrightarrow{\varepsilon_B} B$ such that (GB, ε_B) is cofree over B and the ε_B form the components of a natural transformation $\varepsilon: FG \to 1_{\mathbf{D}}$. These data determine an adjunction $F \dashv G$ of which the ε_B form the components of the counit.

We now consider the situation in which we are given just *one* of the functors F, G. Thus, suppose we are given a functor $G: \mathbf{D} \to \mathbf{C}$ and, for each $\mathbf{C}$-object A a $\mathbf{D}$-object A^* together with an arrow $A \xrightarrow{\eta_A} GA^*$ such that (A^*, η_A) is G-free over A. Then there is a (unique) functor $F: \mathbf{C} \to \mathbf{D}$ such that $F \dashv G$, $FA = A^*$ for all $\mathbf{C}$-objects A, and the η_A form the components of the unit of the adjunction. To see this, given a $\mathbf{C}$-arrow $A \xrightarrow{f} A'$, let $f^*: A^* \to (A')^*$ be the unique $\mathbf{D}$-arrow such that the diagram

$$
\begin{array}{ccc}
A & \xrightarrow{\eta_A} & GA^* \\
& {}_{\eta_{A'}\circ f}\searrow & \big\downarrow{\scriptstyle Gf^*} \\
& & G((A')^*)
\end{array}
\qquad (*)
$$

commutes. If we define $F:\mathbf{C}\rightarrow\mathbf{D}$ by $FA = A^*$, $Ff = f^*$, then F is a functor. For the diagram (*) may now be written

$$
\begin{array}{ccc}
A & \xrightarrow{\eta_A} & GFA \\
{\scriptstyle f}\big\downarrow & & \big\downarrow{\scriptstyle GFf} \\
A' & \xrightarrow{\eta_A} & GFA'
\end{array}
\qquad (**)
$$

where Ff is the *unique* arrow making the diagram commute. Taking $f = 1_A$, since the diagram

$$
\begin{array}{ccc}
A & \xrightarrow{\eta_A} & GFA \\
{\scriptstyle 1_A}\big\downarrow & & \big\downarrow{\scriptstyle 1_{GFA} = G1_{FA}} \\
A & \xrightarrow{\eta_A} & GFA
\end{array}
$$

commutes, uniqueness gives $F1_A = 1_{FA}$. Moreover, for any $A' \xrightarrow{g} A''$, the diagram

$$
\begin{array}{ccc}
A & \xrightarrow{\eta_A} & GFA \\
{\scriptstyle g\circ f}\big\downarrow & & \big\downarrow{\scriptstyle G(Fg\circ Ff)} \\
A'' & \xrightarrow{\eta_{A''}} & GFA''
\end{array}
$$

commutes, so uniqueness implies $F(g \circ f) = Fg \circ Ff$. Therefore F is a functor. Finally, diagram (**) shows that the η_A form the components of a natural transformation $\eta : 1_{\mathbf{C}} \rightarrow GF$. Accordingly, we have verified the conditions just shown to be sufficient for F to be a left adjoint for G, and for η to be the unit of the adjunction.

Dually, suppose we are given a functor $F:\mathbf{C}\rightarrow\mathbf{D}$ and, for each **D**-object B a **C**-object B^+ together with an arrow $FB^+ \xrightarrow{\varepsilon_B} B$ such that (B^+, ε_B) is F-cofree over B. Then there is a (unique) functor $G:\mathbf{D}\rightarrow\mathbf{C}$ such that $F \dashv G$, $GB = B^+$ for all **D**-objects B and the ε_A form the components of the counit of the adjunction.

REMARK. Adjunctions (F, G, ϕ) are often presented and analysed in a 'proof-theoretic' notation: we write

$$
\frac{FA \rightarrow B \text{ (in } \mathbf{D})}{A \rightarrow GB \text{ (in } \mathbf{C})}
$$

to depict the bijective correspondence ϕ between the 'numerator' arrows $FA \rightarrow B$ and the 'denominator' arrows $A \rightarrow GB$. The naturality of this bijection is then succinctly expressed in this notation: for every $A' \rightarrow$

$A, B \to B'$ we have

$$\frac{FA' \to FA \to B \to B'}{A' \to A \to GB \to GB'}.$$

The unit and counit of the adjunction are easily obtained by considering the appropriate identity arrows:

$$\frac{FA \xrightarrow{1_{FA}} FA}{A \xrightarrow{\eta_A} GFA} \qquad \frac{FGB \xrightarrow{\varepsilon_B} B}{GB \xrightarrow{1_{GB}} GB}.$$

Uniqueness of adjoints

1.25 PROPOSITION. A functor has at most one left adjoint (and dually, at most one right adjoint) up to natural isomorphism.

Proof. Let $G : \mathbf{D} \to \mathbf{C}$ and let $F, F' : \mathbf{C} \to \mathbf{D}$ be left adjoints of G. Let η, η' be the units of the adjunctions $F \dashv G$, $F' \dashv G$ respectively, and let $\hat{}$ denote transposition across $F \dashv G$. Then for each $\mathbf{C}$-object A there are unique arrows $\theta_A : FA \to F'A$, $\theta'_A : F'A \to FA$ such that

$$\eta'_A = G(\theta_A) \circ \eta_A, \qquad \eta_A = G(\theta'_A) \circ \eta'_A. \tag{1}$$

We claim that *each θ_A is an isomorphism*. This is because

$$\begin{aligned} G(\theta'_A \circ \theta_A) \circ \eta_A &= G(\theta'_A) \circ G(\theta_A) \circ \eta_A \\ &= G(\theta'_A) \circ \eta'_A \\ &= \eta_A \qquad \text{(by (1))}. \end{aligned}$$

In other words,

$$(\theta'_A \circ \theta_A)\hat{} = (1_{FA})\hat{}.$$

Hence $\theta'_A \circ \theta_A = 1_{FA}$. Similarly $\theta_A \circ \theta'_A = 1_{F'A}$. Therefore θ_A is an isomorphism.

Each θ_A is a natural transformation. We have to show that

$$\begin{array}{ccc} FA & \xrightarrow{\theta_A} & F'A \\ {\scriptstyle Ff}\downarrow & & \downarrow{\scriptstyle F'f} \\ FB & \xrightarrow{\theta} & F'B \end{array} \tag{2}$$

commutes for every $A \xrightarrow{f} B$ in $\mathbf{C}$. We have

$$\begin{aligned} (F'f \circ \theta_A)\hat{} &= GF'f \circ G(\theta_A) \circ \eta_A \\ &= GF'f \circ \eta'_A \qquad \text{(by (1))} \\ &= \eta'_B \circ f, \end{aligned}$$

since η' is a natural transformation. Also,

$$\begin{aligned}(\theta_B \circ Ff)\hat{} &= G(\theta_B) \circ GFf \circ \eta_A \\ &= G(\theta_B) \circ \eta_B \circ f \qquad \text{(since } \eta \text{ is a natural transformation)} \\ &= \eta'_B \circ f \qquad \text{(by (1)).}\end{aligned}$$

Therefore $F'f \circ \theta_A$ and $\theta_{A'} \circ Ff$ have the same transpose across $F \dashv G$, and so are equal. Thus diagram (2) commutes and therefore each θ_A is a natural transformation.

It follows that $\theta = (\theta_A)_{A \in \mathrm{Ob}(\mathbf{C})}$ is a natural isomorphism between F and F'. □

Preservation of limits and colimits

Let $\mathbf{J}$ be a diagram with vertices $\{J_i : i \in I\}$ in a category $\mathbf{C}$. Given a cone

$$K = \{J_i \xrightarrow{f_i} A : i \in I\}$$

under $\mathbf{J}$, and a functor $F : \mathbf{C} \to \mathbf{D}$ to a category $\mathbf{D}$, let

$$FK = \{FJ_i \xrightarrow{Ff_i} FA : i \in I\}$$

and let $F\mathbf{J}$ be the diagram in $\mathbf{D}$ with vertices $\{FJ_i : i \in I\}$ and arrows all Fd with d an arrow of $\mathbf{J}$. Since any diagram

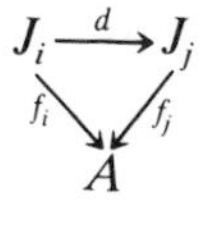

commutes, so does the corresponding diagram

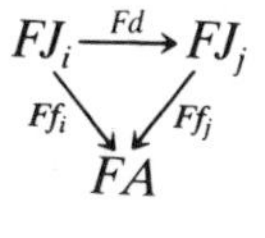

In other words, FK is a cone under $F\mathbf{J}$.

We say that F *preserves the colimit* of $\mathbf{J}$ if whenever K is a colimit for $\mathbf{J}$ in $\mathbf{C}$, FK is a colimit for $F\mathbf{J}$ in $\mathbf{D}$. And we say that F *preserves* (finite) *colimits* if F preserves the colimit of every (finite) diagram in $\mathbf{C}$. (Dually, of course, for limits.) A functor that preserves finite colimits (limits) is called *right* (*left*) *exact.* A functor which is both left and right exact is called *exact.*

We now give some sufficient conditions for a functor to preserve limits or colimits.

1.26 PROPOSITION. (i) Suppose that **C** is complete. Then $F:\mathbf{C}\to\mathbf{D}$ preserves limits iff it preserves products and equalizers.

(ii) Suppose that **C** is finitely complete. Then $F:\mathbf{C}\to\mathbf{D}$ is left exact iff it preserves binary products, equalizers, and terminal objects.

Proof. Both parts follow from the proof of (1.5), in which any limit is constructed from products and equalizers. □

1.27 THEOREM. Any functor $F:\mathbf{C}\to\mathbf{D}$ with a right adjoint preserves colimits. Dually, any functor with a left adjoint preserves limits.

Proof. Let $G:\mathbf{D}\to\mathbf{C}$ be a right adjoint for F, and let ε be the counit of the adjunction $F\dashv G$. Let **J** be a diagram in **C** with vertices $\{J_i : i\in I\}$ and suppose **J** has a colimit $\{J_i \xrightarrow{f_i} A : i\in I\}$ in **C**. We have to show that the cone $\{FJ_i \xrightarrow{Ff_i} FA : i\in I\}$ is a colimit for $F\mathbf{J}$. That is, given any cone under $F\mathbf{J}$,

$$K=\{FJ_i \xrightarrow{g_i} B : i\in I\}\,,$$

we have to show that there is a *unique* $k:FA\to B$ such that

$$g_i = k\circ Ff_i \qquad (*)$$

for all $i\in I$.

Let $\check{\ },\hat{\ }$ denote transposition across $F\dashv G$ as usual. Since K is a cone under $F\mathbf{J}$, we have

$$g_j\circ Fd = g_i$$

for any $i,j\in I$ and $J_i\xrightarrow{d}J_j$ in **J**. Applying (1.23) to this, we get

$$\hat{g}_i=(g_j\circ Fd)\hat{\ }=\hat{g}_j\circ d,$$

so that $\{J_i \xrightarrow{\hat{g}_i} GB : i\in I\}$ is a cone under **J**. Since $\{J_i \xrightarrow{f_i} A : i\in I\}$ is a colimit diagram, there is a unique $A\xrightarrow{h}GB$ such that

$$\hat{g}_i = h\circ f_i. \qquad (**)$$

Putting $k=\check{h}:FA\to B$, we have

$$\begin{aligned} k\circ Ff_i=\check{h}\circ Ff_i &= (h\circ f_i)\check{\ } && \text{(by (1.23))}\\ &=(\hat{g}_i)\check{\ } && \text{(by (**))}\\ &= g_i. && \text{(by (1.20))}\end{aligned}$$

Therefore k satisfies (*).

To establish the uniqueness of k, we note that, if k satisfies (*), then by (1.23)

$$\hat{g}_i=(k\circ Ff_i)\hat{\ }=\hat{k}\circ f_i,$$

that is, $\hat{k}$ satisfies (**). Therefore $\hat{k}$ is uniquely determined, and so, accordingly, is k. □

1.28 THEOREM. For any small category $\mathbf{C}$, the Yoneda embedding $Y:\mathbf{C}\to\mathbf{Set}^{\mathbf{C}^{\mathrm{op}}}$ preserves limits.

Proof. Let $\mathbf{D}$ be a diagram in $\mathbf{C}$ with vertices $\{D_i : i\in I\}$ and suppose that $\{A \xrightarrow{f_i} D_i : i\in I\}$ is a limit diagram for $\mathbf{D}$. We have to show that $\{YA \xrightarrow{Yf_i} YD_i : i\in I\}$ is a limit diagram for $Y\mathbf{D}$.

Let $\{F \xrightarrow{\eta_i} YD_i : i\in I\}$ be a cone over $Y\mathbf{D}$ in $\mathbf{Set}^{\mathbf{C}^{\mathrm{op}}}$. By (1.10) there is a diagram $\mathbf{E}$ in $\mathbf{Set}^{\mathbf{C}^{\mathrm{op}}}$ with vertices $\{YB_j : j\in J\}$ and arrows $\{YB_j \xrightarrow{\beta_j} F : j\in J\}$ such that $F=\operatorname{colim}\mathbf{E}$. Consider the composition

$$YB_j \xrightarrow{\beta_j} F \xrightarrow{\eta_i} YD_i.$$

Then, since Y is full, for each $i\in I$, $j\in J$ there is an arrow $B_j \xrightarrow{f_{ij}} D_i$ such that

$$Yf_{ij} = \eta_i \circ \beta_j \tag{1}$$

Since $A=\lim\mathbf{D}$ there is for each $j\in J$ a unique $B_j \xrightarrow{g_j} A$ such that

$$f_i \circ g_j = f_{ij} \tag{2}$$

for all $i\in I$. It is now easy to verify that $\{YB_j \xrightarrow{Yg_j} YA : j\in J\}$ is a cone under $\mathbf{E}$. Since $F=\operatorname{colim}\mathbf{E}$ there is a unique $F \xrightarrow{\alpha} YA$ such that

$$\alpha \circ \beta_j = Yg_j \tag{3}$$

for all $j\in J$. It follows from (1) and (2) that for any $i\in I$, $j\in J$,

$$\begin{aligned}\eta_i \circ \beta_j &= Yf_{ij} = Y(f_i\circ g_j)\\ &= Y(f_i)\circ Y(g_j)\\ &= Yf_i\circ\alpha\circ\beta_j.\end{aligned} \tag{4}$$

Now $\{YB_j \xrightarrow{\eta_i\circ\beta_j} YD_i : j\in J\}$ is a cone under $\mathbf{E}$ for each $i\in I$, so by the uniqueness property of colimits and (4) we have

$$\eta_i = Yf_i \circ \alpha \tag{5}$$

for all $i\in I$.

To show that $\{YA \xrightarrow{Yf_i} YD_i : i\in I\}$ is a limit diagram for $Y\mathbf{D}$, it remains to show that the arrow $F \xrightarrow{\alpha} YA$ satisfying (5) is unique. Suppose, accordingly, that $F \xrightarrow{\beta} YA$ satisfies $\eta_i = Yf_i\circ\beta$ for all $i\in I$. Since Y is full, for each $j\in J$ there is $B_j \xrightarrow{h_j} A$ such that $\beta\circ\beta_j = Yh_j$. Then, using (1),

$$Yf_{ij} = \eta_i\circ\beta_j = Yf_i\circ\beta\circ\beta_j = Yf_i\circ Yh_j = Y(f_i\circ h_j).$$

Since Y is faithful, it follows that $f_{ij} = f_i\circ h_j$ for all $i\in I$. Now the uniqueness of g_j satisfying (2) implies that $h_j = g_j$, so that

$$\beta\circ\beta_j = Yh_j = Yg_j$$

for all $j \in J$. Since α is the unique arrow satisfying (3), we conclude that $\alpha = \beta$. The proof is complete. $\square$

Cartesian closed categories

Define the *diagonal functor* $\Delta; \mathbf{C} \to \mathbf{C} \times \mathbf{C}$ on a category $\mathbf{C}$ by $\Delta(A) = (A,A)$ and $\Delta(f) = (f,f)$ for $A \xrightarrow{f} A'$. It is easy to see that if $\mathbf{C}$ has binary products, then the Cartesian product bifunctor $\cdot \times \cdot : \mathbf{C} \times \mathbf{C} \to \mathbf{C}$ is right adjoint to Δ. Thus, $\mathbf{C}$ has binary products if and only if Δ has a right adjoint.

The concept of *exponentiation* of objects in a category is also introduced via adjunctions. Let $\mathbf{C}$ be a category with binary products. Given a $\mathbf{C}$-object A, define the functor $A \times \cdot : \mathbf{C} \to \mathbf{C}$ by

$$(A \times \cdot)(B) = A \times B, \qquad (A \times \cdot)(f) = 1_A \times f.$$

Suppose that $A \times \cdot$ has a right adjoint $G : \mathbf{C} \to \mathbf{C}$; let ε be the counit of the adjunction $A \times \cdot \dashv G$. Then for any arrow $A \times B \xrightarrow{f} C$ there is a unique arrow $B \xrightarrow{\hat{f}} GC$ (the transpose of f across the adjunction) such that the diagram

$$\begin{array}{ccc} A \times B & & \\ {\scriptstyle 1_A \times \hat{f}} \downarrow & \searrow^{f} & \\ A \times GC & \xrightarrow{\varepsilon_C} & C \end{array}$$

commutes. We usually write C^A for GC and $\mathrm{ev}_{C,A}$ for $\varepsilon_C : C^A$ is called the *exponential* of C by A (or, suggestively, 'C to the A') and $\mathrm{ev}_{C,A}$ the *evaluation arrow*. We say that $\mathbf{C}$ is *Cartesian closed* if it has a terminal object, binary products and the functor $A \times \cdot$ has a right adjoint for any $\mathbf{C}$-object A. Thus $\mathbf{C}$ is Cartesian closed provided it has a terminal object, binary products and, for any $\mathbf{C}$-objects A, C there is a $\mathbf{C}$-object C^A and an arrow $A \times C^A \xrightarrow{\mathrm{ev}_{C,A}} C$ such that, for any arrow $A \times B \xrightarrow{f} C$, there is a unique arrow $B \xrightarrow{\hat{f}} C^A$ such that the diagram

$$\begin{array}{ccc} A \times B & & \\ {\scriptstyle 1_A \times \hat{f}} \downarrow & \searrow^{f} & \\ A \times C^A & \xrightarrow{\mathrm{ev}_{C,A}} & C \end{array}$$

commutes. $\hat{f}$ is called the *exponential transpose* of f.

In a Cartesian closed category we thus have, for any objects A, B, C, a natural bijection $(A \times B, C) \cong (B, C^A)$ or, more graphically

$$\frac{A \times B \to C}{B \to C^A}.$$

It is readily shown that the property of being Cartesian closed is essentially categorical.

The standard example of a Cartesian closed category is **Set**, in which the exponential C^A is the set of all maps from A to C and $\mathrm{ev}_{C,A}$ is the evaluation map

$$\mathrm{ev}_{C,A}:(a,g)\mapsto g(a)$$

for $a \in A$, $g \in C^A$.

Our next proposition assembles some basic facts about Cartesian closed categories.

1.29 PROPOSITION. Let $\mathbf{C}$ be a Cartesian closed category with an initial object 0. Then, for any objects A, B, C of $\mathbf{C}$,

(i)
$$A \times 0 \cong 0$$
$$A \times (B + C) \cong (A \times B) + (A \times C)$$
$$A \times \coprod_{i \in I} B_i \cong \coprod_{i \in I} A \times B_i.$$

(ii)
$$1^A \cong 1$$
$$(B \times C)^A \cong B^A \times C^A$$

(iii)
$$A^0 \cong 1$$
$$A^1 \cong A$$

(iv) if there exists an arrow $A \to 0$, then $A \cong 0$.

(v) there exists an arrow $1 \to 0$ iff $\mathbf{C}$ is equivalent to the degenerate category **1**.

Proof. (i) The functor $A \times \cdot$ has a right adjoint (namely $\cdot^A$); hence it preserves colimits and, in particular, initial objects and coproducts.

(ii) The functor $\cdot^A$ has a left adjoint (namely $A \times \cdot$); hence it preserves limits and, in particular, terminal objects and products.

(iii) For any object B, we have a natural bijection

$$\frac{B \to A^0}{0 \cong B \times 0 \to A}$$

But there is a unique arrow $0 \to A$, so for any B there is a unique arrow $B \to A^0$. In other words, A^0 is a terminal object, hence $\cong 1$.

To establish the second isomorphism, we note that, since $1 \times B \cong B$, there are natural bijections

$$(B,A^1) \cong (1 \times B, A) \cong (B,A).$$

Therefore both $\cdot^1$ and $1_{\mathbf{C}}$ are right adjoint to $1 \times \cdot$. Since adjoints are unique up to isomorphism, $A^1 \cong A$.

(iv) Suppose $A \xrightarrow{f} 0$; then

$$A \xrightarrow{\langle 1_A, f\rangle} A \times 0 \xrightarrow{\pi_1} A = 1_A.$$

Now, by (i), $A \times 0$ is an initial object, so $1_{A\times 0}$ is the sole arrow $A \times 0 \to A \times 0$. Therefore

$$A \times 0 \xrightarrow{\pi_1} A \xrightarrow{\langle 1_A, f\rangle} A \times 0 = 1_{A\times 0}.$$

Thus $\langle 1_A, f\rangle$ is an isomorphism $A \cong A \times 0$. Since, by (i), $A \times 0 \cong 0$, (iv) follows.

(v) If there exists an arrow $1 \to 0$, then for any object A of $\mathbf{C}$ we have a composition $A \to 1 \to 0$. By (iv), $A \cong 0$, so that $\mathbf{C}$ has a skeleton $\{0, 1_0\} \cong \mathbf{1}$. The converse is trivial. □

We note as an immediate consequence of this proposition that **Grp** is not Cartesian closed. For we have observed that $0 \cong 1$ in **Grp**, but evidently **Grp** is not degenerate.

Now consider a partially ordered set $(P, \leqslant)$. What does it mean to say that $(P, \leqslant)$, considered as a category, is Cartesian closed? First, it must have a terminal object, i.e. a *largest element* 1, and binary products, i.e. an *infimum* $p \wedge q$ for each pair of elements p, q. And finally each pair of elements p, q must have an exponential p^q which, in view of its definition, must satisfy

$$x \leqslant p^q \quad \text{iff} \quad x \wedge q \leqslant p$$

for all $x \in P$. We usually write $q \Rightarrow p$ for p^q, so we have

$$x \leqslant q \Rightarrow p \text{ iff } x \wedge q \leqslant p. \tag{*}$$

That is, $q \Rightarrow p$ is the *largest* element x of P such that $x \wedge q \leqslant p$.

A partially ordered set which, as a category, in addition to being Cartesian closed, also has an initial object and binary coproducts, is called a *Heyting algebra*. To say that a partially ordered set $(P, \leqslant)$ is a Heyting algebra is thus to assert that it is a lattice with least and largest elements together with an operation $\Rightarrow : P \times P \to P$ satisfying (*).

For each element p of a Heyting algebra P the element $p \Rightarrow 0$ is called the *pseudocomplement* of p and is denoted by p^*: clearly p^* is the largest element x of P for which $p \wedge x = 0$. The Heyting algebra P is called a *Boolean algebra* if $p \vee p^* = 1$ for all $p \in P$ (where 1 is the largest element of P); this is easily seen to be equivalent to the condition: $p^{**} = p$ for all $p \in P$.

If a Heyting algebra P is *complete* as a partially ordered set, then, since the map $x \mapsto x \wedge p : P \to P$ for fixed $p \in P$ has a right adjoint, it must preserve colimits. Therefore, in particular, writing $\bigvee X$ for the supremum of any subset X of P, we must have, for any $p \in P$, $X \subseteq P$,

$$\bigvee \{p \wedge x : x \in X\} = p \wedge \bigvee X. \tag{**}$$

Conversely, it is not hard to show that any complete lattice satisfying (**) is a Heyting algebra in which

$$p \Rightarrow q = \bigvee \{x \in P : p \wedge x \leqslant q\}.$$

A complete Heyting algebra is also called a *locale.* The standard example of a locale (and the source of the name) is a *topology* on a set, i.e., a family of subsets closed under arbitrary unions and finite intersections.

Reflective subcategories

When is a right adjoint full or faithful? The answer is supplied in the following theorem.

1.30 THEOREM. Let $F \dashv G$ be an adjunction with counit ε between categories $\mathbf{C}$, $\mathbf{D}$. Then:

(i) G is faithful iff every ε_B is epic;
(ii) G is full iff every ε_B is split monic;
(iii) G is full and faithful iff every ε_B is an isomorphism.

Proof. Let $\hat{\ }$, $\check{\ }$ denote right and left transpose across $F \dashv G$.
(i) We have the following equivalences:

$$\begin{aligned} G \text{ is faithful} &\Leftrightarrow [\text{for all } B, B' \in \mathrm{Ob}(\mathbf{D}),\ B \underset{g'}{\overset{g}{\rightrightarrows}} B',\ Gg = Gg' \Rightarrow g = g'] \\ &\Leftrightarrow [\text{for all } B, B' \in \mathrm{Ob}(\mathbf{D}),\ B \underset{g'}{\overset{g}{\rightrightarrows}} B', \\ &\qquad (g \circ \varepsilon_B)\hat{} = (g' \circ \varepsilon_B)\hat{} \Rightarrow g = g'] \text{ (by (1.22))} \\ &\Leftrightarrow [\text{for all } B, B' \in \mathrm{Ob}(\mathbf{D}),\ B \underset{g}{\overset{g}{\rightrightarrows}} B',\ g \circ \varepsilon_B = g' \circ \varepsilon_B \Rightarrow g = g'] \\ &\Leftrightarrow \text{for all } B \in \mathrm{Ob}(\mathbf{D}),\ \varepsilon_B \text{ is epic.} \end{aligned}$$

This is (i).
(ii) We first observe that

$$\begin{aligned} G \text{ is full} &\Leftrightarrow \text{for all } B, B' \in \mathrm{Ob}(\mathbf{D}), \text{ all } GB \xrightarrow{h} GB' \text{ there is} \\ &\qquad B \xrightarrow{g} B' \text{ such that } h = Gg \\ &\Leftrightarrow \text{for all } B, B' \in \mathrm{Ob}(\mathbf{D}), \text{ all } GB \xrightarrow{h} GB' \text{ there is} \qquad (*) \\ &\qquad B \xrightarrow{g} B' \text{ such that } h = (g \circ \varepsilon_B)\hat{} \text{ (by (1.22)).} \end{aligned}$$

So if G is full, take $B' = FGB$, $h = \eta_{GB}$ in (*). Then there is $g : B \to FGB$ such that $(1_{FGB})\hat{} = \eta_{GB} = (g \circ \varepsilon_B)\hat{}$. Hence $1_{FGB} = g \circ \varepsilon_B$ and so ε_B is split monic. Conversely, suppose ε_B is split monic; let $B \xrightarrow{g} FGB$ satisfy

Table 1.1

C	**D**	L
Esp	**Comp Haus**	Stone–Čech compactification functor
Category of metric spaces with uniformly continuous maps	**Category of complete metric spaces**	'Metric completion' functor
Category of integral domains and injective ring homomorphisms	**Field**	'Field of quotients' functor
Category of partially ordered sets and supremum preserving maps	**Category of complete lattices**	'Order completion' functor
Category of presheaves over a topological space X	**Category of sheaves over X**	'Sheafification' functor (cf. Ch. 5)

$1_{FGB} = g \circ \varepsilon_B$. Given $GB \xrightarrow{h} GB'$, we compute

$$(\tilde{h} \circ g \circ \varepsilon_B)\hat{} = (\tilde{h} \circ 1_{FGB})\hat{} = (\tilde{h})\hat{} = h.$$

So G is full by (*). This proves (ii).

Finally, (iii) is an immediate consequence of (i) and (ii). □

A subcategory **D** of a category **C** is called *reflective* in **C** when the insertion functor $I:\mathbf{D}\to\mathbf{C}$ has a *left* adjoint $L:\mathbf{C}\to\mathbf{D}$. This functor is called a *reflector* and the adjunction $L \dashv I$ a *reflection* of **C** in its subcategory **D**. Since I is always faithful, (1.30) implies that every component ε_B of the counit ε of a reflection is epic. If **D** is a *full* subcategory of **C**, then, by (1.30) again, ε_B is an *isomorphism* $LB \cong B$ for $B \in \mathrm{Ob}(\mathbf{D})$.

A reflection may be described in terms of the concept of freedom: a subcategory **D** of **C** is reflective iff for each **C**-object A there is a **D**-object A^* and an arrow $A \xrightarrow{\eta_A} IA^* = A^*$ such that (A^*, η_A) is I-free over A, that is, every arrow $A \xrightarrow{f} B$ to a **D**-object B has the form $f = g \circ \eta_A$ for a unique **D**-arrow $A^* \xrightarrow{g} B$. If this condition is satisfied, $A \mapsto A^*$ is the (object function of the) reflector.

Some examples of reflections are shown in Table 1.1.

Galois connections

Given an adjunction $F \dashv G$ between categories **C**,**D**, we establish an equivalence between certain subcategories of **C** and **D**. Let η, ε be the unit and counit of $F \dashv G$. Define $\tilde{\mathbf{C}}$ to be the full subcategory of $\mathbf{C}$ whose objects are those $\mathbf{C}$-objects A for which $A \xrightarrow{\eta_A} GFA$ is an isomorphism, and $\tilde{\mathbf{D}}$ to be the full subcategory of $\mathbf{D}$ whose objects are those $\mathbf{D}$-objects B for which $FGB \xrightarrow{\varepsilon_B} B$ is an isomorphism.

1.31 PROPOSITION. F restricts to an equivalence between $\tilde{\mathbf{C}}$ and $\tilde{\mathbf{D}}$ (and G to an equivalence between $\tilde{\mathbf{D}}$ and $\tilde{\mathbf{C}}$).

Proof. First, we observe that F restricts to a functor $\tilde{F}:\tilde{\mathbf{C}} \to \tilde{\mathbf{D}}$. For if A is a $\tilde{\mathbf{C}}$-object, then $A \xrightarrow{\eta_A} GFA$ is an isomorphism, and so therefore is $FA \xrightarrow{F\eta_A} FGFA$. By (1.21), $\varepsilon_{FA} = (F\eta_A)^{-1}$ is also an isomorphism, so that $FA \in \mathrm{Ob}(\tilde{\mathbf{D}})$.

Dually, G restricts to a functor $\tilde{G}$ from $\tilde{\mathbf{D}}$ to $\tilde{\mathbf{C}}$. Evidently $\tilde{F}\tilde{G} \xrightarrow{\varepsilon} 1_{\tilde{\mathbf{D}}}$ and $\tilde{G}\tilde{F} \xrightarrow{\eta} 1_{\tilde{\mathbf{C}}}$ are natural isomorphisms, so $\tilde{F}$ (as well as $\tilde{G}$) is an equivalence. □

As an example, take $\mathbf{C}$ to be **CRng** and $\mathbf{D}$ to be $\mathbf{Top}^{\mathrm{op}}$. Then we have an adjunction $F \dashv G$ between $\mathbf{C}$ and $\mathbf{D}$ given by

$$F(A) = \mathrm{Hom}(A,2) \text{ considered as a subspace of } 2^A,$$

$$G(X) = \text{ring of continuous functions from } X \text{ into } 2.$$

In this case $\tilde{\mathbf{C}}$ is the subcategory of Boolean rings and $\tilde{\mathbf{D}}$ the subcategory of Boolean spaces; F restricts to an equivalence between $\tilde{\mathbf{C}}$ and $\tilde{\mathbf{D}}$ which is the well-known *Stone duality* between Boolean rings and spaces.

The special case in which $\mathbf{C}$ and $\mathbf{D}$ are partially ordered (or preordered) sets is of particular importance. Let $P = (P,\leqslant)$ and $Q = (Q,\leqslant)$ be partially ordered sets. Recall that a functor $F:\mathbf{P} \to \mathbf{Q}$ is then just an order preserving map from P to Q, i.e. such that $p \leqslant p' \Rightarrow Fp \leqslant Fp'$ for all $p,p' \in P$. Then a functor $G:\mathbf{Q} \to \mathbf{P}$ is a *right adjoint* for F if and only if, for all $p \in P$, $q \in Q$,

$$Fp \leqslant q \Leftrightarrow p \leqslant Gq.$$

An adjunction $F \dashv G$ is also known as a (covariant) *Galois connection* between P and Q (the terminology arising from the classical correspondence between Galois extensions of a field and subgroups of the associated Galois group).

Now consider the functors $GF:\mathbf{P} \to \mathbf{P}$ and $FG:\mathbf{Q} \to \mathbf{Q}$. The former is a *closure operation*, i.e. satisfies

$$p \leqslant GFp; \qquad GFGFp = GFp;$$

$$p \leqslant p' \Rightarrow GFp \leqslant GFp'$$

for all $p,p' \in P$. The latter has the dual properties of an *interior operation,* i.e.

$$FGq \leqslant q; \qquad FGq = FGFGq$$

$$q \leqslant q' \Rightarrow FGq \leqslant FGq'.$$

By (1.31), F restricts to an equivalence, i.e. an order isomorphism, between $\tilde{P} = \{p \in P : GFp = p\}$ and $\tilde{Q} = \{q \in Q : FGq = q\}$. This isomorphism is called the associated *Galois isomorphism.*

Examples

(i) Take P and Q both to be the ordered set of natural numbers. Let

$$F(p) = p\text{th prime if } p > 0, \qquad F(0) = 0$$

$$G(q) = \text{number of primes} \leqslant q.$$

Then $GFp = p$ holds for all numbers $p > 0$, while $q \leqslant FGq$ iff q is prime. So the Galois isomorphism here is the bijection between the set of natural numbers and the set of primes.

(ii) Let X be a topological space; take P to be the class $\mathcal{O}(X)$ of open sets in X, and Q the class $\mathcal{C}(X)$ of closed sets in X, both partially ordered by inclusion. Take F to be the closure operation and G the interior operation on X. Then $\tilde{P}$ is the class of *regular open sets* in X, i.e. the $U \in \mathcal{O}(X)$ such that $U = \mathring{\bar{U}}$, and $\tilde{Q}$ the class of *regular closed sets* in X, i.e. the $C \in \mathcal{C}(X)$ such that $C = \bar{\mathring{C}}$. And the Galois isomorphism is the usual bijection between the regular open and regular closed subsets of X.

(iii) Let A and B be sets and let $f : A \to B$. We consider the power sets PA and PB of A and B as partially ordered sets; we have the map $f^{-1} : PB \to PA$ given by

$$f^{-1}(Y) = \{x \in X : f(x) \in Y\}.$$

for $Y \in PB$.

Define the maps $\forall_f : PA \to PB$ and $\exists_f : PA \to PB$ by

$$\forall_f(X) = \{y \in Y : \forall x \in A . f(x) = y \Rightarrow x \in X\},$$

$$\exists_f(X) = \{y \in Y : \exists x \in X . f(x) = y\}$$

for $X \in PA$. It is easy to see that both $\exists_f \dashv f^{-1}$ and $f^{-1} \dashv \forall_f$ are Galois correspondences between PB and PA.

The Galois isomorphism arising from the first of these correspondences has a particularly transparent form: it establishes a bijection between the $Y \in PB$ for which $Y = f[f^{-1}[Y]]$ and $X \in PA$ for which $X = f^{-1}[f[X]]$.

2

Introducing toposes

A *topos* is a category which has certain features in common with the category of sets, features which are sufficiently rich in consequences to justify the claim that a topos is a 'generalization' of **Set**. The first of these features hinges on the notion of a *subobject.*

Subobjects and subobject classifiers

In **Set**, the insertion map $m: B \hookrightarrow A$ of a subset B of a set A is always monic. And conversely, every monic $n: C \rightarrowtail A$ in **Set** is 'almost' an insertion in as much as there is always an isomorphism (i.e., a bijection) $i: C \cong B$ to a subset $B \subseteq A$ such that $m \circ i = n$.

Extending these ideas to an arbitrary category **C**, we call two monics $B \overset{m}{\rightarrowtail} A$ and $C \overset{n}{\rightarrowtail} A$ *equivalent,* and write $m \sim n$, if there is an isomorphism $i: C \cong B$ such that $m \circ i = n$. Clearly $\sim$ is an equivalence relation on the class of all monic arrows in **C** with codomain A; let $[m]$ be the equivalence class of any such monic m. An entity of the form $[m]$ is called a *subobject* of A. When no confusion is likely to occur, we shall identify $[m]$ which its representative m, and sometimes even with the domain B of m. Thus a subobject of an object A may loosely be regarded as a monic with codomain A or as the domain of such a monic. We write Sub(A) for the class of subobjects of A. Given monics $B \overset{m}{\rightarrowtail} A$, $C \overset{n}{\rightarrowtail} A$, we write $m \subseteq n$ and say that m is *included* in n if there is an arrow $B \xrightarrow{f} C$ such that $n \circ f = m$. (Clearly, when m and n are the insertion maps of subsets, inclusion of m in n just means set-theoretic inclusion of B in C.) Evidently $\subseteq$ is a preordering on the class of monics with codomain A. Moreover, it is readily established that $m \subseteq n$ and $n \subseteq m$ iff $m \sim n$. We extend the notion of inclusion to subobjects of A by defining $[m] \subseteq [n]$ iff $m \subseteq n$; it is easy to see that this makes sense, and that $\subseteq$ is actually a *partial ordering* on Sub(A).

Now in **Set** subobjects may be described in another way, namely by *characteristic functions* or *predicates.* The characteristic function of a subset B of a set A is the map $\chi = \chi_B$ from A to the set 2 consisting of two 'truth values' 0 and 1 with $\chi(x) = 1$ when $x \in B$ and $\chi(x) = 0$ when $x \notin B$. More generally, given a monic $m: B \rightarrowtail A$, we define the *characteristic function* of m to be the map $\chi_m: A \to 2$ such that $\chi_m(x) = 1$ when $x \in m[B]$

and $\chi_m(x)=0$ when $x \notin m[B]$. Given another monic $n: C \rightarrowtail A$ it is readily verified that $m \sim n$ iff $\chi_m = \chi_n$; accordingly we may without ambiguity call χ_m the *characteristic function* of the subobject $[m]$. Thus the assignment $[m] \mapsto \chi_m$ correlates each subobject of A with a unique characteristic function. The characteristic function χ_m may also be thought of as the *predicate* defined on A corresponding to the subobject $[m]$ or the monic m.

Conversely, given a map $u: A \to 2$ we may use the pullback technique to correlate a unique subobject of A with u. We take the subset $1 = \{0\} \subseteq 2$ and the monic $1 \rightarrowtail 2$, written $\top$, which sends 0 to 1, and form the pullback in **Set**

$$\begin{array}{ccc} B & \longrightarrow & 1 = \{0\} \\ {\scriptstyle m}\big\downarrow & & \big\downarrow{\scriptstyle \top} \\ A & \xrightarrow{\;u\;} & 2 \end{array}$$

The characteristic function of the subobject $[m]$ of A is then the given map u.

In sum, we have a bijective correspondence between subobjects of a set A and characteristic functions on A.

We can extend this idea to any category **C** with a terminal object. A *subobject classifier* or *truth-value object* in such a category **C** is an object Ω of **C** together with a monic $\top: 1 \to \Omega$, called the *truth arrow*, from the terminal object 1 of **C** such that (i) for each monic $m: B \rightarrowtail A$ there is a *unique* arrow $\chi(m): A \to \Omega$, called the *characteristic arrow of m* (*or* of B), such that the diagram

(2.1)
$$\begin{array}{ccc} B & \longrightarrow & 1 \\ {\scriptstyle m}\big\downarrow & & \big\downarrow{\scriptstyle \top} \\ A & \xrightarrow{\;\chi(m)\;} & \Omega \end{array}$$

is a pullback; and conversely (ii) any diagram of the form $A \xrightarrow{u} \Omega \xleftarrow{\top} 1$ has a pullback. (Note that since any arrow with domain 1 is monic, the pullback of $\top$ along u must also be monic.) We sometimes write $\Omega_{\mathbf{C}}$ for Ω.

By analogy with **Set**, $\chi(m)$ may also be termed the *predicate* on A determined by m.

It is easily checked that a subobject classifier is unique up to isomorphism in the evident sense.

Throughout the rest of this section let **C** be a category with a terminal object 1 and a subobject classifier $(\Omega, \top)$.

For each **C**-arrow $A \xrightarrow{u} \Omega$ choose a monic $\bar{u}$ with codomain A such that

(2.2)
$$\begin{array}{ccc} \mathrm{dom}(\bar{u}) & \longrightarrow & 1 \\ {\scriptstyle \bar{u}}\downarrow & & \downarrow{\scriptstyle \top} \\ A & \xrightarrow{u} & \Omega \end{array}$$

is a pullback. (The existence of such a $\bar{u}$ is guaranteed by condition (ii) in the definition of Ω.) $\bar{u}$ is called the *kernel* of u. It will be convenient to assume—as we evidently may—that $\overline{\chi(1_A)} = 1_A$ for any object A.

2.3 LEMMA. Let A be any object of **C**, let m and n be any **C**-monics with codomain A and let $u: A \to \Omega$ be a **C**-arrow. Then

(i) $u = \chi(\bar{u})$
(ii) $[\overline{\chi(m)}] = [m]$
(iii) $[m] = [n] \Leftrightarrow \chi(m) = \chi(n)$.

Proof. (i) Since (2.1) with $B = \mathrm{dom}(\bar{u})$, $m = \bar{u}$ and (2.2) are both pullback diagrams, it follows that $\chi(\bar{u}) = u$, i.e. (i).

(ii) Similar to (i).

(iii) If $\chi(m) = \chi(n)$ then $[m] = [\overline{\chi(m)}] = [\overline{\chi(n)}] = [n]$ by (ii). Conversely, if $[m] = [n]$, there is an isomorphism $f: \mathrm{dom}(n) \cong \mathrm{dom}(m)$ such that the diagram

$$\begin{array}{ccc} \mathrm{dom}(n) & \xrightarrow{f} & \mathrm{dom}(m) \\ & {\scriptstyle n}\searrow \quad \swarrow{\scriptstyle m} & \\ & A & \end{array}$$

commutes. Combining this with the pullback diagram (2.1), where $B = \mathrm{dom}(m)$, we see that

$$\begin{array}{ccc} \mathrm{dom}(n) & \longrightarrow & 1 \\ {\scriptstyle n}\downarrow & & \downarrow{\scriptstyle \top} \\ A & \xrightarrow{\chi(m)} & \Omega \end{array}$$

is also a pullback diagram. Therefore $\chi(m)$ is equal to the characteristic arrow $\chi(n)$ of n. □

It follows immediately from this lemma that the map $[m] \mapsto \chi(m)$ is a bijection between the class Sub(A) of subobjects of A and the *set* $\mathbf{C}(A, \Omega)$ of **C**-arrows $A \to \Omega$. Therefore, in particular, Sub(A) is a set. (More precisely, Sub(A) has a *set* of representatives.) Moreover, we can use this bijection to transfer the partial ordering $\subseteq$ on Sub(A) to $\mathbf{C}(A, \Omega)$. That is, we define, for $u, v \in \mathbf{C}(A, \Omega)$

$$u \leqslant v \Leftrightarrow [m] \subseteq [n] \quad \text{with} \quad \chi(m) = u, \quad \chi(n) = v$$
$$\Leftrightarrow [\bar{u}] \subseteq [\bar{v}]$$
$$\Leftrightarrow \bar{u} \subseteq \bar{v}$$
$$\Leftrightarrow \text{for some arrow } \operatorname{dom}(\bar{u}) \xrightarrow{f} \operatorname{dom}(\bar{v}), \; \bar{u} = \bar{v} \circ f.$$

Since $\subseteq$ is a partial ordering of $\operatorname{Sub}(A)$, $\leqslant$ is a partial ordering of $\mathbf{C}(A, \Omega)$, and $(\operatorname{Sub}(A), \subseteq) \cong (\mathbf{C}(A, \Omega), \leqslant)$.

For any arrow $A \xrightarrow{u} \Omega$ we call $[\bar{u}]$ (which by (2.3) is uniquely determined by u) the *subobject of A classified* by u.

Let m be a monic with codomain A and let $f : C \to A$. We define the *inverse image* $f^{-1}(m)$ of m under f by

(2.4) $$f^{-1}(m) = \overline{\chi(m) \circ f}.$$

Thus $f^{-1}(m)$ is characterized by the pullback square

(2.5)

$$
\begin{array}{ccc}
\operatorname{dom}(f^{-1}(m)) & \longrightarrow & 1 \\
{\scriptstyle f^{-1}(m)}\downarrow & & \downarrow{\scriptstyle \top} \\
C & \xrightarrow{\chi(m)\circ f} & \Omega
\end{array}
$$

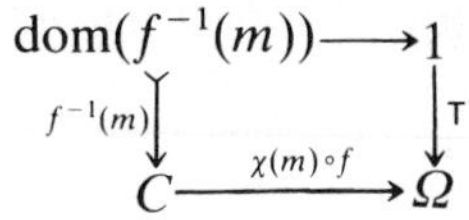

It follows that the diagram

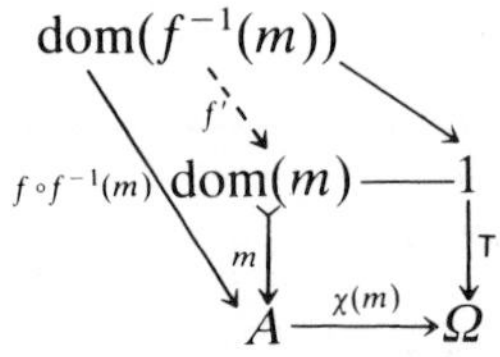

commutes. Since the lower right-hand square is a pullback, there must be a (unique) $f' : \operatorname{dom}(f^{-1}(m)) \to \operatorname{dom}(m)$ such that $f \circ f^{-1}(m) = m \circ f'$, i.e. such that the diagram

(2.6)

$$
\begin{array}{ccc}
\operatorname{dom}(\bar{f}^{1}(m)) & \xrightarrow{f'} & \operatorname{dom}(m) \\
{\scriptstyle f^{-1}(m)}\downarrow & & \downarrow{\scriptstyle m} \\
C & \xrightarrow{f} & A
\end{array}
$$

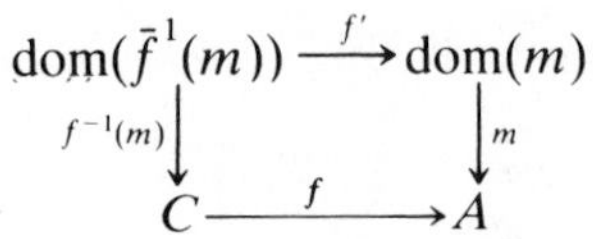

commutes. Thus the pullback square (2.5) may be written as a composition of commutative squares:

$$
\begin{array}{ccccc}
\operatorname{dom}(f^{-1}(m)) & \xrightarrow{f'} & \operatorname{dom}(m) & \longrightarrow & 1 \\
{\scriptstyle f^{-1}(m)}\downarrow & & \downarrow & & \downarrow{\scriptstyle \top} \\
C & \xrightarrow{f} & A & \xrightarrow{\chi(m)} & \Omega
\end{array}
$$

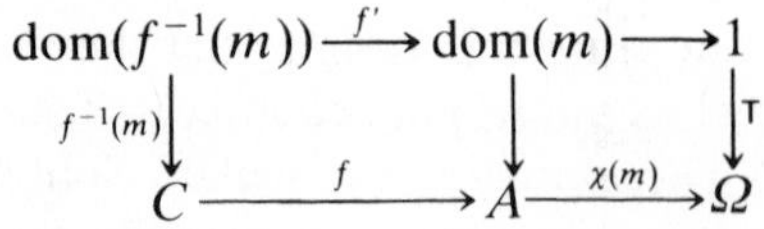

But since the right-hand square here is a pullback, so the left-hand square, i.e. (2.6), is also a pullback. *We have therefore shown that in a category with a subobject classifier, any monic arrow has a pullback, or inverse image, under any arrow.*

We note that equation (2.4) may also be written

(2.7) $$\chi(f^{-1}(m)) = \chi(m)\circ f.$$

Our next proposition follows immediately from (2.7) and the pullback property of (2.6).

2.8 PROPOSITION. For any monics m, n with codomains A, C respectively and any $f: C \to A$, the following are equivalent:

(i) $n \sim f^{-1}(m)$;
(ii) $\chi(n) = \chi(m)\circ f$;
(iii) there is $g: \mathrm{dom}(n) \to \mathrm{dom}(m)$ such that the diagram

$$\begin{array}{ccc} \mathrm{dom}(n) & \xrightarrow{g} & \mathrm{dom}(m) \\ {\scriptstyle n}\downarrow & & \downarrow{\scriptstyle m} \\ C & \xrightarrow{f} & A \end{array}$$

is a pullback. □

Similarly, we also have, under the same conditions as (2.8):

2.9 PROPOSITION. The following are equivalent:

(i) $n \subseteq f^{-1}(m)$;
(ii) there is $g: \mathrm{dom}(n) \to \mathrm{dom}(m)$ such that the diagram

$$\begin{array}{ccc} \mathrm{dom}(n) & \xrightarrow{g} & \mathrm{dom}(m) \\ {\scriptstyle n}\downarrow & & \downarrow{\scriptstyle m} \\ C & \xrightarrow{f} & A \end{array}$$

commutes. □

We also observe that inclusions are preserved by inverse images, i.e.

2.10 PROPOSITION. For any monics m, n with codomain A and any $f: C \to A$, we have

$$m \subseteq n \Rightarrow f^{-1}(m) \subseteq f^{-1}(n).$$

Proof. If $m \subseteq n$, there is $g: \mathrm{dom}(m) \to \mathrm{dom}(n)$ such that $m = n\circ g$. Then the diagram

$$\begin{array}{ccc} \mathrm{dom}(f^{-1}(m)) & \xrightarrow{f'} & \mathrm{dom}(m) \\ {\scriptstyle f^{-1}(m)}\downarrow & & \downarrow{\scriptstyle m = n\circ g} \\ C & \xrightarrow{f} & A \end{array}$$

commutes. Since the square

$$\begin{array}{ccc} \mathrm{dom}(f^{-1}(n)) & \longrightarrow & \mathrm{dom}(n) \\ {\scriptstyle f^{-1}(n)}\downarrow & & \downarrow{\scriptstyle n} \\ C & \xrightarrow{\;\;f\;\;} & A \end{array}$$

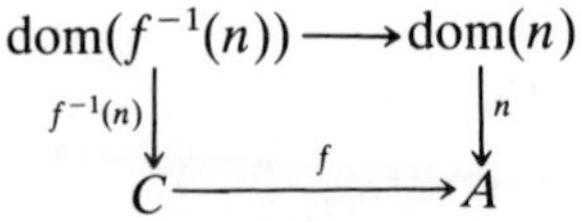

is a pullback, there must accordingly be an $h:\mathrm{dom}(f^{-1}(m))\to \mathrm{dom}(f^{-1}(n))$ making the left-hand triangle in the (commutative) diagram

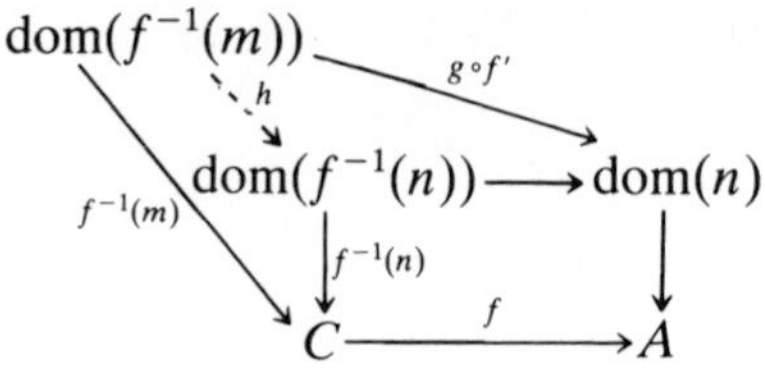

commute. It follows that $f^{-1}(m)\subseteq f^{-1}(n)$. □

Given an object A, we recall that the *diagonal arrow* is given by $\delta_A = \langle 1_A,1_A\rangle:A\to A\times A$. Clearly δ_A is always monic. We have the following fact about δ_A.

2.11 PROPOSITION.   Given a diagram

$$\mathrm{dom}(n) \xrightarrow{\;\;n\;\;} C \xrightarrow[g]{f} A$$

with n monic, then

$$f\circ n = g\circ n \Leftrightarrow n\subseteq \langle f,g\rangle^{-1}(\delta_A).$$

Proof.   By 2.9, the following are equivalent:

(1) $n\subseteq \langle f,g\rangle^{-1}(\delta_A)$
(2)   for some $\mathrm{dom}(n)\xrightarrow{h}A$,

$$\langle f\circ n, g\circ n\rangle = \langle f,g\rangle\circ n = \delta_A\circ h$$
$$= \langle 1_A,1_A\rangle\circ h = \langle h,h\rangle.$$

But (2) is clearly equivalent to

$$f\circ n = g\circ n,$$

as required.   □

We have already observed that pullbacks of monics may be constructed in **C**. In particular, we may construct the *intersection* of two monics with common codomain as follows. Given two monics m,n with common codomain A, we form the pullback square

$$\begin{array}{ccc} \bullet & \xrightarrow{\ n^{-1}(m)\ } & \mathrm{dom}(m) \\ {\scriptstyle m^{-1}(n)}\downarrow & & \downarrow{\scriptstyle n} \\ \mathrm{dom}(n) & \xrightarrow[\ m\ ]{} & A \end{array} \qquad (*)$$

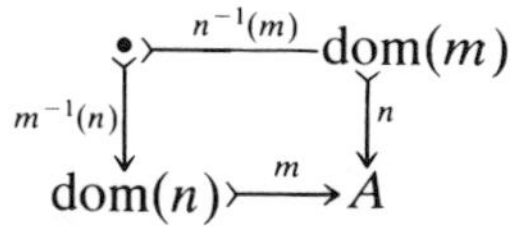

The composition $m \circ m^{-1}(n)$ (or, equivalently, $n \circ n^{-1}(m)$) is then a monic arrow with codomain A; we denote it by $m \cap n$ and call it the *intersection* of m and n. It follows immediately from the pullback property of (*) that, for any monic arrow p with codomain A,

(2.12) $p \subseteq m \cap n \Leftrightarrow p \subseteq m$ and $p \subseteq n$.

This implies that $[m \cap n]$ is the *infimum* of the set of subobjects $\{[m], [n]\}$ w.r.t. $\subseteq$; therefore $(\mathrm{Sub}(A), \subseteq)$ is a *lower semilattice*, that is, a partially ordered set in which every pair of elements has an infimum.

Moreover, intersections are preserved under inverse images, in the sense that, for $f : C \to A$,

(2.13) $f^{-1}(m \cap n) \sim f^{-1}(m) \cap f^{-1}(n)$.

To prove (2.13), we observe that, since $m \cap n \subseteq m$ and $m \cap n \subseteq n$, it follows from 2.10 that $f^{-1}(m \cap n) \subseteq f^{-1}(m)$ and $f^{-1}(m \cap n) \subseteq f^{-1}(n)$, and so $f^{-1}(m \cap n) \subseteq f^{-1}(m) \cap f^{-1}(n)$ by (2.12). To prove the reverse inclusion, consider the commutative diagram

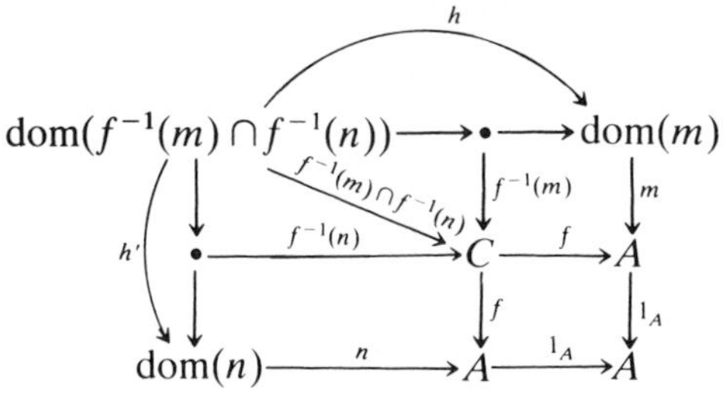

Then $m \circ h = n \circ h'$, so since

$$\begin{array}{ccc} \mathrm{dom}(m \cap n) & \xrightarrow{\ n^{-1}(m)\ } & \mathrm{dom}(n) \\ {\scriptstyle m^{-1}(n)}\downarrow & & \downarrow{\scriptstyle n} \\ \mathrm{dom}(m) & \xrightarrow[\ m\ ]{} & A \end{array}$$

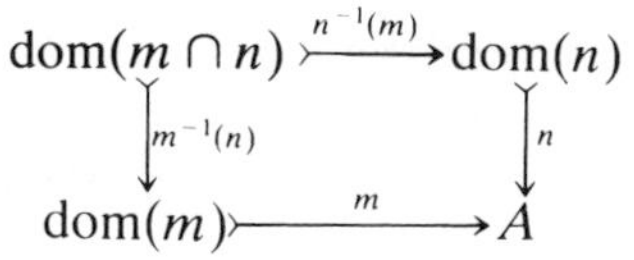

is a pullback square, there is $g : \mathrm{dom}(f^{-1}(m) \cap f^{-1}(n)) \to \mathrm{dom}(m \cap n)$ such that $h = m^{-1}(n) \circ g$. Then we have

$$f \circ (f^{-1}(m) \cap f^{-1}(n)) = m \circ h = m \circ m^{-1}(n) \circ g$$
$$= (m \cap n) \circ g.$$

Hence, by (2.9),

$$f^{-1}(m) \cap f^{-1}(n) \subseteq f^{-1}(m \cap n)$$

and (2.13) follows.

The results so far obtained can all be reformulated in terms of characteristic arrows instead of monics. Thus, for example, consider (2.9). Given $u: C \to \Omega$, $v: A \to \Omega$, $f: C \to A$, putting $n = \bar{u}$, $m = \bar{v}$ in (2.9) gives, using (2.7),

(2.14) $$u \leqslant v \circ f \Leftrightarrow \text{there is } g: \mathrm{dom}(\bar{u}) \to \mathrm{dom}(\bar{v}) \text{ such that } \bar{v} \circ g = f \circ \bar{u}.$$

Next, consider (2.10). Taking characteristic arrows on both sides gives, using (2.3) and the definition of $\leqslant$,

$$\chi(m) \leqslant \chi(n) \Rightarrow \chi(f^{-1}(m)) \leqslant \chi(f^{-1}(n)).$$

Writing $u = \chi(m)$, $v = \chi(n)$, it now follows from (2.7) that for any $u, v: A \to \Omega$ and any $f: C \to A$,

(2.15) $$u \leqslant v \Rightarrow u \circ f \leqslant v \circ f.$$

Turning now to (2.11), let us first define $\mathrm{eq}_A: A \times A \to \Omega$ by

$$\mathrm{eq}_A = \chi(\delta_A).$$

(Where A is understood, we write eq for eq_A.) Given $u: C \to \Omega$ and $f, g: C \to A$, take $n = \bar{u}$ in (2.11) and use (2.7) and (2.3) to obtain

$$\begin{aligned} f \circ \bar{u} = g \circ \bar{u} &\Leftrightarrow \bar{u} \subseteq \langle f, g \rangle^{-1}(\delta_A) \\ &\Leftrightarrow u \leqslant \chi(\langle f, g \rangle^{-1}(\delta_A)) \\ &\qquad = \chi(\delta_A) \circ \langle f, g \rangle \\ &\qquad = \mathrm{eq}_A \circ \langle f, g \rangle. \end{aligned}$$

Hence

(2.16) $$f \circ \bar{u} = g \circ \bar{u} \quad \text{iff} \quad u \leqslant \mathrm{eq}_A \circ \langle f, g \rangle.$$

To obtain the corresponding versions of (2.12) and (2.13), we need to define the operation on characteristic arrows that corresponds to the intersection of their kernels. Thus we define the *meet* $u \wedge v$ of two arrows $u, v: A \to \Omega$ by

$$u \wedge v = \chi(\bar{u} \cap \bar{v}).$$

Then (2.12) gives, for any arrow $w: A \to \Omega$,

(2.17) $$w \leqslant u \wedge v \Leftrightarrow w \leqslant u \quad \text{and} \quad w \leqslant v,$$

and (2.13) becomes, for any $f:C \to A$,

(2.18) $$(u \wedge v) \circ f = u \circ f \wedge v \circ f.$$

It follows from (2.17) that $u \wedge v$ is the *infimum* of the set $\{u, v\}$ w.r.t. $\leqslant$, so that $(\mathbf{C}(A, \Omega), \leqslant)$ is a *lower semilattice*. In particular, $\wedge$ satisfies the *commutative* and *associative* laws:

$$u \wedge v = v \wedge u \quad \text{and} \quad (u \wedge v) \wedge w = u \wedge (v \wedge w).$$

In view of the latter, for any $u_1, \ldots, u_m \in \mathbf{C}(A, \Omega)$ we may write $u_1 \wedge \cdots \wedge u_m$ unambiguously for the infimum of $\{u_1, \ldots, u_m\}$.

To conclude this section we define the *maximal* characteristic arrow T_A on A by

$$T_A = \chi(1_A): A \to \Omega.$$

(Where A is understood, we write just T for T_A.) We note that $\bar{T}_A = 1_A$. T_A is called maximal because it is evidently the case that $u \leqslant T_A$ for any $u: A \to \Omega$. We assemble the basic facts about maximal characteristic arrows in a final proposition.

2.19 PROPOSITION. Let $u, v: A \to \Omega$ and $f, g: C \to A$. Then

(i) $T_C = v \circ f \Leftrightarrow$ there is $g: C \to \mathrm{dom}(\bar{v})$ such that $f = \bar{v} \circ g$.

(ii) $T = v \circ \bar{u} \Leftrightarrow u \leqslant v$.

(iii) $T_C = \mathrm{eq}_A \circ \langle f, g \rangle \Leftrightarrow f = g$.

Proof. (i) is an immediate consequence of (2.8), and (ii) of (i). Finally, using (2.16),

$$\begin{aligned} T_C = \mathrm{eq}_A \circ \langle f, g \rangle &\Leftrightarrow T_C \leqslant \mathrm{eq}_A \circ \langle f, g \rangle \\ &\Leftrightarrow f \circ \bar{T}_C = g \circ \bar{T}_C \\ &\Leftrightarrow f = g. \end{aligned}$$ □

REMARK. We can give an alternative description of subobject classifiers in any category $\mathbf{C}$ in which pullbacks of monic arrows exist, and for which $\mathrm{Sub}(A)$ is a set for each $\mathbf{C}$-object A. Under these conditions we can turn Sub into a contravariant set-valued functor on $\mathbf{C}$: for $A \xrightarrow{f} B$, $\mathrm{Sub}(f): \mathrm{Sub}(B) \to \mathrm{Sub}(A)$ is defined by $\mathrm{Sub}(f)([m]) = [f^{-1}(m)]$ for any monic m with codomain B.. It is now easily seen that a *subobject classifier* Ω *in* $\mathbf{C}$ *is a representing object for* Sub, *that is,* $\mathrm{Sub} \cong H^{\Omega}$. *And conversely, if* Sub *is representable, then* $\mathbf{C}$ *has a subobject classifier.*

For suppose that Ω is a $\mathbf{C}$-object for which there is a natural isomorphism $\eta: H^{\Omega} \to \mathrm{Sub}$. Now the Yoneda lemma (1.7) says that η is determined by $\eta_{\Omega}(1_{\Omega}) \in \mathrm{Sub}(\Omega)$. Let $\mathrm{dom}(m) \overset{m}{\rightarrowtail} \Omega$ be such that $[m] = \eta_{\Omega}(1_{\Omega})$. Then an easy calculation shows that, for any A and any

$f \in \mathbf{C}(A, \Omega)$,

$$\eta_\Omega(f) = \mathrm{Sub}(f)([m]) = [f^{-1}(m)].$$

Thus $f \mapsto [f^{-1}(m)]$ is an isomorphism $\mathbf{C}(A, \Omega) \cong \mathrm{Sub}(A)$. This implies that, for any monic n with codomain A there is a unique $f: A \to \Omega$ for which there is $g: \mathrm{dom}(n) \to \mathrm{dom}(m)$ such that

$$\begin{array}{ccc} \mathrm{dom}(n) & \xrightarrow{g} & \mathrm{dom}(m) \\ {\scriptstyle n}\downarrow & & \downarrow{\scriptstyle m} \\ A & \xrightarrow{f} & \Omega \end{array}$$

is a pullback square. We need finally to show that $\mathrm{dom}(m)$ is a terminal object in $\mathbf{C}$. To see this, note that for any A there is a unique $A \xrightarrow{f} \Omega$ for which there is $A \xrightarrow{g} \mathrm{dom}(m)$ such that

$$\begin{array}{ccc} A & \xrightarrow{g} & \mathrm{dom}(m) \\ {\scriptstyle 1_A}\downarrow & & \downarrow{\scriptstyle m} \\ A & \xrightarrow{f} & \Omega \end{array}$$

is a pullback square. Now, given any $A \xrightarrow{h} \mathrm{dom}(m)$, it is easy to see that

$$\begin{array}{ccc} A & \xrightarrow{h} & \mathrm{dom}(m) \\ {\scriptstyle 1_A}\downarrow & & \downarrow{\scriptstyle m} \\ A & \xrightarrow{m \circ h} & \Omega \end{array}$$

is a pullback square. Therefore, by uniqueness of f, $m \circ h = f = m \circ g$, so $h = g$ by the monicity of m. Therefore g is the unique arrow $A \to \mathrm{dom}(m)$ and $\mathrm{dom}(m)$ is a terminal object.

Power objects: the concept of topos

We next describe the second feature possessed by the category of sets to which we will devote special attention and which will complete the formulation of the concept of topos, viz., the presence of *power objects*.

Given a set A, we may form its *power set* PA, the set of all the subsets of A. The correspondence $Y \mapsto \chi_Y$ between subsets of A and characteristic functions on A establishes a bijection between PA and the exponential 2^A.

We recall that 2^A is characterized by the following property. For any map $f:A\times B\to 2$, there is a unique $\hat{f}:B\to 2^A$ such that

$$\begin{array}{ccc} A\times B & & \\ {\scriptstyle 1_A\times\hat{f}}\downarrow & \searrow^{f} & \\ A\times 2^A & \xrightarrow{\mathrm{ev}_A} & 2 \end{array}$$

commutes, where ev_A is the evaluation map. Since $PA\cong 2^A$, it follows that PA satisfies the same condition, that is, for any set A there is a map $e_A:A\times PA\to 2$ such that, corresponding to any map $f:A\times B\to 2$ there is a unique $\hat{f}:B\to PA$ such that the diagram

$$\begin{array}{ccc} A\times B & & \\ {\scriptstyle 1_A\times\hat{f}}\downarrow & \searrow^{f} & \\ A\times PA & \xrightarrow{e_A} & 2 \end{array}$$

commutes. (Explicitly, e_A is given by $e_A(a,X)=1$ iff $a\in X$ and $\hat{f}$ by $\hat{f}(b)=\{a\in A:f(a,b)=1\}$.)

We now extend these ideas to more general categories. Let **C** be a category with finite products and a subobject classifier Ω. Given a **C**-object A, a *power object* for A is a pair (PA, e_A) consisting of a **C**-object A and an arrow $e_A:A\times PA\to\Omega$ (called the *evaluation arrow* of A) such that corresponding to any $A\times B\xrightarrow{f}\Omega$ there is a unique $B\xrightarrow{\hat{f}}PA$ (called the *power transpose* of f) such that the diagram

$$\begin{array}{ccc} A\times B & & \\ {\scriptstyle 1_A\times\hat{f}}\downarrow & \searrow^{f} & \\ A\times PA & \xrightarrow{e_A} & \Omega \end{array}$$

commutes. (By abuse of language, we also call PA the power object of A.) Since PA, if it exists, is just the exponential Ω^A, it is unique up to isomorphism in the usual sense. **C** is said to *have power objects* if (PA, e_A) exists for each **C**-object A.

We note that, for any $g:B\to PA$, we have

(2.20) $$g=(e_A\circ(1_A\times g))\hat{}\,.$$

For, setting $h=e_A\circ(1_A\times g))$, it follows from the uniqueness condition on $\hat{h}$ that $g=\hat{h}$, i.e.. (2.20). Conversely, it is easy to see that if (2.20) holds for every $g:B\to PA$, then $\hat{f}$ is uniquely determined for every $f:A\times B\to\Omega$.

We note that PA, as the exponential Ω^A, may be equivalently described by asserting that, for any B, there is an isomorphism $\mathbf{C}(A\times B,\Omega)\cong\mathbf{C}(B,PA)$. But since $\mathbf{C}(A\times B,\Omega)\cong\mathrm{Sub}(A\times B)$, we get an isomorphism

$$\mathrm{Sub}(A\times B)\cong\mathbf{C}(B,PA).$$

It is natural to call a subobject of $A \times B$ a *relation* between A and B. Accordingly, relations between A and B correspond bijectively to arrows $B \to PA$. In particular, taking $B = 1$, and noting that relations between A and 1 correspond bijectively to subobjects of A, it follows that subobjects of A correspond bijectively to arrows $1 \to PA$, i.e. to **C**-elements of PA.

We are now in a position to introduce the central concept of this book.

DEFINITION. A *topos* is a category with finite products (equivalently, with binary products and a terminal object), a subobject classifier and power objects.

It is readily seen that the property of being a topos is essentially categorical. Evidently, also, **Set** is a topos. Moreover, the degenerate category **1** is a topos, which we naturally call the *degenerate topos*; more generally, we term *degenerate* any topos equivalent to the degenerate topos.

We are eventually going to show that any topos has, in addition to its defining properties, many other important properties possessed by **Set**, for example, Cartesian closedness and finite completeness and cocompleteness. We shall not, however, establish these facts by category-theoretic means (which would be somewhat difficult) but instead by employing a technique that combines proof-theoretic and model-theoretic elements in a way to be described in the next chapter.

We turn instead to establishing the existence of a rich class of toposes.

$\mathbf{Set}^{\mathbf{C}}$ as a topos

Let **C** be a small category. We show that *the functor category* $\mathbf{Set}^{\mathbf{C}}$ *is a topos.*

First, the *terminal object* of $\mathbf{Set}^{\mathbf{C}}$ is the constant functor $1 : \mathbf{C} \to \mathbf{Set}$ which assigns to each **C**-object A the value $1 = \{0\}$.

Next, *products* in $\mathbf{Set}^{\mathbf{C}}$ are constructed 'pointwise': given objects $F, G \in \mathbf{Set}^{\mathbf{C}}$, $F \times G \in \mathbf{Set}^{\mathbf{G}}$ is defined by

$$(F \times G)A = FA \times GA$$

for any **C**-object A, and for $A \xrightarrow{f} B$, the arrow $(F \times G)A \xrightarrow{(F \times G)f} (F \times G)B$ is given by

$$(F \times G)f((x,y)) = ((Ff)(x), (Gf)(y)).$$

Before identifying the subobject classifier in $\mathbf{Set}^{\mathbf{C}}$, we remark that, if $F \in \mathbf{Set}^{\mathbf{C}}$ and $G \overset{m}{\rightarrowtail} F$ is any monic in $\mathbf{Set}^{\mathbf{C}}$, then it is easy to see that each $m_A : GA \to FA$ is monic, and so, just as in **Set**, G is isomorphic to its 'image' under m, which in this case is the functor $G' : \mathbf{C} \to \mathbf{Set}$ given by $G'A = m_A[GA] \subseteq FA$. Thus $[G] = [G']$ as subobjects of F. The functor G' is in an obvious sense a *subfunctor* of F; formally, a functor $H : \mathbf{C} \to \mathbf{Set}$ is called a *subfunctor* of F, and we write $H \subseteq F$, if $HA \subseteq FA$ for all $\mathbf{C}$-objects A and, for any $A \xrightarrow{f} B$, the diagram

$$\begin{array}{ccc} HA & \xrightarrow{Hf} & HB \\ \downarrow & & \downarrow \\ FA & \xrightarrow{Ff} & FB \end{array}$$

commutes, i.e. Hf is the restriction of Ff to HA. Our remarks above then show that in $\mathbf{Set}^{\mathbf{C}}$ *there is a bijection between subobjects and subfunctors.*

Now we can proceed to identify the *subobject classifier* of $\mathbf{Set}^{\mathbf{C}}$. By the Yoneda lemma (1.7), if $\Omega \in \mathbf{Set}^{\mathbf{C}}$ *were* a subobject classifier in $\mathbf{Set}^{\mathbf{C}}$, then for any $\mathbf{C}$-object A,

$$\Omega(A) \cong \mathrm{Nat}(H_A, \Omega) \cong \mathrm{Sub}(H_A). \qquad (*)$$

As we have already observed, members of $\mathrm{Sub}(H_A)$ may be identified with subfunctors of H_A. Now it is easily seen that, for any subfunctor F of H_A, the set $\bigcup \{FB : B \in \mathrm{Ob}(\mathbf{C})\}$ is a collection of arrows with domain A closed under composition on the left. We call such a set a *sieve* on A; thus a *sieve* on A is a set S of arrows of $\mathbf{C}$ with domain A such that, for any $f \in S$ and any $\mathbf{C}$-arrow g,

$$\mathrm{dom}(g) = \mathrm{cod}(f) \Rightarrow g \circ f \in S.$$

So every subfunctor of H_A gives rise to a sieve on A; conversely, any sieve S on A determines a subfunctor of H_A, namely the functor whose object function is

$$B \mapsto \{f \in S : \mathrm{cod}(f) = B\}.$$

Accordingly, *subfunctors of* H_A *may be identified with sieves on* A. This, together with the isomorphism (*) above, suggests that we take the object functor of the subobject classifier Ω in $\mathbf{Set}^{\mathbf{C}}$ to be defined by:

$$\Omega(A) = \text{set of all sieves on } A.$$

We turn Ω into a functor in the natural way by 'pulling back': for $A \xrightarrow{f} B$ and $S \in \Omega(A)$,

$$(\Omega f)S = \{g \in \mathrm{Arr}(\mathbf{C}) : \mathrm{dom}(g) = B \quad \text{and} \quad g \circ f \in S\}.$$

EXERCISE. Let $A \xrightarrow{f} B$ and let $f^* : H_B \to H_A$ be the natural transformation induced by f via the Yoneda lemma. Given $S \in \Omega(A)$, regard S as a subfunctor of H_A. Show that $(\Omega f)S$, regarded as a subfunctor of H_B, is the pullback of S under f^* in $\mathbf{Set}^{\mathbf{C}}$.

For each **C**-object A let $\max_A$ be the *maximal sieve* on A, i.e.

$$\max_A = \{f \in \mathrm{Arr}(\mathbf{C}) : \mathrm{dom}(f) = A\}.$$

We define $1 \xrightarrow{\top} \Omega$ in $\mathbf{Set}^{\mathbf{C}}$ by

$$\top_A(0) = \max_A$$

for any **C**-object A.

To show that $(\Omega, \top)$ as just defined is a subobject classifier in $\mathbf{Set}^{\mathbf{C}}$, take any monic $G \rightarrowtail F$ in $\mathbf{Set}^{\mathbf{C}}$. According to our remarks above, we may without loss of generality assume that G is a subfunctor of F. We define the *characteristic arrow* $\chi_G : F \to \Omega$ of G by putting, for each **C**-object A, $x \in FA$,

$$(\chi_G)_A(x) = \{A \xrightarrow{f} B : (Ff)x \in GB\}.$$

It is readily verified that χ_G is a natural transformation $F \to \Omega$, and hence an arrow in $\mathbf{Set}^{\mathbf{C}}$. Moreover, it follows from the definition of χ_G that for any A,

$$GA = ((\chi_G)_A)^{-1}(\max_A)$$

and hence that the diagram

$$\begin{array}{ccc} G & \longrightarrow & 1 \\ \downarrow & & \downarrow \top \\ F & \xrightarrow{\chi_G} & \Omega \end{array}$$

is a pullback. The uniqueness of χ is also clear.

Conversely, given any arrow $F \xrightarrow{\eta} \Omega$ in $\mathbf{Set}^{\mathbf{C}}$, we let G be the subfunctor of F defined by

$$GA = \eta^{-1}(\max_A)$$

and, for $f : A \to B$,

$$Gf = Ff \restriction GA.$$

It is now easily verified that the diagram

$$\begin{array}{ccc} G & \longrightarrow & 1 \\ \downarrow & & \downarrow \top \\ F & \xrightarrow{\eta} & \Omega \end{array}$$

is a pullback. Accordingly, $(\Omega, \top)$ is a subobject classifier in $\mathbf{Set}^{\mathbf{C}}$ as claimed.

Finally, we construct *power objects* in $\mathbf{Set}^{\mathbf{C}}$. Given $F \in \mathbf{Set}^{\mathbf{C}}$, if PF *were* a power object for F, then by the Yoneda lemma (1.7), for any A,

$$(PF)A \cong \mathrm{Nat}(H_A, PF) \cong \mathrm{Nat}(F \times H_A, \Omega) \cong \mathrm{Sub}(F \times H_A).$$

This suggests that we take

$$(PF)A = \text{set of subfunctors of } F \times H_A.$$

We turn PF into a functor $\mathbf{C} \to \mathbf{Set}$ as follows. Given $A \xrightarrow{f} B$, we define $(PF)f : (PF)A \to (PF)B$ by setting, for each $G \in (PF)A$, $\mathbf{C} \in \mathrm{Ob}(C)$,

$$((PF)f)(G)(C) = \{(x,g) \in FC \times H_B(C) : (x, g \circ f) \in GC\}.$$

We define the *evaluation arrow* $e_F : F \times PF \to \Omega$ by

$$(e_F)_A((x,G)) = \{A \xrightarrow{f} B : ((Ff)x,f) \in GB\}$$

for $A \in \mathrm{Ob}(\mathbf{C})$, $x \in FA$, $G \in (PF)A$. (It is readily checked that e_F is a natural transformation.)

Now, given an arrow $F \times G \xrightarrow{\eta} \Omega$ in $\mathbf{Set}^{\mathbf{C}}$, we define $\hat{\eta} : G \to PF$ by setting, for each $A, B \in \mathrm{Ob}(\mathbf{C})$, $y \in GA$,

$$\hat{\eta}_A(y)(B) = \{(z,f) \in FB \times H_A(B) : \eta_B((z,(Gf)y)) = \max_B\}.$$

One checks that $\hat{\eta}$ is a natural transformation. Also the diagram

$$\begin{array}{ccc} F \times G & & \\ {\scriptstyle 1_F \times \hat{\eta}} \downarrow & \searrow^{\eta} & \\ F \times PF & \xrightarrow{e_F} & \Omega \end{array}$$

commutes, since given $A \xrightarrow{f} B$, $x \in FA$, $y \in GA$,

$$\begin{aligned} f \in (e_F)_A((x,\hat{\eta}_A(y)) &\Leftrightarrow ((Ff)x,f) \in \hat{\eta}_A(y)(B) \\ &\Leftrightarrow \eta_B(((Ff)x,(Gf)y)) = \max_B \\ &\Leftrightarrow \eta_B((F \times G)f)(x,y)) = \max_B \\ &\Leftrightarrow 1_B \in \eta_B((F \times G)f)(x,y) \\ &\Leftrightarrow f \in \eta_A((x,y)). \end{aligned}$$

Finally, $\hat{\eta}$ is the unique transformation making the above diagram commute, since, for any natural transformation $\sigma : G \to PF$, we have

$$\sigma = (e_F \circ (1_F \times \sigma))^\wedge.$$

To see this, observe that, for $A \xrightarrow{f} B$, $y \in GA$, $z \in FB$,

$$\begin{aligned} &(z,f) \in (e_F \circ (1_F \times \sigma)^\wedge)(y)(B) \\ &\quad \Leftrightarrow (e_F)_B((z,\sigma_B((Gf)y))) = \max_B \\ &\quad \Leftrightarrow 1_B \in (e_F)_B((z,\sigma_B((Gf)y))) \\ &\quad \Leftrightarrow (z,1_B) \in \sigma_B((Gf)y)(B) = ((PF)f)(\sigma_A(y))(B) \\ &\quad \Leftrightarrow (z,f) \in \sigma_A(y)(B). \end{aligned}$$

This completes the proof that $\mathbf{Set}^{\mathbf{C}}$ is a topos.

It is helpful to think of any object in $\mathbf{Set}^{\mathbf{C}}$, i.e. any functor $F:\mathbf{C}\to\mathbf{Set}$, as *a set varying over the category* $\mathbf{C}$. The idea here is that the objects A of $\mathbf{C}$ are to be taken as stages marking the 'development' of the varying set F, and FA as the 'set of elements' or 'content' of F at stage A. Any $\mathbf{C}$-arrow $A\xrightarrow{f}B$ then engenders a *transition map* $FA\xrightarrow{Ff}FB$ from the elements of F at stage A to those at stage B.

For these reasons $\mathbf{Set}^{\mathbf{C}}$ is called the *topos of sets varying over* $\mathbf{C}$.

Let us now discuss some special cases. To begin with, consider the case in which $\mathbf{C}$ is the category $\mathbf{P}$ associated with a partially ordered set P. An *object* in $\mathbf{Set}^{\mathbf{P}}$ is then given by: a map F which assigns to each $p\in P$ a set $F(p)$, and to each pair $p,q\in P$ for which $p\leq q$ a map $F_{pq}:F(p)\to F(q)$ such that, if $p\leqslant q\leqslant r$, then $F_{pr}=F_{qr}\circ F_{pq}$, and F_{pp} is the identity on $F(p)$. An *arrow* $\eta:F\to G$ in $\mathbf{Set}^{\mathbf{P}}$ is determined by a collection of maps $\eta_p:F(p)\to G(p)$ for $p\in P$ such that, whenever $p\leqslant q$, we have

$$\eta_q\circ F_{pq}=G_{pq}\circ\eta_p.$$

A *sieve* in P on an element $p\in P$ is just what is often called a *filter over* p, that is, a subset U of the set $O_p=\{q\in P:p\leqslant q\}$ such that $q\in U$ and $q\leqslant r$ imply $r\in U$. The *maximal sieve* on p is O_p. The *subobject classifier* Ω in $\mathbf{Set}^{\mathbf{P}}$ then has

$$\Omega(p)=\text{set of all filters over } p,$$

and for $U\in\Omega(p)$, $p\leqslant q$

$$\Omega_{pq}(U)=U\cap O_q.$$

Next, suppose that $\mathbf{C}$ is the category $\mathbf{M}$ associated with a *monoid* $(M,\cdot)$. It is readily seen that any *object* in $\mathbf{Set}^{\mathbf{M}}$ may be identified with a set *acted on by* M, or an *M-set* for short, that is, a pair $(X,\cdot)$ in which X is a set and $\cdot$ is a map from $M\times X$ to X satisfying

$$\alpha\cdot(\beta\cdot x)=(\alpha\beta)\cdot x,\qquad 1\cdot x=x$$

for all $\alpha,\beta\in M$, $x\in X$. An *arrow* $f:(X,\cdot)\to(Y,\cdot)$ between two M-sets is an *equivariant* map $f:X\to Y$, i.e. such that $f(\alpha\cdot x)=\alpha\cdot f(x)$ for all $x\in X$, $\alpha\in M$. The *subobject classifier* Ω in $\mathbf{Set}^{\mathbf{M}}$ is the collection of all *left ideals* of M, where a subset I of M is a *left ideal* if it is closed under multiplication on the left:

$$\alpha\in I,\ \beta\in M\Rightarrow\beta\alpha\in I.$$

The action of M on Ω is *division:*

$$\alpha\cdot I=\{\beta\in M:\beta\alpha\in I\}.$$

The truth arrow $1\xrightarrow{\top}\Omega$ is the function with value M, the unit ideal.

$\mathbf{Set}^{\mathbf{M}}$ is called the *topos of M-sets.*

Let X be a topological space, and $\mathcal{O}(X)$ its family of open sets partially ordered by inclusion. A set varying over the partially ordered set $\mathcal{O}(X)^{\mathrm{op}}$ is classically termed a *presheaf on X*. In consonance with this, a set varying over $\mathbf{C}^{\mathrm{op}}$, where $\mathbf{C}$ is a small category, is called a *presheaf on* $\mathbf{C}$. The topos $\mathbf{Set}^{\mathbf{C}^{\mathrm{op}}}$ is accordingly called the *topos of presheaves* on $\mathbf{C}$. We note that, in $\mathbf{Set}^{\mathbf{C}^{\mathrm{op}}}$, the subobject classifier Ω has, for any $\mathbf{C}$-object A,

$$\Omega(A) = \text{set of all } \textit{cosieves} \text{ on } A,$$

where, of course, a *cosieve* on A is the dual of a sieve on A, that is to say, a set of $\mathbf{C}$-arrows with *codomain* A closed under composition on the *right*.

REMARK. Let F be any presheaf on $\mathbf{C}$ and let A be any $\mathbf{C}$-object. If $Y: \mathbf{C} \to \mathbf{Set}^{\mathbf{C}^{\mathrm{op}}}$ is the Yoneda embedding, then by the Yoneda lemma (1.7) there is a bijective correspondence between the elements of FA, i.e. the elements of F at stage A, and the arrows $YA \to F$ in $\mathbf{Set}^{\mathbf{C}^{\mathrm{op}}}$. If $\mathbf{C}$ has a terminal object 1, it is readily seen that $Y1$ is a terminal object in $\mathbf{Set}^{\mathbf{C}^{\mathrm{op}}}$, and so the elements of F at stage 1 correlate exactly with the arrows $1 \to F$ in $\mathbf{Set}^{\mathbf{C}^{\mathrm{op}}}$, in other words with the $\mathbf{Set}^{\mathbf{C}^{\mathrm{op}}}$-*elements* of F.

Geometric morphisms

For each set I we define the *constant presheaf* $\hat{I}$ *on* $\mathbf{C}$ *with value* I to be the functor $\hat{I} \in \mathbf{Set}^{\mathbf{C}^{\mathrm{op}}}$ satisfying for any $A \in \mathrm{Ob}(\mathbf{C})$ and $A \xrightarrow{f} B$ in $\mathbf{C}$,

$$\hat{I}(A) = I, \qquad \hat{I}(f) = 1_I.$$

$\hat{I}$ may be thought of as the *representative* of the set I in $\mathbf{Set}^{\mathbf{C}^{\mathrm{op}}}$. Since I is the I-fold copower of 1 in $\mathbf{Set}$, and colimits are determined pointwise in $\mathbf{Set}^{\mathbf{C}^{\mathrm{op}}}$ (by 1.11), it follows that

$$\hat{I} = \coprod_I 1$$

in $\mathbf{Set}^{\mathbf{C}^{\mathrm{op}}}$.

The assignment $I \mapsto \hat{I}$ can be turned into a functor $\gamma^*: \mathbf{Set} \to \mathbf{Set}^{\mathbf{C}^{\mathrm{op}}}$ by defining

$$\gamma^*(I) = \hat{I}$$

and, for $I \xrightarrow{f} J$ in $\mathbf{Set}$,

$$(\gamma^*(f))_A = f$$

for $A \in \mathrm{Ob}(\mathbf{C})$.

Notice that γ^* is *left exact*. To prove this, it suffices to show that γ^* preserves 1, binary products and equalizers. Preservation of 1 is obvious.

For products, we note that since the product in $\mathbf{Set}^{\mathbf{C}^{\mathrm{op}}}$ is determined 'pointwise', we have

$$\begin{aligned}\gamma^*(I \times J)(A) &= (I \times J)\hat{}(A)\\ &= I \times J\\ &= \hat{I}(A) \times \hat{J}(A)\\ &= (\hat{I} \times \hat{J})(A)\\ &= (\gamma^*(I) \times \gamma^*(J))(A).\end{aligned}$$

The preservation of equalizers is established similarly.

Moreover, γ^* has a *right adjoint* which we will denote by $\gamma_*: \mathbf{Set}^{\mathbf{C}^{\mathrm{op}}} \to \mathbf{Set}$. What form must γ_* have? To determine this, observe that, if $\gamma^* \dashv \gamma_*$, then for any set I and any $F \in \mathbf{Set}^{\mathbf{C}^{\mathrm{op}}}$ we must have

$$\mathbf{Set}^{\mathbf{C}^{\mathrm{op}}}(\hat{I}, F) = \mathbf{Set}^{\mathbf{C}^{\mathrm{op}}}(\gamma^* I, F) \cong \mathbf{Set}(I, \gamma_* F).$$

Taking $I = 1$, we see that, since $\hat{1} \cong 1$,

$$\mathbf{Set}^{\mathbf{C}^{\mathrm{op}}}(1, F) \cong \mathbf{Set}(1, \gamma_* F).$$

Now $\mathbf{Set}(1, \gamma_* F)$ is essentially the set of elements of $\gamma_* F$ and so may be identified with $\gamma_* F$ itself. Thus we are led to define

$$\gamma_* F = \mathbf{Set}^{\mathbf{C}^{\mathrm{op}}}(1, F) = \text{set of } \mathbf{Set}^{\mathbf{C}^{\mathrm{op}}}\text{-elements of } F. \qquad (*)$$

We can, however, give an explicit description of the $\mathbf{Set}^{\mathbf{C}^{\mathrm{op}}}$-elements of F. For any arrow $1 \xrightarrow{\xi} F$ in $\mathbf{Set}^{\mathbf{C}^{\mathrm{op}}}$, each $\xi_A : 1 \to F$ may be identified with its image $x_A = \xi_A(0) \in FA$. The x_A then satisfy the condition: for any $A \xrightarrow{f} B$,

$$(Ff)(x_B) = x_A. \qquad (**)$$

A set $\{x_A : A \in \mathrm{Ob}(\mathbf{C})\}$ such that $x_A \in FA$ for all $A \in \mathrm{Ob}(\mathbf{C})$ and satisfying (**) is called an *F-matching family*. Thus any arrow $1 \xrightarrow{\xi} F$ in $\mathbf{Set}^{\mathbf{C}^{\mathrm{op}}}$ gives rise to an F-matching family; conversely it is easy to see that any F-matching family engenders an arrow $1 \to F$ in $\mathbf{Set}^{\mathbf{C}^{\mathrm{op}}}$. Accordingly, $\mathbf{Set}^{\mathbf{C}^{\mathrm{op}}}$-elements of F may be identified with F-matching families.

The pair (γ_*, γ^*) is an example of what is called a *geometric*† *morphism of toposes* which play a particularly important role in the theory. Explicitly, if $\mathbf{E}$ and $\mathbf{E}'$ are toposes, a *geometric morphism* $\mu: \mathbf{E} \to \mathbf{E}'$ consists of a pair of functors $\mu_*: \mathbf{E} \to \mathbf{E}'$, $\mu^*: \mathbf{E}' \to \mathbf{E}$ such that $\mu^* \dashv \mu_*$ and μ^* is left exact. In particular, the pair $\gamma = (\gamma_*, \gamma^*)$ described above is a geometric morphism $\mathbf{Set}^{\mathbf{C}^{\mathrm{op}}} \to \mathbf{Set}$.

It turns out that the particular structure of the geometric morphism $\gamma: \mathbf{Set}^{\mathbf{C}^{\mathrm{op}}} \to \mathbf{Set}$ is actually typical of geometric morphisms to $\mathbf{Set}$, for we have the following proposition.

† For an explanation of the origin of this term, see the end of Chapter 5.

2.21 PROPOSITION. Let **E** be a topos. If there exists a geometric morphism $\mu:\mathbf{E}\to\mathbf{Set}$, then it is uniquely determined up to isomorphism; in fact $\mu_*\cong H_1$ and, for any set I, $\mu^*(I)\cong\coprod_I 1$, the I-fold copower of 1 in **E** (which accordingly exists).

Proof. Let $\mu:\mathbf{E}\to\mathbf{Set}$ be a geometric morphism. Since μ^* is left exact, it preserves terminal objects, i.e. $\mu^*(1)\cong 1$. Then for each **E**-object A we have natural bijections

$$\begin{aligned}\mathbf{Set}(1,\mu_*A)\cong\mathbf{E}(\mu^*1,A)\cong\mathbf{E}(1,A)\\ =H_1(A).\end{aligned}$$

So $\mu_*\cong H_1$. Moreover, for any set I we have natural bijections

$$\mathbf{E}(\mu^*I,A)\cong\mathbf{Set}(I,\mu_*A)\cong\mathbf{Set}(I,\mathbf{E}(1,A)).$$

Thus **E**-arrows $\mu^*I\to A$ correspond bijectively to maps $I\to\mathbf{E}(1,A)$, which in turn correspond to I-indexed sets of arrows $\{1\xrightarrow{f_i}A:i\in I\}$. But this is precisely the condition describing $\coprod_I 1$ in **E**; it follows that $\mu^*I\cong\coprod_I 1$. □

In Chapter 4 we will show, conversely, that if **E** has arbitrary set-fold copowers of 1, i.e. if $\coprod_I 1$ exists in **E** for any set I, then the pair $\mu=(\mu_*,\mu^*)$ defined by $\mu_*(A)=H_1(A)$, $\mu^*(I)=\coprod_I 1$ is a geometric morphism $\mathbf{E}\to\mathbf{Set}$. *A topos having arbitrary set-fold copowers of* 1 *is said to be* defined over **Set**: it follows from these remarks that *a topos* **E** *is defined over* **Set** *if and only if there is a (unique) geometric morphism* $\mathbf{E}\to\mathbf{Set}$.

3

Local set theories

The prime example of a topos is the category of sets. The fact that it is a topos is, of course, a consequence of the axioms of classical set theory. In this chapter we formulate a generalization of the system of classical set theory—*local set theory*— within which the construction of a corresponding 'category of sets' can still be carried out, and shown to be a topos—exactly as in the classical case. We will show that *any* topos is obtainable (up to equivalence of categories) as the category of sets within some local set theory, thus revealing the precise sense in which toposes are to be regarded as generalizations of the category of sets.

We will also show that toposes are in a natural sense the *models* or *interpretations* of local set theories. Introducing the concept of *validity* of an assertion of a local set theory under an interpretation, we will establish that such interpretations are *sound* in the sense that any theorem of a local set theory is valid under any interpretation validating its axioms and *complete* in the sense that, conversely, any assertion of a local set theory valid under every interpretation validating its axioms is itself a theorem. The basic axioms and rules of local set theories will be chosen in such a way as to yield as theorems precisely those of (higher-order) *intuitionistic logic*. These basic theorems will then coincide with those assertions that are valid under *every* interpretation.

In a local set theory the set concept, as a primitive, is replaced by the notion of *type*. Types may be thought of as *natural kinds* or *species* from which sets are extracted as subspecies. The resulting theory of sets is *local* in the sense that, for example, the inclusion relation will only obtain among sets which have the same type, i.e. are subspecies of the same species.

A local set theory, then, is a type-theoretic system built on the same primitive symbols $=$, $\in$, $\{:\}$ as classical set theory, in which the set-theoretic operations of forming products and powers of types can be performed, and which in addition contains a 'truth value' type acting as the range of values of 'characteristic functions' on types. A local set theory is determined by specifying a collection of *axioms* formulated within a *local language* defined as follows.

Local languages and local set theories

A *local language* $\mathscr{L}$ is determined by first specifying the following classes of symbols:

(S1) symbols $\mathbf{1},\mathbf{\Omega}$ called the *unity type symbol* and the *truth value type symbol,* respectively;

(S2) a (possibly empty) collection of symbols $\mathbf{A},\mathbf{B},\mathbf{C}, \ldots$ called *ground type symbols*;

(S3) a (possibly empty) collection of symbols $\mathbf{f},\mathbf{g},\mathbf{h}, \ldots$ called *function symbols.*

The *type symbols* of $\mathscr{L}$ are now defined recursively as follows.

(TS1) $\mathbf{1},\mathbf{\Omega}$ are type symbols;
(TS2) any ground type symbol is a type symbol;
(TS3) if $\mathbf{A}_1, \ldots, \mathbf{A}_n$ are type symbols, so as $\mathbf{A}_1 \times \cdots \times \mathbf{A}_n$, with the proviso that, if $n = 1$, then $\mathbf{A}_1 \times \cdots \times \mathbf{A}_n$ is $\mathbf{A}_1$, and if $n = 0$, then $\mathbf{A}_1 \times \cdots \times \mathbf{A}_n$ is $\mathbf{1}$;
(TS4) if $\mathbf{A}$ is a type symbol, so is $\mathbf{PA}$.

The type symbols $\mathbf{A}_1 \times \cdots \times \mathbf{A}_n$ and $\mathbf{PA}$ are called the *product* of $\mathbf{A}_1, \ldots, \mathbf{A}_n$ and the *power* of $\mathbf{A}$, respectively. A type symbol of the form $\mathbf{PA}$ is called a *power type symbol.*

For each type symbol $\mathbf{A}$ we assume that $\mathscr{L}$ contains a set of symbols $x_{\mathbf{A}},y_{\mathbf{A}},z_{\mathbf{A}}, \ldots$ called *variables* of type $\mathbf{A}$. In addition we assume that $\mathscr{L}$ contains the symbol $*$.

Each function symbol of $\mathscr{L}$ is assumed to be assigned a *signature* of the form $\mathbf{A} \to \mathbf{B}$, where $\mathbf{A},\mathbf{B}$ are type symbols. We assume that the collection of function symbols of any given signature forms a *set.*

Now we can define the *terms* of $\mathscr{L}$ and their associated *types* recursively as follows.

(T1) $*$ is a term of type $\mathbf{1}$;

(T2) for each type symbol $\mathbf{A}$, variables $x_{\mathbf{A}},y_{\mathbf{A}},z_{\mathbf{A}}, \ldots$ are terms of type $\mathbf{A}$;

(T3) if $\mathbf{f}$ is a function symbol of signature $\mathbf{A} \to \mathbf{B}$, and τ is a term of type $\mathbf{A}$, $\mathbf{f}(\tau)$ is a term of type $\mathbf{B}$;

(T4) if $\tau_1, \ldots, \tau_n$ are terms of types $\mathbf{A}_1, \ldots, \mathbf{A}_n$, $\langle \tau_1, \ldots, \tau_n \rangle$ is a term of type $\mathbf{A}_1 \times \cdots \times \mathbf{A}_n$, with the proviso that, if $n = 1$, then $\langle \tau_1, \ldots, \tau_n \rangle$ is τ_1, while if $n = 0$, then $\langle \tau_1, \ldots, \tau_n \rangle$ is $*$;

(T5) if τ is a term of type $\mathbf{A}_1 \times \cdots \times \mathbf{A}_n$, and $1 \leqslant i \leqslant n$, $(\tau)_i$ is a term of type $\mathbf{A}_i$;

(T6) if α is a term of type $\mathbf{\Omega}$, and $x_{\mathbf{A}}$ is a variable of type $\mathbf{A}$, $\{x_{\mathbf{A}} : \alpha\}$ is a term of type $\mathbf{PA}$;

(T7) if σ, τ are terms of the same type, $\sigma = \tau$ is a term of type $\mathbf{\Omega}$;

(T8) if σ, τ are terms of types $\mathbf{A}, \mathbf{PA}$ respectively, $\sigma \in \tau$ is a term of type $\mathbf{\Omega}$.

In (T4), $\langle \tau_1, \ldots, \tau_n \rangle$ is the *ordered n-tuple* of $\tau_1, \ldots, \tau_n$; in (T5) $(\boldsymbol{\tau})_i$ is the ith *coordinate of* τ; and in (T6) $\{x_A : \alpha\}$ is *the set of all* x_A *such that* α.

A term of type $\mathbf{\Omega}$ is called a *formula*.

Arbitrary terms will be denoted by Greek letters $\sigma, \tau, \ldots$. Symbols $\omega, \omega', \omega'', \ldots$ will be reserved for variables of type $\mathbf{\Omega}$ and α, β, γ for formulae. We shall usually suppress the type subscripts attached to variables when confusion is unlikely to arise: thus variables will usually appear as $x, y, z, \ldots$.

Sometimes a term will be introduced as $\tau(x)$; this is done solely to draw attention to the variable x and does not imply that x occurs in τ.

An occurrence of a variable x in a term τ is *bound* if it appears within a context of the form $\{x : \alpha\}$; otherwise the occurrence is *free*. A term with no free variables is said to be *closed*; a closed formula is called a *sentence*. For any terms τ, σ and any variable x of the same type as σ we write $\tau(x/\sigma)$ (or sometimes just $\tau(\sigma)$) for the term obtained from τ by substituting σ for each free occurrence of x. The term σ is, as usual, said to be *free for x in τ* if, for any free variable y in σ, any occurrence of y in $\tau(x/\sigma)$ not already in τ is a free occurrence. If Γ is any set of formulae, we write $\Gamma(x/\sigma)$ for the collection of formulae of the form $\tau(x/\sigma)$ with τ in Γ, and say that σ is *free for x in Γ* if σ is free for x in every formula in Γ. However, we also say that x is *free in Γ* if it is free in *some* member of Γ.

Similarly, for any variables $x_1, \ldots, x_n$ and terms $\sigma_1, \ldots, \sigma_n$ of the appropriate types, we write $\tau(x_1/\sigma_1, \ldots, x_n/\sigma_n)$ or briefly $\tau(\mathbf{x}/\boldsymbol{\sigma})$ for the result of substituting σ_i for x_i in τ for $1 \leqslant i \leqslant n$. This notation extends naturally to $\Gamma(x_1/\sigma_1, \ldots, x_n/\sigma_n)$ or $\Gamma(\mathbf{x}/\boldsymbol{\sigma})$.

Logical operations, axioms and inference rules. The reader will observe that the local language $\mathscr{L}$ does not have any logical operations among its primitive symbols. The reason for this is that in a typed language such as $\mathscr{L}$ the logical operations can be *defined* as follows. We write

(L1)	$\alpha \Leftrightarrow \beta$	for	$\alpha = \beta$
(L2)	*true*	for	$* = *$
(L3)	$\alpha \wedge \beta$	for	$\langle \alpha, \beta \rangle = \langle true, true \rangle$
(L4)	$\alpha \Rightarrow \beta$	for	$(\alpha \wedge \beta) \Leftrightarrow \alpha$
(L5)†	$\forall x \alpha$	for	$\{x : \alpha\} = \{x : true\}$
(L6)	*false*	for	$\forall \omega . \omega$
(L7)	$\neg \alpha$	for	$\alpha \Rightarrow false$
(L8)	$\alpha \vee \beta$	for	$\forall \omega[(\alpha \Rightarrow \omega \wedge \beta \Rightarrow \omega) \Rightarrow \omega]$
(L9)†	$\exists x \alpha$	for	$\forall \omega[\forall x(\alpha \Rightarrow \omega) \Rightarrow \omega]$

† For clarity we shall occasionally write $\forall x.\alpha$, $\exists x.\alpha$ for $\forall x \alpha$, $\exists x \alpha$.

In (L8) and (L9) ω is assumed to be a variable of type $\mathbf{\Omega}$ not occurring in α or β.

Once we have specified the axioms and inference rules of local set theories, we will see that the logical operations introduced in (L1)–(L9) satisfy the laws of intuitionistic logic.

We therefore turn now to the axioms and inference rules for a local set theory. These will be given in terms of *sequents*.

A *sequent* (in $\mathscr{L}$) is an expression of the form

$$\Gamma : \alpha$$

where α is a formula and Γ is a (possibly empty) *finite* set of formulae. We write:

$$\begin{array}{rcl} \Gamma, \Delta : \alpha & \text{for} & \Gamma \cup \Delta : \alpha \\ \beta, \Gamma : \alpha \ \text{ or } \ \Gamma, \beta : \alpha & \text{for} & \Gamma \cup \{\beta\} : \alpha \\ \beta_1, \ldots, \beta_n : \alpha & \text{for} & \{\beta_1, \ldots, \beta_n\} : \alpha \\ : \alpha & \text{for} & \varnothing : \alpha. \end{array}$$

The *basic axioms* for a local set theory are the following sequents:

Tautology	$\alpha : \alpha$
Unity	$: x_1 = *$
Equality	$x = y, \alpha(z/x) : \alpha(z/y)$ with x and y free for z in α
Products	$: (\langle x_1, \ldots, x_n \rangle)_i = x_i$
	$: x = \langle (x)_1, \ldots, (x)_n \rangle$
Comprehension	$: x \in \{x : \alpha\} \Leftrightarrow \alpha$

The *rules of inference* for a local set theory are the following.

Thinning $$\frac{\Gamma : \alpha}{\beta, \Gamma : \alpha}$$

Cut $$\frac{\Gamma : \alpha \quad \alpha, \Gamma : \beta}{\Gamma : \beta} \text{(any free variable of } \alpha \text{ free in } \Gamma \text{ or } \beta)$$

Substitution $$\frac{\Gamma : \alpha}{\Gamma(x/\tau) : \alpha(x/\tau)} (\tau \text{ free for } x \text{ in } \Gamma \text{ and } \alpha)$$

Extensionality $$\frac{\Gamma : x \in \sigma \Leftrightarrow x \in \tau}{\Gamma : \sigma = \tau} (x \text{ not free in } \Gamma, \sigma, \tau)$$

Equivalence $$\frac{\alpha, \Gamma : \beta \quad \beta, \Gamma : \alpha}{\Gamma : \alpha \Leftrightarrow \beta}$$

If S is any collection of sequents, we define a *proof* from S to be a finite tree with its vertex at the bottom whose nodes are correlated with sequents in such a way that (i) any sequent associated with a node which has nodes above it is a direct consequence by one of the rules of inference of the sequent(s) associated with the node(s) immediately above it and (ii) every topmost node is correlated with either a basic axiom or a member of S. The sequent assigned to the vertex of the tree is called the *conclusion* of the proof.

We say that the sequent $\Gamma : \alpha$ is *derivable* from S, and write

$$\Gamma \vdash_S \alpha$$

provided there is a proof from S of which the sequent $\Gamma : \alpha$ is the conclusion. We also write

$$\Gamma \vdash \alpha$$

for $\Gamma \vdash_\varnothing \alpha$ and say that $\Gamma : \alpha$ is a *valid* sequent. When $\varnothing \vdash_S \alpha$ we write $\vdash_S \alpha$ and say that α is *provable* from S. We note that if $S \subseteq S'$ then $\vdash_S \alpha$ implies $\vdash_{S'} \alpha$.

Successive applications of the cut rule will be written in the form

$$\Gamma \vdash_S \alpha_1 \vdash_S \alpha_2 \vdash_S \cdots \vdash_S \alpha_n$$

yielding

$$\Gamma \vdash_S \alpha_n.$$

A *local set theory*, or just a *theory*, in $\mathscr{L}$ is now formally defined to be a collection S of sequents which is closed under derivability. That is, S is a local set theory if, for any sequent $\Gamma : \alpha$,

$$\Gamma \vdash_S \alpha \quad \text{iff} \quad (\Gamma : \alpha) \text{ is in } S.$$

Clearly any collection S of sequents *generates* a local set theory $\bar{S}$ given by

$$(\Gamma : \alpha) \text{ is in } \bar{S} \quad \text{iff} \quad \Gamma \vdash_S \alpha.$$

S is said to be a *set of axioms* for a local set theory S' if $\bar{S} = S'$.

Given two local set theories S, S' in $\mathscr{L}$, S' is said to be an *extension* of S if for any sequent $\Gamma : \alpha$ of $\mathscr{L}$, $\Gamma \vdash_S \alpha$ implies $\Gamma \vdash_{S'} \alpha$.

A local set theory S is said to be *consistent* if it is *not* the case that $\vdash_S$ *false*.

The local set theory in $\mathscr{L}$ generated by the empty set of axioms is called the *pure local set theory* in $\mathscr{L}$ and is denoted by L.

The local language with no ground types or function symbols is called the *pure local language* and is denoted by $\mathscr{L}_0$. The pure local set theory in $\mathscr{L}_0$ is denoted by L_0.

Logic in a local set theory

We now verify that the logical operations on formulae defined above obey the rules customary in intuitionistic logic, except for an additional restriction on variables in the quantifier rules.

In displaying proofs we shall often omit steps which are evident or have been previously established: sometimes such omitted steps will be indicated by a series of dots:. . . .

We write

$$\frac{\Gamma_1 : \alpha_1, \ldots, \Gamma_n : \alpha_n}{\Delta : \beta} \qquad (*)$$

for 'there exists a proof from the set of sequents $\{\Gamma_1 : \alpha_1, \ldots, \Gamma_n : \alpha_n\}$ with the sequent $\Delta : \beta$ as conclusion'. Once an assertion of the form (*) has been established we shall feel free to treat it as a derived rule of inference and thence to employ it without further comment in subsequent proofs.

Now to the proofs themselves.

3.1 CONJUNCTION

(3.1.1) $\quad x = x', y = y' \vdash \langle x, y \rangle = \langle x', y' \rangle.$ □

(3.1.2) (i) $\langle x, y \rangle = \langle x', y' \rangle \vdash x = x'.$ □

(ii) $\langle x, y \rangle = \langle x', y' \rangle \vdash y = y'.$ □

(3.1.3) $\quad \alpha \vdash \alpha.$ □

(3.1.4) $\quad \vdash \mathit{true}.$ □

(3.1.5) (i) $\alpha \vdash \alpha = \mathit{true}$

(ii) $\alpha = \mathit{true} \vdash \alpha.$

Proof. (i)

$$\frac{\mathit{true}, \alpha : \alpha \qquad \overline{\alpha, \alpha : \mathit{true}}}{\alpha : \alpha = \mathit{true}}$$

(ii)

$$\frac{\dfrac{\omega = \omega', \omega' : \omega}{\alpha = \mathit{true}, \mathit{true} : \alpha} \qquad \overline{: \mathit{true}}}{\alpha = \mathit{true} : \alpha} \qquad \square$$

(3.1.6)

$$\frac{\Gamma : \alpha \qquad \Gamma : \beta}{\Gamma : \alpha \wedge \beta}$$

Proof.

$$\dfrac{\Gamma:\beta \qquad \dfrac{\Gamma:\alpha \qquad \dfrac{\overline{\beta:\beta=true} \qquad \dfrac{\overline{\alpha:\alpha=true} \qquad \overline{\alpha=true,\ \beta=true:\alpha\wedge\beta}}{\alpha,\beta=true:\alpha\wedge\beta}}{\alpha,\beta:\alpha\wedge\beta}}{\Gamma,\beta:\alpha\wedge\beta}}{\Gamma:\alpha\wedge\beta}$$ □

(3.1.7) (i) $\dfrac{\alpha,\Gamma:\gamma}{\alpha\wedge\beta,\ \Gamma:\gamma}$ (ii) $\dfrac{\beta,\Gamma:\gamma}{\alpha\wedge\beta,\ \Gamma:\gamma}$

Proof. (i)

$$\dfrac{\dfrac{\overline{\alpha\wedge\beta:\alpha=true} \qquad \overline{\alpha=true:\alpha}}{\alpha\wedge\beta:\alpha} \qquad \alpha,\Gamma:\gamma}{\alpha\wedge\beta,\Gamma:\gamma}$$

(ii) is similar. □

3.2 IMPLICATION

(3.2.1)

$$\dfrac{\alpha,\Gamma:\beta}{\Gamma:\alpha\Rightarrow\beta}$$

Proof.

$$\dfrac{\dfrac{\dfrac{\alpha:\alpha}{\vdots}}{\alpha\wedge\beta,\Gamma:\alpha} \qquad \dfrac{\dfrac{\alpha,\Gamma:\alpha \qquad \alpha,\Gamma:\beta}{\vdots}}{\alpha,\Gamma:\alpha\wedge\beta}}{\Gamma:\alpha\Rightarrow\beta}$$ □

(3.2.2) (i)

$$\dfrac{\Gamma:\alpha \qquad \beta,\Gamma:\gamma}{\alpha\Leftrightarrow\beta,\Gamma:\gamma}$$

(ii)

$$\dfrac{\Gamma:\alpha \qquad \beta,\Gamma:\gamma}{\beta\Leftrightarrow\alpha,\Gamma:\gamma}$$

Proof. (i)

$$\dfrac{\dfrac{\Gamma:\alpha \qquad \alpha\Leftrightarrow\beta,\alpha:\beta}{\alpha\Leftrightarrow\beta,\Gamma:\beta} \qquad \beta,\Gamma:\gamma}{\alpha\Leftrightarrow\beta,\Gamma:\gamma}$$

(ii) is similar. □

(3.2.3) (i) $\dfrac{\Gamma:\alpha \qquad \beta,\Gamma:\gamma}{\alpha\Rightarrow\beta,\ \Gamma:\gamma}$ (ii) $\dfrac{\Gamma:\alpha\Rightarrow\beta}{\Gamma,\alpha:\beta}$.

Proof. (i)

$$\frac{\Gamma:\alpha \qquad \dfrac{\beta,\Gamma:\gamma}{\alpha\wedge\beta,\Gamma:\gamma}}{\alpha\Rightarrow\beta,\ \Gamma:\gamma}$$

(ii)

$$\frac{\dfrac{\Gamma:\alpha\Rightarrow\beta}{\Gamma,\alpha:\alpha\Rightarrow\beta} \qquad \dfrac{\Gamma,\alpha:\alpha \qquad \Gamma,\alpha,\beta:\beta}{\Gamma,\alpha,\alpha\Rightarrow\beta:\beta}}{\Gamma,\alpha:\beta} \qquad \square$$

3.3 CONJUNCTION (continued)

(3.3.1) (i) $\dfrac{\alpha,\beta:\gamma}{\alpha\wedge\beta:\gamma}$ (ii) $\dfrac{\alpha\wedge\beta:\gamma}{\alpha,\beta:\gamma}$

Proof. (i)

$$\frac{\alpha\wedge\beta:\alpha \qquad \dfrac{\dfrac{\dfrac{\dfrac{}{\alpha\wedge\beta:\beta} \qquad \dfrac{\alpha,\beta:\gamma}{\beta:\alpha\Rightarrow\gamma} \qquad \vdots}{\alpha\wedge\beta:\alpha\Rightarrow\gamma \qquad \alpha\Rightarrow\gamma,\alpha:\gamma}}{\vdots}}{\alpha\wedge\beta,\alpha:\gamma}}{\alpha\wedge\beta:\gamma}$$

(ii)

$$\frac{\dfrac{\alpha,\beta:\alpha \qquad \alpha,\beta:\beta}{\alpha,\beta:\alpha\wedge\beta} \qquad \alpha\wedge\beta:\gamma}{\alpha,\beta:\gamma} \qquad \square$$

If $\Gamma=\{\alpha_1,\ldots,\alpha_n\}$, write $\bigwedge\Gamma$ or $\alpha_1\wedge\cdots\wedge\alpha_n$ for $(\alpha_1\wedge(\alpha_2\wedge(\cdots\wedge(\alpha_{n-1}\wedge\alpha_n)\cdots)$.

(3.3.2) (i) $\dfrac{\Gamma:\alpha}{\bigwedge\Gamma:\alpha}$ (ii) $\dfrac{\bigwedge\Gamma:\alpha}{\Gamma:\alpha}$

Proof. (i) Induction on the cardinality of Γ, using (3.2.3) and (3.3.1). Suppose the assertion true for all Γ with less than n members. Then, writing Γ' for $\{\alpha_2,\ldots,\alpha_{n-1}\}$,

$$\cfrac{\cfrac{\cfrac{\cfrac{\cfrac{\cfrac{\Gamma:\alpha}{\alpha_1,\ \Gamma':\alpha}}{\Gamma':\alpha_1\Rightarrow\alpha}}{\bigwedge\Gamma':\alpha_1\Rightarrow\alpha}}{\alpha_1,\ \bigwedge\Gamma':\alpha}}{\alpha_1\wedge\bigwedge\Gamma':\alpha}}{\bigwedge\Gamma:\alpha}$$

(ii) Same as (i), only turn the proof tree upside down. $\square$

3.4 UNIVERSAL QUANTIFICATION

(3.4.1)
$$\frac{\Gamma:\alpha\Leftrightarrow\beta}{\Gamma:\{x:\alpha\}=\{x:\beta\}}$$

provided x is not free in Γ.

Proof.

$$\dfrac{\dfrac{\dfrac{\dfrac{\Gamma:\alpha\Leftrightarrow\beta}{\vdots}}{\alpha,\Gamma:\beta}\;\vdots}{x\in\{x:\alpha\},\,\Gamma:x\in\{x:\beta\}}\quad\dfrac{\dfrac{\dfrac{\Gamma:\alpha\Leftrightarrow\beta}{\vdots}}{\beta,\Gamma:\alpha}\;\vdots}{x\in\{x:\beta\},\,\Gamma:x\in\{x:\alpha\}}}{\dfrac{\Gamma:x\in\{x:\alpha\}\Leftrightarrow x\in\{x:\beta\}}{\Gamma:\{x:\alpha\}=\{x:\beta\}.}}$$ □

(3.4.2)
$$\frac{\Gamma:\alpha}{\Gamma:\forall x\alpha}$$

provided either (i) x is not free in Γ or (ii) x is not free in α.

Proof. (i)

$$\dfrac{\Gamma:\alpha\qquad\overline{\alpha:\alpha\Leftrightarrow \mathit{true}}}{\dfrac{\dfrac{\Gamma:\alpha\Leftrightarrow \mathit{true}}{\vdots}}{\Gamma:\forall x\alpha}}$$

(ii) Let v be a new variable. Then

$$\dfrac{\dfrac{\dfrac{\Gamma:\alpha}{\Gamma(x/v):\alpha}}{\Gamma(x/v):\forall x\alpha}}{\Gamma:\Sigma\forall x\alpha}$$ □

(3.4.3)
$$\frac{\Gamma:\{x:\alpha\}=\{x:\beta\}}{\Gamma:\alpha\Leftrightarrow\beta}$$

provided that x is free in at least one of α,β.

Proof. This is shown as Fig. 3.1. □

(3.4.4)
$$\forall x\alpha\vdash\alpha$$

provided x is free in α.

$$\dfrac{\Gamma:\{x:\alpha\}=\{x:\beta\} \qquad \dfrac{\dfrac{\dfrac{}{\alpha:x\in\{x:\alpha\}} \qquad \dfrac{\{x:\alpha\}=\{x:\beta\},\, x\in\{x:\alpha\}:x\in\{x:\beta\} \qquad \dfrac{}{x\in\{x:\beta\}:\beta}}{\{x:\alpha\}=\{x:\beta\},\, x\in\{x:\alpha\}:\beta}}{\{x:\alpha\}=\{x:\beta\},\, \alpha:\beta} \qquad \overline{\{x:\alpha\}=\{x:\beta\},\, \beta:\alpha}}{\{x:\alpha\}=\{x:\beta\}:\alpha\Leftrightarrow\beta}}{\Gamma:\alpha\Leftrightarrow\beta}$$

Figure 3.1 Proof of 3.4.3

Proof.

$$\dfrac{\dfrac{\forall x\alpha:\forall x\alpha}{\forall x\alpha:\alpha=true} \qquad \dfrac{}{\alpha=true:\alpha}}{\forall x\alpha:\alpha} \quad \square$$

(3.4.5) $$\forall u\alpha(x/u)\Leftrightarrow\forall x\alpha$$

provided u is free for x and not free in α.

Proof.

$$\dfrac{\dfrac{\dfrac{\dfrac{\dfrac{:u\in\{u:\alpha(x/u)\}\Leftrightarrow\alpha(x/u)}{:x\in\{u:\alpha(x/u)\}\Leftrightarrow\alpha}}{\vdots}}{:x\in\{u:\alpha(x/u)\}\Leftrightarrow x\in\{x:\alpha\}}}{:\{u:\alpha(x/u)\}=\{x:\alpha\}}}{\dfrac{\vdots}{:\forall u\alpha(x/u)\Leftrightarrow\forall x\alpha}} \quad \square$$

(3.4.6) $$\dfrac{\Gamma:\alpha(x/u)}{\Gamma:\forall x\alpha}$$

provided that either (i) u is free for x in α and not free in Γ or $\forall x\alpha$ or (ii) x is not free in α.

Proof.

$$\dfrac{\dfrac{\Gamma:\alpha(x/u)}{\Gamma:\forall u\alpha(x/u)} \qquad \dfrac{}{\forall u\alpha(x/u):\forall x\alpha}}{\Gamma:\forall x\alpha} \quad \square$$

(3.4.7) $$\dfrac{\alpha(x/\tau),\, \Gamma:\beta}{\forall x\alpha,\, \Gamma:\beta}$$

provided that τ is free for x in α, x is free in α and any free variable of τ is free in $\forall x\alpha$, Γ, or β.

Proof.

$$\dfrac{\dfrac{\overline{\forall x\alpha : \alpha}}{\forall x\alpha : \alpha(x/\tau)} \qquad \alpha(x/\tau),\, \Gamma : \beta}{\forall x\alpha,\, \Gamma : \beta} \qquad \square$$

3.5 NEGATION

(3.5.1) $$\textit{false} \vdash \alpha$$

Proof.

$$\dfrac{\overline{\textit{false} : \omega}}{\textit{false} : \alpha} \quad \square$$

(3.5.2) $$\dfrac{\alpha, \Gamma : \textit{false}}{\Gamma : \neg\alpha} \quad \square$$

(3.5.3) $$\dfrac{\Gamma : \alpha}{\neg\alpha, \Gamma : \textit{false}} \quad \square$$

3.6 DISJUNCTION

(3.6.1) $$\dfrac{\alpha, \Gamma : \gamma \qquad \beta, \Gamma : \gamma}{\alpha \vee \beta, \Gamma : \gamma}$$

Proof.

$$\dfrac{\dfrac{\dfrac{\dfrac{\alpha,\Gamma : \gamma}{\Gamma : \alpha \Rightarrow \gamma} \qquad \dfrac{\beta,\Gamma : \gamma}{\Gamma : \beta \Rightarrow \gamma}}{\Gamma : (\alpha \Rightarrow \gamma) \wedge (\beta \Rightarrow \gamma)} \qquad \gamma : \gamma}{(\alpha \Rightarrow \gamma \wedge \beta \Rightarrow \gamma) \Rightarrow \gamma,\, \Gamma \vdash \gamma}}{\alpha \vee \beta, \Gamma : \gamma} \qquad \square$$

(3.6.2) (i) $\dfrac{\Gamma : \alpha}{\Gamma : \alpha \vee \beta}$ (ii) $\dfrac{\Gamma : \beta}{\Gamma : \alpha \vee \beta}$

Proof. (i)

$$\dfrac{\dfrac{\dfrac{\dfrac{\Gamma : \alpha \qquad \omega : \omega}{\alpha \Rightarrow \omega,\, \Gamma : \omega}}{\alpha \Rightarrow \omega \wedge \beta \Rightarrow \omega, \Gamma : \omega}}{\Gamma : (\alpha \Rightarrow \omega \wedge \beta \Rightarrow \omega) \Rightarrow \omega}}{\Gamma : \alpha \vee \beta}$$

(ii) is similar. $\square$

3.7 EXISTENTIAL QUANTIFICATION

(3.7.1) $$\alpha \vdash \exists x\alpha$$

provided x is free in α.

Proof.

$$\dfrac{\dfrac{\dfrac{\dfrac{\alpha:\alpha \qquad \omega:\omega}{\alpha\Rightarrow\omega,\alpha:\omega}}{\forall x(\alpha\Rightarrow\omega),\ \alpha:\omega}}{\alpha:\forall x(\alpha\Rightarrow\omega)\Rightarrow\omega}}{\alpha:\exists x\alpha} \qquad \square$$

(3.7.2) $$\dfrac{\alpha,\Gamma:\beta}{\exists x\alpha,\Gamma:\beta}$$

provided that either (i) x is not free in Γ or β or (ii) x is not free in α.

Proof. (i)

$$\dfrac{\dfrac{\dfrac{\dfrac{\alpha,\Gamma:\beta}{\Gamma:\alpha\Rightarrow\beta}}{\Gamma:\forall x(\alpha\Rightarrow\beta)} \qquad \beta:\beta}{\forall x(\alpha\Rightarrow\beta)\Rightarrow\beta,\Gamma:\beta}}{\exists x\alpha,\ \Gamma:\beta}$$

(ii) Let u be a new variable. Then

$$\dfrac{\dfrac{\dfrac{\alpha,\Gamma:\beta}{\alpha,\Gamma(x/u):\beta(x/u)}}{\exists x\alpha,\Gamma(x/u):\beta(x/u)}}{\exists x\alpha,\Gamma:\beta} \qquad \square$$

(3.7.3) $$\dfrac{\alpha(x/u),\ \Gamma:\beta}{\exists x\alpha,\ \Gamma:\beta}$$

provided that either (i) u is free for x in α and not free in $\exists x\alpha$, Γ or β, or (ii) x is not free in α.

Proof.

$$\dfrac{\exists x\alpha:\exists x\alpha(x/u) \qquad \dfrac{\alpha(x/u),\ \Gamma:\beta}{\exists u\alpha(x/u),\ \Gamma:\beta}}{\exists x\alpha,\ \Gamma:\beta} \qquad \square$$

(3.7.4) $$\dfrac{\Gamma:\alpha(x/\tau)}{\Gamma:\exists x\alpha}$$

provided that τ is free for x in α, x is free in α and any free variable of τ is free in Γ or $\exists x\alpha$.

Proof.

$$\dfrac{\Gamma:\alpha(x/\tau) \qquad \dfrac{\dfrac{\vdots}{\alpha:\exists x\alpha}}{\alpha(x/\tau):\exists x\alpha}}{\Gamma:\exists x\alpha} \qquad \square$$

(3.7.5) $$\vdash \exists u\alpha(x/u) \Leftrightarrow \exists x\alpha$$

provided that u is free for x but not free in α.

Proof.

$$\dfrac{\dfrac{\alpha(x/u):\alpha(x/u)}{\alpha(x/u):\exists x\alpha}}{\exists u\alpha(x/u):\exists x\alpha}$$

Noting that $\alpha(x/u)(u/x)$ is just α,

$$\dfrac{\dfrac{\alpha:\alpha(x/u)(u/x)}{\alpha:\exists u\alpha(x/u)}}{\exists x\alpha:\exists u\alpha(x/u)} \qquad \square$$

(3.7.6) $$\vdash \exists x(\alpha\wedge\beta) \Leftrightarrow \alpha\wedge\exists x\beta$$

provided x is free in β but not α.

Proof

$$\dfrac{\dfrac{\dfrac{}{\alpha\wedge\beta:\alpha}}{\exists x(\alpha\wedge\beta):\alpha} \qquad \dfrac{\dfrac{\dfrac{}{\alpha\wedge\beta:\beta \qquad \beta:\exists x\beta}}{\alpha\wedge\beta:\exists x\beta}}{\exists x(\alpha\wedge\beta):\exists x\beta}}{\exists x(\alpha\wedge\beta):\alpha\wedge\exists x\beta}$$

and

$$\dfrac{\dfrac{\dfrac{\dfrac{}{\alpha,\beta:\alpha\wedge\beta}}{\alpha,\beta:\exists x(\alpha\wedge\beta)}}{\alpha,\exists x\beta:\exists x(\alpha\wedge\beta)}}{\alpha\wedge\exists x\beta:\exists x(\alpha\wedge\beta)} \qquad \square$$

(3.7.7) $$\dfrac{\Gamma:\alpha \qquad \alpha,\Gamma:\beta}{\exists x_1(x_1=x_1),\ldots,\exists x_n(x_n=x_n),\Gamma:\beta}$$

where $x_1, \ldots, x_n$ are the free variables of α which have no free occurrences in Γ or in β.

Proof.

$$\dfrac{\dfrac{\dfrac{\Gamma:\alpha}{x_1=x_1, \ldots, x_n=x_n, \Gamma:\alpha} \qquad \dfrac{\alpha,\Gamma:\beta}{x_1=x_1, \ldots, x_n=x_n, \alpha, \Gamma:\beta}}{x_1=x_1, \ldots, x_n=x_n, \Gamma:\beta}}{\exists x_1(x_1=x_1), \ldots, \exists x_n(x_n=x_n), \Gamma:\beta} \qquad \square$$

This yields another version of the cut rule

(3.7.8)
$$\frac{\Gamma:\alpha \qquad \alpha,\Gamma:\beta}{\Gamma:\beta}$$

provided that, whenever $\mathbf{A}$ is the type of a free variable of α with no free occurrences in Γ or β, there is a closed term of type $\mathbf{A}$.

Proof. Let $x_1, \ldots, x_n$ be the free variables of α with no free occurrences in Γ or β, and let $\mathbf{A}_1, \ldots, \mathbf{A}_n$ be their respective types. Let $\tau_1, \ldots, \tau_n$ be closed terms of types $\mathbf{A}_1, \ldots, \mathbf{A}_n$. Then

$$\dfrac{\dfrac{\overline{:\tau_1=\tau_1, \ldots, :\tau_n=\tau_n}}{:\exists x_1(x_1=x_1), \ldots, :\exists x_n(x_n=x_n)} \qquad \dfrac{\Gamma:\alpha \qquad \alpha,\Gamma:\beta}{\exists x_1(x_1=x_1), \ldots, \exists x_n(x_n=x_n), \Gamma:\beta}}{\Gamma:\beta}$$

$\square$

This concludes our verification of the rules satisfied by the logical operators in local set theories.

Our final task in this section is to show that local set theories satisfy *eliminability of descriptions for formulae* and *for terms of power type.* To explain this, define the *unique existential quantifier* $\exists!$ by writing

$$\exists! x\alpha \qquad \text{for} \qquad \exists x(\alpha \wedge \forall y(\alpha(x/y) \Rightarrow x=y)),$$

where y is different from x and not free in α. Then, if $\alpha(x)$ is any formula and x is any variable either of type $\mathbf{\Omega}$ or of power type, we will exhibit a term τ of the same type as x, for which we have

$$\exists! x\alpha \vdash \alpha(x/\tau).$$

τ is thus an explicit term satisfying the description: 'the unique x such that α'.

3.8 PROPOSITION (Eliminability of descriptions for formulae).

$$\exists! \omega\alpha \vdash \alpha(\omega/\alpha(\omega/\mathit{true})).$$

Proof. We have

$$\exists!\omega\alpha,\ \alpha(\omega/true),\ \alpha \vdash \omega = true \vdash \omega.$$

Also

$$\omega, \alpha \vdash \omega = true \wedge \alpha \vdash \alpha(\omega/true).$$

Therefore

$$\exists!\omega\alpha, \alpha \vdash \omega = \alpha(\omega/true),$$

whence

$$\exists!\omega\alpha, \alpha \vdash \alpha(\omega/\alpha(\omega/true)),$$

and so

$$\exists!\omega\alpha, \exists\omega\alpha \vdash \alpha(\omega/\alpha(\omega/true)).$$

But clearly $\exists!\omega\alpha \vdash \exists\omega\alpha$ and so the cut rule yields the conclusion. □

3.9 PROPOSITION (Eliminability of descriptions for terms of power type). Let α be a formula and let w be a variable of power type. Let x be a variable not occurring in α and define

$$a = \{x : \exists w(x \in w \wedge \alpha)\}.$$

Then

$$\exists!w\alpha \vdash \alpha(w/a).$$

Proof. First, we have

$$\begin{aligned} \alpha, x \in w &\vdash \exists w(x \in w \wedge \alpha) \\ &\vdash x \in a \end{aligned}$$

whence

$$\alpha \vdash x \in w \Rightarrow x \in a. \quad (1)$$

Next, if z is a new variable,

$$\begin{aligned} \exists!w\alpha,\ \alpha,\ x \in a &\vdash \exists!w\alpha \wedge \alpha \wedge \exists z(x \in z \wedge \alpha(w/z)) \\ &\vdash \exists z(\exists!w\alpha \wedge \alpha \wedge \alpha(w/z) \wedge x \in z) \\ &\vdash \exists z(w = z \wedge x \in z) \\ &\vdash x \in w. \end{aligned}$$

Hence

$$\exists!w\alpha, \alpha \vdash x \in a \Rightarrow x \in w,$$

which together with (1) gives

$$\exists!w\alpha, \alpha \vdash a = w \wedge \alpha \vdash \alpha(w/a).$$

Therefore $\exists!w\alpha, \exists w\alpha \vdash \alpha(w/a)$. But clearly $\exists!w\alpha \vdash \exists w\alpha$ and the conclusion follows. □

REMARK. It can be shown (cf. (4.9)) that $\mathbf{\Omega}$ is in a natural sense 'isomorphic' to **P1** and so is 'virtually' a power type symbol. So (3.8) is really a special case of (3.9).

Set theory in a local language

If X is a term of power type (i.e. of type **PA** for some type symbol **A**) in a local language $\mathscr{L}$, then assertions of the form $x \in X$ are meaningful (for any variable X of type **A**), so that X may be held to be capable of possessing *elements*. Moreover, in view of the axioms of extensionality and comprehension, X may be regarded as being *determined* by its elements, and so it is natural to consider X as being a *set*. This leads to the following

DEFINITION. The *set-like terms* of a local language $\mathscr{L}$ are the terms of power type. A *closed* set-like term will be called an *$\mathscr{L}$-set* or simply a *set*.

We use upper case letters $A, B, \ldots, Z$ for sets. We introduce the following abbreviations:

$\forall x \in X.\alpha$	for	$\forall x(x \in X \Rightarrow \alpha)$
$\exists x \in X.\alpha$	for	$\exists x(x \in X \wedge \alpha)$
$\exists! x \in X.\alpha$	for	$\exists! x(x \in X \wedge \alpha)$
$\{x \in X : \alpha\}$	for	$\{x : x \in X \wedge \alpha\}$.

We can now define the usual set-theoretic operations and relations in $\mathscr{L}$.

3.10 SET-THEORETIC DEFINITIONS IN $\mathscr{L}$.

(i)	$X \subseteq Y$	for	$\forall x \in X. x \in Y$
(ii)	$X \cap Y$	for	$\{x : x \in X \wedge x \in Y\}$
(iii)	$X \cup Y$	for	$\{x : x \in X \vee x \in Y\}$
(iv)	$U_{\mathbf{A}}$ or A	for	$\{x_{\mathbf{A}} : true\}$
(v)	$\varnothing_{\mathbf{A}}$ or $\varnothing$	for	$\{x_{\mathbf{A}} : false\}$
(vi)	$-X$	for	$\{x : \neg(x \in X)\}$
(vii)	PX	for	$\{u : u \subseteq X\}$
(viii)	$\bigcap U$	for	$\{x : \forall u \in U. x \in u\}$
(ix)	$\bigcup U$	for	$\{x : \exists u \in U. x \in u\}$

(x)	$\bigcap_{i \in I} X_i$	for	$\{x : \forall i \in I. x \in X_i\}$
(xi)	$\bigcup_{i \in I} X_i$	for	$\{x : \exists i \in I. x \in X_i\}$
(xii)	$\{\tau\}$	for	$\{x : x = \tau\}$
(xiii)	$\{\sigma, \tau\}$	for	$\{x : x = \sigma \vee x = \tau\}$
(xiv)	$\{\tau : \alpha\}$	for	$\{z : \exists x_1 \cdots \exists x_n (z = \tau \wedge \alpha)\}$
(xv)	$X \times Y$	for	$\{\langle x,y \rangle : x \in X \wedge y \in Y\}$
(xvi)	$X + Y$	for	$\{\langle \{x\}, \varnothing \rangle : x \in X\} \cup \{\langle \varnothing, \{y\} \rangle : y \in Y\}$.
(xvii)	X^Y	for	$\{u : u \subseteq Y \times X \wedge \forall y \in Y \exists! x \in X. \langle y,x \rangle \in u\}$
(xviii)	$\prod_{i \in I} X_i$	for	$\{u \in U^I_{\mathbf{A}} : \forall i \in I. \{x : \langle i,x \rangle \in u\} \subseteq X_i\}$
(xix)	$\coprod_{i \in I} X_i$	for	$\{\langle i,x \rangle : i \in I \wedge x \in X_i\}$.

REMARKS. In (i), (ii), and (iii) X and Y are required to be of the same type, while in (xvi), (xvii), and (xviii) their types may be different. If X, Y are of types **PA**, **PB** respectively, then $X \times Y$, $X + Y$, X^Y are of types **P(A × B)**, **P(PA × PB)**, **PP(B × A)** respectively.

In (viii) and (ix) U is of type **PPA**.

In (xii) and (xiii) x is required to be not free in τ.

In (xiv) $x_1, \ldots, x_n$ are the free variables of τ and τ is required not to be a variable.

In (x), (xi), (xviii), (xix) X_i is a term of type **PA** which may or may not have the variable i as a free variable.

By abuse of terminology, an $\mathscr{L}$-set of the form $U_{\mathbf{A}}$ will be called a *type*.

The following facts concerning the set-theoretic operations and relations are now established as straightforward consequences of their definitions; we omit the simple proofs.

3.11 PROPOSITION

(i) $\vdash X = Y \Leftrightarrow \forall x (x \in X \Leftrightarrow x \in Y)$

(ii) (a) $\vdash X \subseteq X$

(b) $\vdash (X \subseteq Y \wedge Y \subseteq X) \Rightarrow X = Y$

(c) $\vdash (X \subseteq Y \wedge Y \subseteq Z) \Rightarrow X \subseteq Z$

(iii) $\vdash Z \subseteq X \cap Y \Leftrightarrow Z \subseteq X \wedge Z \subseteq Y$

(iv) $\vdash X \cup Y \subseteq Z \Leftrightarrow X \subseteq Z \wedge Y \subseteq Z$

(v) $\vdash x_{\mathbf{A}} \in U_{\mathbf{A}}$

(vi) $\vdash \neg(x_{\mathbf{A}} \in \varnothing_{\mathbf{A}})$

(vii) $\vdash X \in PY \Leftrightarrow X \subseteq Y$

(viii) $\vdash X \subseteq \bigcap U \Leftrightarrow \forall u \in U . X \subseteq u$

(ix) $\vdash \bigcup U \subseteq X \Leftrightarrow \forall u \in U . u \subseteq X$

(x) $\vdash x \in \{y\} \Leftrightarrow x = y$

(xi) $\vdash \alpha \Rightarrow \tau \in \{\tau : \alpha\}$. □

Here (i) is the *axiom of extensionality,* (iv) the *axiom of binary unions,* (vi) the *axiom of the empty set,* (vii) the *power set axiom,* (ix) the *axiom of unions* and (x) the *axiom of singletons.* These, together with the comprehension axiom, form the core axioms for set theory in $\mathscr{L}$. The set theory is *local* because some of the set-theoretic operations, e.g. intersection and union, may only be performed on sets of the same type, i.e. 'locally'. Moreover, variables are constrained to range only over given types—locally—in contrast with the situation in classical set theory where they are permitted to range globally over a putative universe of discourse.

The category of sets determined by a local set theory

We now show that any local set theory S determines a category $\mathbf{C}(S)$ in essentially the same way as classical set theory determines the category **Set** of sets. As in the classical case, the associated category $\mathbf{C}(S)$ turns out to be a topos.

Let S be a local set theory in a local language $\mathscr{L}$. We define the relation $\sim_S$ on the collection of $\mathscr{L}$-sets by

$$X \sim_S Y \qquad \text{iff} \qquad \vdash_S X = Y.$$

It is readily checked that $\sim_S$ is an equivalence relation. An S-*set* is defined to be an equivalence class $[X]_S$ of $\mathscr{L}$-sets under the relation $\sim_S$. We shall usually *identify* an $\mathscr{L}$-set X with the corresponding S-set $[X]_S$ and denote the latter simply by X. (Thus an S-set will often just be called a *set.*) Notice that then $X = Y$ (i.e., X and Y denote the same S-set) iff $\vdash_S X = Y$.

An S-*map* $X \xrightarrow{f} Y$ is a triple of S-sets (f, X, Y) such that

$$\vdash_S f \in Y^X.$$

We shall usually simply write f for (f, X, Y) and call f an S-map. X and Y are, respectively, the *domain* $\operatorname{dom}(f)$ and *codomain* $\operatorname{cod}(f)$ of f. An S-map will sometimes just be called a *map.*

We now show that the collection of S-sets and maps form a category.

3.12 LEMMA. Let $X \xrightarrow{f} Y$, $X \xrightarrow{g} Y$. Then

$$f = g \quad \text{iff} \quad x \in X \vdash_S \langle x,y \rangle \in f \Leftrightarrow \langle x,y \rangle \in g.$$

Proof. Rule of extensionality. □

If $X \xrightarrow{f} Y \xrightarrow{g} Z$, define the composition $g \circ f$ by

$$g \circ f = \{\langle x,z \rangle : \exists y(\langle x,y \rangle \in f \wedge \langle y,z \rangle \in g)\}.$$

3.13 LEMMA

(i) $X \xrightarrow{g \circ f} Z$
(ii) $\circ$ is associative.

Proof. As for composition of maps in classical set theory, using (3.12). □

Given an S-set X, define

$$\Delta_X = \{\langle x,x \rangle : x \in X\}$$

and

$$1_X = (\Delta_X, X, X).$$

Δ_X is the *diagonal* on X and 1_X the *identity map* on X.

3.14 LEMMA

(i) $X \xrightarrow{1_X} X$,
(ii) for any $X \xrightarrow{f} Y$, $Z \xrightarrow{g} X$, $f \circ 1_X = f$, $1_X \circ g = g$.

Proof. As for classical set theory. □

Lemmas 3.13 and 3.14 show that the collection of S-sets and maps forms a category†. We denote this category by $\mathbf{C}(S)$, and call it the *category of S-sets* (*and maps*), or the *category associated with S*.

Our next task is to show that $\mathbf{C}(S)$ is a topos.

To do this we first show that every term in $\mathscr{L}$ gives rise to a map (i.e., a set) in $\mathbf{C}(S)$: this is another respect in which the set theory is *local*, in contrast with classical set theory in which terms yield globally defined operations which are not, in general, sets.

Suppose that τ is a term such that

$$\langle x_1, \ldots, x_n \rangle \in X \vdash_S \tau \in Y;$$

† Strictly speaking, for this to be the case we need also show that the collection of S-maps between any pair of S-sets forms an (intuitive) set. Actually, this follows easily from the assumption made initially that the collection of function symbols of $\mathscr{L}$ of a given signature forms a set: we leave the proof to the reader.

we write

$$X \xrightarrow{\langle x_1,\ldots,x_n\rangle \mapsto \tau} Y \qquad \text{or} \qquad (\langle x_1, \ldots, x_n\rangle \mapsto \tau) \qquad \text{or} \qquad (\mathbf{x} \mapsto \tau)$$

for

$$\{\langle\langle x_1, \ldots, x_n\rangle, \tau\rangle : \langle x_1, \ldots, x_n\rangle \in X\}.$$

It is easy to see that if $x_1, \ldots, x_n$ includes all the free variables of τ and X and Y are S-sets, then $X \xrightarrow{\langle x_1,\ldots,x_n\rangle \mapsto \tau} Y$ is an S-map $X \to Y$. In this case, when there is no risk of confusion, we shall sometimes write $X \xrightarrow{\tau} Y$ for this map.

If $\mathbf{f}$ is a function symbol, we sometimes write f for the map $x \mapsto \mathbf{f}(x)$.

3.15 LEMMA

$$(\langle y_1, \ldots, y_m\rangle \mapsto \tau) \circ (\langle x_1, \ldots, x_n\rangle \mapsto \langle \sigma_1, \ldots, \sigma_m\rangle) = \langle x_1, \ldots, x_n\rangle \mapsto \tau(\mathbf{y}/\boldsymbol{\sigma})$$

provided σ_i is free for y_i in τ for each i, $1 \leq i \leq n$. □

We can now establish

3.16 THEOREM. For any local set theory S, the category $\mathbf{C}(S)$ is a topos.

The proof proceeds by verifying that $\mathbf{C}(S)$ has the appropriate properties.

(3.16.1) $\mathbf{C}(S)$ *has a terminal object.*

Proof. Let 1 be $U_{\mathbf{1}}$. For any S-set X, consider

$$X \xrightarrow{x \mapsto *} 1.$$

If $X \xrightarrow{f} 1$ then

$$x \in X \vdash_S \langle x, *\rangle \in f$$

and it follows from (3.12) that

$$f = (x \mapsto *).$$

Hence for any X there is a unique S-map $X \to 1$, so that 1 is a terminal object in $\mathbf{C}(S)$. □

(3.16.2) $\mathbf{C}(S)$ *has binary products.*

Proof. Let X and Y be S-sets. Define

$$X \xleftarrow{\pi_1} X \times Y \xrightarrow{\pi_2} Y$$

by $\pi_1 = (\langle x,y\rangle \mapsto x)$, $\pi_2 = (\langle x,y\rangle \mapsto y)$. Then if $X \xleftarrow{f} Z \xrightarrow{g} Y$, define

$Z \xrightarrow{\langle f,g\rangle} X \times Y$ by

$$\langle f,g\rangle = \{\langle z, \langle x,y\rangle\rangle : \langle z,x\rangle \in f \wedge \langle z,y\rangle \in g\}.$$

Clearly $\pi_1 \circ \langle f,g\rangle = f$ and $\pi_2 \circ \langle f,g\rangle = g$. If $Z \xrightarrow{h} X \times Y$ satisfies $\pi_1 \circ h = f$ and $\pi_2 \circ h = g$, then

$$z \in Z \vdash_S \langle z, \langle x,y\rangle\rangle \in h \Leftrightarrow (\langle z,x\rangle \in f \wedge \langle z,y\rangle \in g),$$

so that $h = \langle f,g\rangle$.

Therefore $X \times Y$ is a product of X and Y. □

Next, we need to characterize the monic arrows in $\mathbf{C}(S)$.

(3.16.3) *An S-map f is monic iff*

$$\langle x,z\rangle \in f, \langle y,z\rangle \in f \vdash_S x = y. \qquad (*)$$

Proof. Suppose that $Y \xrightarrow{f} Z$ is monic. Let R be the S-set

$$\{\langle x,y\rangle : \exists z(\langle x,z\rangle \in f \wedge \langle y,z\rangle \in f\}.$$

Define $R \xrightarrow{g} Y$, $R \xrightarrow{h} Y$ by

$$g = (\langle x,y\rangle \mapsto x)$$
$$h = (\langle x,y\rangle \mapsto y).$$

Then we have

$$\exists z(\langle x,z\rangle \in f \wedge \langle y,z\rangle \in f) \vdash_S \langle \langle x,y\rangle, z\rangle \in f \circ g \Leftrightarrow \langle \langle x,y\rangle, z\rangle \in f \circ h.$$

Therefore $f \circ g = f \circ h$, and so, since f is monic, $g = h$. (*) follows.

Conversely suppose that (*) is satisfied and that $X \xrightarrow{g} Y$, $X \xrightarrow{h} Y$ are such that $f \circ g = f \circ h$. Then

$$x \in X \vdash_S \exists u\, \exists v(\langle x,u\rangle \in g \wedge \langle x,v\rangle \in h \\ \wedge \langle u,z\rangle \in f \wedge \langle v,z\rangle \in f).$$

Since (*) holds, it follows that

$$x \in X \vdash_S \exists u(\langle x,u\rangle \in g \wedge \langle x,u\rangle \in h)$$

and hence $g = h$. Therefore f is monic. □

Given an S-map $X \xrightarrow{f} U_{\Omega}$, write

$$f^*(x) \qquad \text{for} \qquad \langle x, \mathit{true}\rangle \in f.$$

Then since $\vdash_S \exists! \omega . \langle x, \omega\rangle \in f$, it follows from (3.8) that

$$x \in X \vdash \langle x, f^*(x)\rangle \in f.$$

Accordingly $f^*(x)$ may be thought of as the *value* of f at x. The following proposition is then an immediate consequence.

(3.16.4) (i) $X \xrightarrow{x \mapsto f^*(x)} U_\Omega = f$,

(ii) $\vdash_S (\langle x_1, \ldots, x_n \rangle \mapsto \alpha)^*(\langle x_1, \ldots, x_n \rangle) \Leftrightarrow \alpha$. □

Now write Ω for U_Ω and

$$1 \xrightarrow{\top} \Omega \qquad \text{for} \qquad 1 \xrightarrow{x \mapsto true} \Omega.$$

Then we have

(3.16.5) *A diagram in* $\mathbf{C}(S)$

$$\begin{array}{ccc} Y & \longrightarrow & 1 \\ {\scriptstyle m}\big\downarrow & & \big\downarrow{\scriptstyle \top} \\ X & \xrightarrow{h} & \Omega \end{array} \qquad (m \text{ monic})$$

is a pullback if and only if

$$h = (x \mapsto \exists y.\langle y,x \rangle \in m). \qquad (**)$$

Proof. First, assume (**). Then we have

$$\begin{aligned} \vdash_S \langle z,\omega \rangle \in h \circ m \\ &\Leftrightarrow \exists x(\langle z,x \rangle \in m \wedge \omega = \exists y.\langle y,x \rangle \in m) \\ &\Leftrightarrow \omega = true. \end{aligned}$$

It follows that (*) commutes.

To see that (*) is a pullback, suppose that $Z \xrightarrow{f} X$ is such that the diagram

$$\begin{array}{ccc} Z & \longrightarrow & 1 \\ {\scriptstyle f}\big\downarrow & & \big\downarrow{\scriptstyle \top} \\ X & \xrightarrow{h} & \Omega \end{array}$$

commutes. Define

$$g = \{\langle z,y \rangle : \exists x(\langle z,x \rangle \in f \wedge \langle y,x \rangle \in m)\}.$$

Using the monicity of f, it is easily checked that g is a map $Z \xrightarrow{g} Y$. Moreover $m \circ g = f$. Also, if $m \circ h = f$, then $h = g$ since m is monic. Therefore (*) is a pullback.

Conversely, suppose that (*) is a pullback. Since (*) commutes, we have

$$\vdash_S \langle x,\omega \rangle \in h \circ m \Leftrightarrow \omega = true.$$

Therefore

$$\langle y,x \rangle \in m \wedge \langle x,\omega \rangle \in h \vdash_S \omega = true,$$

whence

$$\langle x,y\rangle \in m \vdash_S \langle m,\omega\rangle \in h \Rightarrow \omega = true$$

so that

$$\langle y,x\rangle \in m \vdash_S \langle x, true\rangle \in h \vdash_S h^*(x)$$

and

$$\exists y.\langle y,x\rangle \in m \vdash_S h^*(x). \tag{1}$$

Now let $Z = \{x : h^*(x)\}$ and define $Z \xrightarrow{\bar{h}} X$ by $\bar{h} = (x \mapsto x)$. Then clearly the diagram

$$\begin{array}{ccc} Z & \longrightarrow & 1 \\ {\scriptstyle \bar{h}}\downarrow & & \downarrow{\scriptstyle \mathsf{T}} \\ X & \xrightarrow{h} & \Omega \end{array}$$

commutes. Since (*) is a pullback, there is $Z \xrightarrow{f} Y$ such that $\bar{h} = m \circ f$. It follows that

$$\begin{aligned} h^*(x) &\vdash_S \exists y(\langle x,y\rangle \in f \wedge \langle y,x\rangle \in m) \\ &\vdash_S \exists y.\langle y,x\rangle \in m. \end{aligned}$$

This, together with (1), gives

$$\vdash_S h^*(x) = (\exists y.\langle y,x\rangle \in m),$$

which, by (3.16.4), gives (**). □

Now we can prove

(3.16.6) **C**(S) *has a subobject classifier, namely* (Ω, T).

Proof. Suppose we are given a monic arrow $Y \xrightarrow{m} X$ in **C**(S). Define $X \xrightarrow{\chi(m)} \Omega$ by

$$\chi(m) = (x \mapsto \exists y.\langle y,x\rangle \in m).$$

Then, by (3.16.5), $\chi(m)$ is the unique arrow $X \to \Omega$ such that

$$\begin{array}{ccc} Y & \longrightarrow & 1 \\ {\scriptstyle m}\downarrow & & \downarrow{\scriptstyle \mathsf{T}} \\ X & \xrightarrow{\chi(m)} & \Omega \end{array}$$

is a pullback.

Conversely, given $X \xrightarrow{h} \Omega$ define $Z = \{x : h^*(x)\}$ and define $Z \xrightarrow{\bar{h}} X$ by $\bar{h} = (x \mapsto x)$. Then we have

$$\begin{aligned} \chi(\bar{h}) &= (x \mapsto \exists y.\langle y,x\rangle \in \bar{h}) \\ &= x \mapsto h^*(x) \\ &= h \end{aligned}$$

by (3.16.4).

Thus (Ω, T) is a subobject classifier in $\mathbf{C}(S)$, as claimed. □

(3.16.7) $\mathbf{C}(S)$ *has power objects.*

Proof. Given an S-set X, we claim that PX is a power object for X. First, define $X \times PX \xrightarrow{e_x} \Omega$ by

$$e_X = (\langle x,z \rangle \mapsto x \in z).$$

Then if $X \times Y \xrightarrow{f} \Omega$, define $Y \xrightarrow{\hat{f}} PX$ by

$$\hat{f} = (y \mapsto \{x : f^*(\langle x,y \rangle)\}).$$

Then we have

$$e_X \circ (1_X \times \hat{f}) = (\langle x,y \rangle \mapsto f^*(\langle x,y \rangle)) \\ = f$$

by (3.16.4).

It is easily verified that if $Y \xrightarrow{g} PX$ satisfies $e_X \circ (1_X \times g) = f$, then

$$\vdash_S f^*(\langle x,y \rangle) = \exists w \in PX(x \in w \wedge \langle y,w \rangle \in g).$$

Also

$$y \in Y \vdash_S \langle y,z \rangle \in \hat{f} \Leftrightarrow z = \{x : f^*(\langle x,y \rangle)\}.$$

Hence

$$y \in Y \vdash_S \langle y,z \rangle \in \hat{f} \Leftrightarrow z = \{x : \exists w \in PX(x \in w \wedge \langle y,w \rangle \in g)\}. \qquad (1)$$

But

$$y \in Y \vdash_S \exists! w \in PX. \langle y,w \rangle \in g.$$

So by (3.9) and (1)

$$y \in Y \vdash_S \langle y,z \rangle \in \hat{f} \Leftrightarrow \langle y,z \rangle \in g.$$

It follows that $g = \hat{f}$. □

This completes the proof of Theorem 3.16. A topos of the form $\mathbf{C}(S)$ will be called a *linguistic topos*, to point up its derivation from a *language*.

Interpreting a local language in a topos: the soundness theorem

We next show how local languages may be *interpreted* in arbitrary toposes. This gives rise to a notion of *validity* for formulae for which we will establish the property of soundness: derivability in pure local set theory implies validity in every topos.

Let $\mathscr{L}$ be a local language, and **E** a topos. Assume that products, power objects, the terminal object and the subobject classifier in **E** have been specified.

An *interpretation* I of $\mathscr{L}$ in **E** is an assignment

(i) to each type **A** of an **E**-object $\mathbf{A}_I$ such that

$$(\mathbf{A}_1 \times \cdots \times \mathbf{A}_n)_I = (\mathbf{A}_1)_I \times \cdots \times (\mathbf{A}_n)_I$$
$$(\mathbf{PA})_I = P(\mathbf{A}_I)$$
$$\mathbf{1}_I = 1, \text{ the terminal object of } \mathbf{E}$$
$$\mathbf{\Omega}_I = \Omega_{\mathbf{E}},$$

(ii) to each function symbol **f** with signature $\mathbf{A} \to \mathbf{B}$ of an **E**-arrow

$$\mathbf{A}_I \xrightarrow{\mathbf{f}_I} \mathbf{B}_I.$$

More generally, an *interpretation* of $\mathscr{L}$ is a pair (**E**,I) consisting of a topos **E** and an interpretation I of $\mathscr{L}$ in **E**.

We shall often write $\mathbf{A}_E$ *or just* A *for* $\mathbf{A}_I$.

Given an interpretation (**E**,I) of $\mathscr{L}$, we shall extend I to all the terms of $\mathscr{L}$. Let τ be a term of type **B** and let $x_1, \ldots, x_n$ be distinct variables of types $\mathbf{A}_1, \ldots, \mathbf{A}_n$ including all the free variables of τ. Write **x** for the sequence $(x_1, \ldots, x_n)$. We define an **E**-arrow

$$A_1 \times \cdots \times A_n \xrightarrow{[\![\tau]\!]_{I,x_1,\ldots,x_n}} B$$

often written $[\![\tau]\!]_{I,\mathbf{x}}$ or $[\![\tau]\!]_{\mathbf{x}}$ recursively as follows.

(I1) $[\![*]\!]_{\mathbf{x}} =$ unique arrow $A_1 \times \cdots \times A_n \to 1$ in **E**
(I2) $[\![x_i]\!]_{\mathbf{x}} = \pi_i$, the projection $A_1 \times \cdots \times A_n \to A_i$
(I3) $[\![\mathbf{f}(\tau)]\!]_{\mathbf{x}} = \mathbf{f}_I \circ [\![\tau]\!]_{\mathbf{x}}$
(I4) $[\![\langle \tau_1, \ldots, \tau_n \rangle]\!]_{\mathbf{x}} = \langle [\![\tau_1]\!]_{\mathbf{x}}, \ldots, [\![\tau_n]\!]_{\mathbf{x}} \rangle$
(I5) $[\![(\tau)_i]\!]_{\mathbf{x}} = \pi_i \circ [\![\tau]\!]_{\mathbf{x}}$
(I6) $[\![\{y : \alpha\}]\!]_{\mathbf{x}} = ([\![\alpha(y/u)]\!]_{u\mathbf{x}} \circ \text{can})\hat{}$

where u is not one of $x_1, \ldots, x_n$ but is free for y in α,y is of type **C** (so that **B** is **PC**), 'can' is the canonical isomorphism

$$C \times (A_1 \times \cdots \times A_n) \cong C \times A_1 \times \cdots \times A_n$$

and $\hat{f}$ is the power transpose of f (defined just before (2.20)).

(I7) $[\![\sigma = \tau]\!]_{\mathbf{x}} = \text{eq}_C \circ [\![\langle \sigma, \tau \rangle]\!]_{\mathbf{x}}$

where σ,τ are of type **C**. Here eq_C is as defined just after (2.15).

(I8) $[\![\sigma \in \tau]\!]_{\mathbf{x}} = e_C \circ [\![\langle \sigma, \tau \rangle]\!]_{\mathbf{x}}$

where σ is of type **C**. Here e_C is the evaluation arrow of A defined just before (2.20).

We note that if τ is a *closed* term (of type **B**), then $\mathbf{x}$ may be taken to be the empty sequence $\varnothing$. Writing $[\![\tau]\!]$ for $[\![\tau]\!]_{I,\varnothing}$, it is evident that $[\![\tau]\!]$ is an **E**-element of B. In particular, if τ is a closed *set-like* term $\{y:\alpha\}$ of type **PC**, then $[\![\{y:\alpha\}]\!]$ is an **E**-element of PC, which in turn corresponds (via power transposition) to a *subobject* of C, viz., the subobject of C classified by $[\![\alpha]\!]_y$.

It is readily verified that

(3.17) $$[\![true]\!]_{\mathbf{x}} = T.$$

To establish our next results we shall need the following technical lemma whose proof (a simple diagram chase) we leave to the reader.

3.18 LEMMA. Let **C** be a category with finite products. Let $B, A_1, \ldots, A_n, B_1, \ldots, B_m$ be **C**-objects. Write 'can', 'can*' for the canonical isomorphisms

$$B \times (A_1 \times \cdots \times A_n) \cong B \times A_1 \times \cdots \times A_n,$$
$$B \times (B_1 \times \cdots \times B_m) \cong B \times B_1 \times \cdots \times B_m,$$

write 'proj' for the composition

$$B \times A_1 \times \cdots \times A_n \xrightarrow{\text{can}^{-1}} B \times (A_1 \times \cdots \times A_n) \xrightarrow{\pi_2} A_1 \times \cdots \times A_n$$

and π_1 for the projection $B \times A_1 \times \cdots \times A_n \to B$. Then for any arrows

$$A_1 \times \cdots \times A_n \xrightarrow{f_1} B_1, \ldots, A_1 \times \cdots \times A_n \xrightarrow{f_m} B_m,$$

we have

$$\langle \pi_1, f_1 \circ \text{proj}, \ldots, f_m \circ \text{proj} \rangle \circ \text{can} = \text{can}^* \circ (1_B \times \langle f_1, \ldots, f_m \rangle). \quad \square$$

We use this in the proof of the

3.19 LEMMA ON SUPERFLUOUS VARIABLES. Let τ be a term with free variables among $x_1, \ldots, x_n$. Suppose that $1 < p_1 < \cdots < p_m < n$ and $x_{p_1}, \ldots, x_{p_m}$ includes all the free variables of τ. Then, writing $(x_1, \ldots, x_n) = \mathbf{x}$, $(x_{p_1}, \ldots, x_{p_m}) = \mathbf{x}'$, we have, for any interpretation of $\mathscr{L}$ in a topos **E**,

$$[\![\tau]\!]_{\mathbf{x}} = [\![\tau]\!]_{\mathbf{x}'} \circ [\![\langle x_{p_1}, \ldots, x_{p_m} \rangle]\!]_{\mathbf{x}}.$$

Proof. By induction on the formation of τ. The only difficult induction step arises when τ is of the form $\{y:\alpha\}$. In this case we may assume that y is of type **B** and τ is of type **PB**. If $x_1, \ldots, x_n$ are of types $\mathbf{A}_1, \ldots, \mathbf{A}_n$, then, using the notation of (3.18), we have

$$\begin{aligned}
[\![\{y:\alpha\}]\!]_{\mathbf{x}} &= ([\![\alpha(y/u)]\!]_{u\mathbf{x}}\circ \mathrm{can})\hat{} \\
&= [[\![\alpha(y/u)]\!]_{u\mathbf{x}'}\circ[\![\langle u,x_{p_1},\ldots,x_{p_m}\rangle]\!]_{u\mathbf{x}}\circ \mathrm{can}]\hat{} \qquad \text{(induction hypothesis)}\\
&= [[\![\alpha(y/u)]\!]_{u\mathbf{x}'}\circ\langle \pi_1,[\![x_{p_1}]\!]_{u\mathbf{x}},\ldots,[\![x_{p_1}]\!]_{u\mathbf{x}}\rangle\circ \mathrm{can}]\hat{}\\
&= [[\![\alpha(y/u)]\!]_{u\mathbf{x}'}\circ\langle \pi_1,[\![x_{p_1}]\!]_{\mathbf{x}}\circ\mathrm{proj},\ldots,[\![x_{p_m}]\!]_{\mathbf{x}}\circ\mathrm{proj}\rangle\circ \mathrm{can}]\hat{}\\
&= [[\![\alpha(y/u)]\!]_{u\mathbf{x}'}\circ \mathrm{can}^*\circ(1_B\times[\![\langle x_{p_1},\ldots,x_{p_m}\rangle]\!]_{\mathbf{x}})]\hat{} \qquad \text{(by 3.18)}\\
&= [e_B\circ[1_B\times([\![\alpha(y/u)]\!]_{u\mathbf{x}'}\circ\mathrm{can}^*)\hat{}]\circ(1_B\times[\![\langle x_{p_1},\ldots,x_{p_m}\rangle]\!]_{\mathbf{x}})]\hat{}\\
&= [e_B\circ 1_B\times[\![\{y:\alpha\}]\!]_{\mathbf{x}'}\circ[\![\langle x_{p_1},\ldots,x_{p_m}\rangle]\!]_{\mathbf{x}}]\hat{}\\
&= [\![\{y:\alpha\}]\!]_{\mathbf{x}'}\circ[\![\langle x_{p_1},\ldots,x_{p_m}\rangle]\!]_{\mathbf{x}} \qquad \text{(by (2.20))}.
\end{aligned}$$

This completes the induction step. □

Now we can prove the

3.20 SUBSTITUTION LEMMA. Let τ be a term with free variables among $z_1,\ldots,z_m$ and let $\sigma_1,\ldots,\sigma_m$ be terms such that σ_i is free for z_i in τ for each $i=1,\ldots,m$. Then for any interpretation of $\mathscr{L}$ in a topos $\mathbf{E}$, we have

$$[\![\tau(\mathbf{z}/\boldsymbol{\sigma})]\!]_{\mathbf{x}} = [\![\tau]\!]_{\mathbf{z}}\circ[\![\langle\sigma_1,\ldots,\sigma_m\rangle]\!]_{\mathbf{x}}$$

(where $\mathbf{x}$ includes all free variables of $\sigma_1,\ldots,\sigma_m$).

Proof. Again the proof is by induction on the formation of τ, the only difficult induction step arising when τ is of the form $\{y:\alpha\}$. In this case we may assume that y is of type $\mathbf{B}$ and τ is of type $\mathbf{PB}$. Let $x_1,\ldots,x_n$ be of types $\mathbf{A}_1,\ldots,\mathbf{A}_n$. Now we have two subcases:

(i) y is not among $z_1,\ldots,z_m$. Then

$$\begin{aligned}
[\![\{y:\alpha\}(\mathbf{z}/\boldsymbol{\sigma})]\!]_{\mathbf{x}} &= [\![\{y:\alpha(\mathbf{z}/\boldsymbol{\sigma})\}]\!]_{\mathbf{x}}\\
&= [[\![\alpha(\mathbf{z}/\boldsymbol{\sigma})(y/u)]\!]_{u\mathbf{x}}\circ\mathrm{can}]\hat{}\\
&= [[\![\alpha(y/u)(\mathbf{z}/\boldsymbol{\sigma})]\!]_{u\mathbf{x}}\circ\mathrm{can}]\hat{}.
\end{aligned}$$

(ii) y is among $z_1,\ldots,z_m$. In this case we may write $\mathbf{z}=y\mathbf{z}'$, $\boldsymbol{\sigma}=\rho\boldsymbol{\sigma}'$ and we then have

$$\begin{aligned}
[\![\{y:\alpha\}(y\mathbf{z}'/\rho\boldsymbol{\sigma}')]\!]_{\mathbf{x}} &= [\![\{y:\alpha(\mathbf{z}/\boldsymbol{\sigma})\}]\!]_{\mathbf{x}}\\
&= [[\![\alpha(\mathbf{x}/\boldsymbol{\sigma})(y/u)]\!]_{u\mathbf{x}}\circ\mathrm{can}]\hat{}\\
&= [[\![\alpha(y/u)(y\mathbf{z}'/\rho\boldsymbol{\sigma}')]\!]_{u\mathbf{x}}\circ\mathrm{can}]\hat{}.
\end{aligned}$$

Accordingly in both cases, i.e. for arbitrary $\mathbf{z}$, we have, using the induction hypothesis, and adopting the notation of (3.18),

$$\begin{aligned}
[\![\{y:\alpha\}(\mathbf{z}/\boldsymbol{\sigma})]\!]_{\mathbf{x}} &= [[\![\alpha(y/u)]\!]_{u\mathbf{z}} \circ [\![\langle u, \sigma_1, \ldots, \sigma_m\rangle]\!]_{u\mathbf{x}} \circ \mathrm{can}]^\wedge \\
&= [[\![\alpha(y/u)]\!]_{u\mathbf{z}} \circ \langle \pi_1, [\![\sigma_1]\!]_{u\mathbf{x}}, \ldots, [\![\sigma_m]\!]_{u\mathbf{x}}\rangle \circ \mathrm{can}]^\wedge \\
&= [[\![\alpha(y/u)]\!]_{u\mathbf{z}} \circ \langle \pi_1, [\![\sigma_1]\!]_{\mathbf{x}} \circ [\![\langle x_1, \ldots, x_n\rangle]\!]_{u\mathbf{x}}, \ldots, \\
&\qquad [\![\sigma_m]\!]_{\mathbf{x}} \circ [\![\langle x_1, \ldots, x_n\rangle]\!]_{u\mathbf{x}}\rangle \circ \mathrm{can}]^\wedge \qquad \text{(by 3.19)} \\
&= [[\![\alpha(y/u)]\!]_{u\mathbf{z}} \circ \langle \pi_1, [\![\sigma_1]\!]_{\mathbf{x}} \circ \mathrm{proj}, \ldots, [\![\sigma_m]\!]_{\mathbf{x}} \circ \mathrm{proj}\rangle \circ \mathrm{can}]^\wedge \\
&= [[\![\alpha(y/u)]\!]_{u\mathbf{z}} \circ \mathrm{can}^* \circ (1_B \times [\![\langle \sigma_1, \ldots, \sigma_m\rangle]\!]_{\mathbf{x}})]^\wedge \qquad \text{(by 3.18)} \\
&= [e_B \circ (1_B \times [\![\alpha(y/u)]\!]_{u\mathbf{z}} \circ \mathrm{can}^*)^\wedge \circ (1_B \times [\![\langle \sigma_1, \ldots, \sigma_m\rangle]\!]_{\mathbf{x}})]^\wedge \\
&= [e_B \circ (1_B \times [\![\{y:\alpha\}]\!]_{\mathbf{z}}) \circ (1_B \times [\![\langle \sigma_1, \ldots, \sigma_m\rangle]\!]_{\mathbf{x}})]^\wedge \\
&= [e_B \circ 1_B \times [\![\{y:\alpha\}]\!]_{\mathbf{z}} \circ [\![\langle \sigma_1, \ldots, \sigma_m\rangle]\!]_{\mathbf{x}}]^\wedge \\
&= [\![\{y:\alpha\}]\!]_{\mathbf{z}} \circ [\![\langle \sigma_1, \ldots, \sigma_m\rangle]\!]_{\mathbf{x}}. \qquad \text{(by (2.20))}
\end{aligned}$$

This completes the induction step. □

An immediate consequence of the substitution lemma is the

3.21 INDEPENDENCE LEMMA. If u is free for x and not free in a term τ, then, for any interpretation of $\mathscr{L}$,

$$[\![\tau(x/u)]\!]_{u\mathbf{y}} = [\![\tau]\!]_{x\mathbf{y}}. \quad \square$$

We can now finally introduce the notion of *validity* of a formula in a topos. Given an interpretation I of $\mathscr{L}$ in a topos $\mathbf{E}$, for any finite set $\Gamma = \{\alpha_1, \ldots, \alpha_m\}$ of formulae write

$$[\![\Gamma]\!]_{I,\mathbf{x}} \text{ for } \begin{cases} [\![\alpha_1]\!]_{I,\mathbf{x}} \wedge \cdots \wedge [\![\alpha_m]\!]_{I,\mathbf{x}} & \text{if } m > 0 \\ T & \text{if } m = 0. \end{cases}$$

(Note that the order and grouping of the terms $[\![\alpha_1]\!]_{I,\mathbf{x}}, \ldots, [\![\alpha_m]\!]_{I,\mathbf{x}}$ in $[\![\Gamma]\!]_{I,\mathbf{x}}$ is irrelevant.) Given a formula β, let $(x_1, \ldots, x_n) = \mathbf{x}$ list all the free variables in $\Gamma \cup \{\beta\}$; we write

(3.22) $$\Gamma \models_I \beta \quad \text{or} \quad \Gamma \models_E \beta \qquad \text{for} \qquad [\![\Gamma]\!]_{I,\mathbf{x}} \leq [\![\beta]\!]_{I,\mathbf{x}}.$$

If $\Gamma \models_I \beta$, we say that the sequent $\Gamma : \beta$ is *valid* under the interpretation I (or in the topos $\mathbf{E}$).

We have the following basic facts concerning the concept of validity.

(3.23.1) $$\models_I \alpha \quad \text{iff} \quad [\![\alpha]\!]_{\mathbf{x}} = T. \quad \square$$

It follows immediately from (3.23.1) that, *if I is an interpretation in a degenerate topos, then $\models_I \alpha$ for every formula α.*

Recall that, for any arrow u with codomain Ω in a topos, $\bar{u}$ denotes the *kernel* of u (defined immediately after (2.2)).

(3.23.2) $$\Gamma \models_I \alpha \quad \text{iff} \quad [\![\alpha]\!]_{\mathbf{x}} \circ \overline{[\![\Gamma]\!]_{\mathbf{x}}} = T.$$

Proof. By (2.19(ii)). □

(3.23.3) $$\Gamma \models_I \sigma = \tau \quad \text{iff} \quad [\![\sigma]\!]_{\mathbf{x}} \circ \overline{[\![\Gamma]\!]_{\mathbf{x}}} = [\![\tau]\!]_{\mathbf{x}} \circ \overline{[\![\Gamma]\!]_{\mathbf{x}}}.$$

Proof. We have

$$\begin{aligned} \Gamma \models_I \sigma = \tau \quad &\text{iff} \quad [\![\sigma = \tau]\!]_{\mathbf{x}} \circ \overline{[\![\Gamma]\!]_{\mathbf{x}}} = T \qquad \text{(by (3.23.2))} \\ &\text{iff} \quad \mathrm{eq}_B \circ \langle [\![\sigma]\!]_{\mathbf{x}}, [\![\tau]\!]_{\mathbf{x}} \rangle \circ \overline{[\![\Gamma]\!]_{\mathbf{x}}} = T \\ &\text{iff} \quad \mathrm{eq}_B \circ \langle [\![\sigma]\!]_{\mathbf{x}} \circ \overline{[\![\Gamma]\!]_{\mathbf{x}}}, [\![\tau]\!]_{\mathbf{x}} \circ \overline{[\![\Gamma]\!]_{\mathbf{x}}} \rangle = T \\ &\text{iff} \quad [\![\sigma]\!]_{\mathbf{x}} \circ \overline{[\![\Gamma]\!]_{\mathbf{x}}} = [\![\tau]\!]_{\mathbf{x}} \circ \overline{[\![\Gamma]\!]_{\mathbf{x}}}. \qquad \text{(by (2.19(iii))} \end{aligned}$$ □

We shall also need to establish the following facts concerning the behaviour of superfluous variables under an interpretation.

Assume throughout (3.24) that x is not free in Γ or α.

(3.24.1) $$[\![\Gamma]\!]_{x\mathbf{y}} = [\![\Gamma]\!]_{\mathbf{y}} \circ \pi_2 \circ \mathrm{can}^{-1}.$$

Proof. By (3.19) and (2.18),

$$\begin{aligned} [\![\Gamma]\!]_{x\mathbf{y}} &= [\![\Gamma]\!]_{\mathbf{y}} \circ [\![\langle y_1, \ldots, y_n \rangle]\!]_{x\mathbf{y}} \\ &= [\![\Gamma]\!]_{\mathbf{y}} \circ \pi_2 \circ \mathrm{can}^{-1}. \end{aligned}$$ □

(3.24.2) $$[\![\Gamma]\!]_{\mathbf{y}} \leqslant [\![\alpha]\!]_{\mathbf{y}} \quad \text{implies} \quad [\![\Gamma]\!]_{x\mathbf{y}} \leqslant [\![\alpha]\!]_{x\mathbf{y}}.$$

Proof. This follows immediately from (3.24.1) and (2.15). □

(3.24.3) $\overline{[\![\Gamma]\!]_{x\mathbf{y}}} \sim \mathrm{can} \circ (1 \times \overline{[\![\Gamma]\!]_{\mathbf{y}}})$, where $\sim$ denotes equivalence of monics in the sense of Chapter 2.

Proof. Consider the commutative diagram

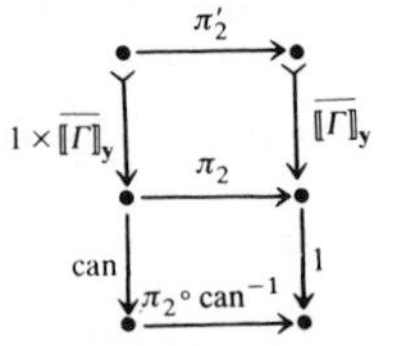

Clearly both constituent squares are pullbacks, and so therefore is the whole diagram. It now follows from (2.8) that

$$\begin{aligned} \chi((\mathrm{can} \circ (1 \times \overline{[\![\Gamma]\!]_{\mathbf{y}}})) &= \chi(\overline{[\![\Gamma]\!]_{\mathbf{y}}}) \circ \pi_2 \circ \mathrm{can}^{-1} \\ &= [\![\Gamma]\!]_{\mathbf{y}} \circ \pi_2 \circ \mathrm{can}^{-1} \\ &= [\![\Gamma]\!]_{x\mathbf{y}} \end{aligned}$$

by (3.24.1). Applying (2.3), we have our result. □

We are now in a position to prove the central *soundness theorem,* the core of which is the assertion that the theorems of any pure local set theory are valid under any interpretation.

Let us write

$$\Gamma \models \alpha$$

for

$$\Gamma \models_I \alpha \quad \text{for any interpretation } I \text{ of } \mathscr{L};$$

$$\frac{\Gamma_1 \models_I \alpha_1, \ldots, \Gamma_n \models_I \alpha_n}{\Delta \models_I \beta}$$

for

$$\Gamma_1 \models_I \alpha_1 \text{ and } \cdots \text{ and } \Gamma_n \models_I \alpha_n \text{ imply } \Delta \models_I \beta;$$

$$\frac{\Gamma_1 \models \alpha_1, \ldots, \Gamma_n \models \alpha_n}{\Delta \models \beta}$$

for

$$\frac{\Gamma_1 \models_I \alpha_1, \ldots, \Gamma_n \models_I \alpha_n}{\Delta \models_I \beta}$$

for all interpretations I.

For any local set theory S in $\mathscr{L}$, an interpretation I of $\mathscr{L}$ (in a topos $\mathbf{E}$) is called a *model* of S (in $\mathbf{E}$) if every axiom of S is valid under I. We write

$$\Gamma \models_S \alpha$$

if $\Gamma \models_I \alpha$ for any model I of S.

Note that if $\mathbf{E}$ is degenerate, then by (3.23.1) any interpretation in $\mathbf{E}$ is a model of the inconsistent theory.

3.25 SOUNDNESS THEOREM.

(i) $\Gamma \vdash \alpha$ implies $\Gamma \models \alpha$.

(ii) $\dfrac{\Gamma_1 : \alpha_1, \ldots, \Gamma_n : \alpha_n}{\Delta : \beta}$ implies $\dfrac{\Gamma_1 \models \alpha_1, \ldots, \Gamma_n \models \alpha_n}{\Delta \models \beta}$

(iii) $\Gamma \vdash_S \alpha$ implies $\Gamma \models_S \alpha$.

To prove (i) and (ii) of the soundness theorem, it is enough to establish the validity under any interpretation I of the axioms and rules of inference of pure local set theory. This we presently proceed to do. To get (iii) of the soundness theorem, we then merely observe that if $\Gamma \vdash_S \alpha$, then

$$\frac{\Gamma_1 : \alpha_1, \ldots, \Gamma_n : \alpha_n}{\Gamma : \alpha}$$

for some axioms $\Gamma_1 : \alpha_1, \ldots, \Gamma_n : \alpha_n$ of S, and apply (ii) of the soundness theorem.

We now verify the validity of the axioms and rules of pure local set theory.

(3.25.1) (i) *Tautology.* $\alpha \models_I \alpha$. □
(ii) *Unity.* $\models_I x_1 = *$. □

(3.25.2) *Equality.*
(i) $\models_I x = x$,
(ii) $x = y, \alpha(z/x) \models_I \alpha(z/y)$,

provided that x and y are free for z in α.

Proof. (i) We have

$$[\![x = x]\!]_x = \text{eq} \circ \langle 1, 1 \rangle = T. \quad \square$$

(ii) Assume that x, y and z are distinct and that z, but neither x nor y, occur free in α. The remaining cases are either proved similarly or are trivial.

Let $z, v_1, \ldots, v_n$ be the free variables of α, and suppose that

$$u \leqslant [\![x = y]\!]_{xy\mathbf{v}} \wedge [\![\alpha(z/x)]\!]_{xy\mathbf{v}}.$$

Then by (2.17)

$$u \leqslant \text{eq} \circ \langle [\![x]\!]_{xy\mathbf{v}}, [\![y]\!]_{xy\mathbf{v}} \rangle \tag{1}$$

and by (2.17) and (3.20),

$$u \leqslant [\![\alpha]\!]_{z\mathbf{v}} \circ [\![\langle x, v_1, \ldots, v_n \rangle]\!]_{xy\mathbf{v}}. \tag{2}$$

By (1) and (2.16), we have

$$[\![x]\!]_{xy\mathbf{v}} \circ \bar{u} = [\![y]\!]_{xy\mathbf{v}} \circ \bar{u} \tag{3}$$

and by (2) and (2.14) there is an arrow h in $\mathbf{E}$ such that

$$\begin{aligned} \overline{[\![\alpha]\!]_{z\mathbf{v}}} \circ h &= [\![\langle x, v_1, \ldots, v_n \rangle]\!]_{xy\mathbf{v}} \circ \bar{u} \\ &= \langle [\![x]\!]_{xy\mathbf{v}} \circ \bar{u}, \ldots, [\![v_n]\!]_{xy\mathbf{v}} \circ \bar{u} \rangle \\ &= \langle [\![y]\!]_{xy\mathbf{v}} \circ \bar{u}, \ldots, [\![v_n]\!]_{xy\mathbf{v}} \circ \bar{u} \rangle \qquad \text{(by (3))} \\ &= [\![\langle y, v_1, \ldots, v_n \rangle]\!]_{xy\mathbf{v}} \circ \bar{u}. \end{aligned}$$

Hence, using (2.14) again,

$$\begin{aligned} u &\leqslant [\![\alpha]\!]_{z\mathbf{v}} \circ [\![\langle y, v_1, \ldots, v_n \rangle]\!]_{xy\mathbf{v}} \\ &= [\![\alpha(z/y)]\!]_{xy\mathbf{v}} \end{aligned}$$

by (3.20). Therefore, since u was arbitrary,

$$[\![x = y]\!]_{xy\mathbf{v}} \wedge [\![\alpha(z/x)]\!]_{xy\mathbf{v}} \leqslant [\![\alpha(z/y)]\!]_{xy\mathbf{v}}$$

and the result follows. □

(3.25.3) *Products.*

(i) $\models_I (\langle x_1, \ldots, x_n \rangle)_i = x_i$. □

(ii) $\models_I x = \langle (x)_1, \ldots, (x)_n \rangle$. □

(3.25.4) *Comprehension.* $\models_I x \in \{x : \alpha\} \Leftrightarrow \alpha$.

Proof. The verification here is a bit tedious: we first require a

LEMMA. Let can′ be the canonical isomorphism between $A \times (A \times A_1 \times \cdots \times A_n)$ and $A \times A \times A_1 \times \cdots \times A_n$ and let

$$\eta = \langle \pi_1, \pi_3, \ldots, \pi_n \rangle \circ \text{can}' : A \times (A \times A_1 \times \cdots \times A_n) \to A \times A_1 \times \cdots \times A_n.$$

Also let

$$\xi = \langle \pi_2, \pi_3, \ldots, \pi_n \rangle : A \times A_1 \times \cdots \times A_n \to A_1 \times \cdots \times A_n.$$

Then for any $g : A = A_1 \times \cdots \times A_n \to \Omega$ we have

$$(g \circ \eta)\hat{} = (g \circ \text{can})\hat{} \circ \xi,$$

where 'can' is, as usual, the canonical isomorphism between

$$A \times (A_1 \times \cdots \times A_n) \quad \text{and} \quad A \times A_1 \times \cdots \times A_n.$$

Proof of lemma. Given any $A_1 \times \cdots \times A_n \xrightarrow{h} X$ it is readily checked that

$$(1_A \times h) \circ \text{can}^{-1} \circ \eta = 1_A \times (h \circ \xi).$$

Therefore, taking $h = (g \circ \text{can})\hat{}$, we get

$$(1_A \times (g \circ \text{can})\hat{}) \circ \text{can}^{-1} \circ \eta = 1_A \times (g \circ \text{can})\hat{} \circ \xi.$$

Hence

$$\begin{aligned} & e_A \circ (1_A = (g \circ \text{can})\hat{} \circ \xi) \\ &= e_A \circ (1_A \times (g \circ \text{can})\hat{}) \circ \text{can}^{-1} \circ \eta \\ &= g \circ \text{can} \circ \text{can}^{-1} \circ \eta \\ &= g \circ \eta. \end{aligned}$$

The lemma now follows from (2.20). □

Now we can establish (3.25.4). For if x is of type **A** and $v_1, \ldots, v_n$ are of types $\mathbf{A}_1, \ldots, \mathbf{A}_n$, then

$$\begin{aligned}
&[\![x \in \{x : \alpha\}]\!]_{x\mathbf{v}} \\
&\quad = e_A \circ \langle [\![x]\!]_{x\mathbf{v}}, ([\![\alpha(x/u)]\!]_{ux\mathbf{v}} \circ \mathrm{can}')\hat{}\,\rangle \\
&\quad = e_A \circ \langle \pi_1, ([\![\alpha]\!]_{x\mathbf{v}} \circ [\![\langle u, v_1, \ldots, v_n\rangle]\!]_{ux\mathbf{v}} \circ \mathrm{can}')\hat{}\,\rangle \qquad \text{(by 3.20)} \\
&\quad = e_A \circ \langle \pi_1, ([\![\alpha]\!]_{x\mathbf{v}} \circ \eta)\hat{}\,\rangle \\
&\quad = e_A \circ \langle \pi_1, ([\![\alpha]\!]_{x\mathbf{v}} \circ \mathrm{can})\hat{}\circ \xi\rangle \qquad \text{(by the Lemma)} \\
&\quad = e_A \circ (1_A \times ([\![\alpha]\!]_{x\mathbf{v}} \circ \mathrm{can})\hat{}\,) \circ \mathrm{can}^{-1} \\
&\quad = [\![\alpha]\!]_{x\mathbf{v}} \circ \mathrm{can} \circ \mathrm{can}^{-1} \\
&\quad = [\![\alpha]\!]_{x\mathbf{v}}.
\end{aligned}$$

The result now follows from (3.23.3). □

(3.25.5) *Thinning.*

$$\frac{\Gamma \models_I \alpha}{\beta, \Gamma \models_I \alpha}$$

Proof. If $[\![\Gamma]\!]_{\mathbf{x}} \leq [\![\alpha]\!]_{\mathbf{x}}$, then by (3.24.2), $[\![\Gamma]\!]_{\mathbf{xy}} \leq [\![\alpha]\!]_{\mathbf{xy}}$, where $y_1, \ldots, y_n$ are the additional free variables of β. Therefore $[\![\beta]\!]_{\mathbf{xy}} \wedge [\![\Gamma]\!]_{\mathbf{xy}} \leq [\![\alpha]\!]_{\mathbf{xy}}$, and the result follows. □

(3.25.6) *Cut.*

$$\frac{\Gamma \models_I \alpha \quad \alpha, \Gamma \models_I \beta}{\Gamma \models_I \beta}$$

provided each of the free variables of α is free in Γ or β.

Proof. Suppose $[\![\Gamma]\!]_{\mathbf{x}} \leq [\![\alpha]\!]_{\mathbf{x}}$ and $[\![\alpha]\!]_{\mathbf{xy}} \wedge [\![\Gamma]\!]_{\mathbf{xy}} \leq [\![\beta]\!]_{\mathbf{xy}}$ where $y_1, \ldots, y_n$ are the additional free variables of β. By (3.24.2), $[\![\Gamma]\!]_{\mathbf{xy}} \leq [\![\alpha]\!]_{\mathbf{xy}}$. Hence

$$[\![\Gamma]\!]_{\mathbf{xy}} = [\![\Gamma]\!]_{\mathbf{xy}} \wedge [\![\alpha]\!]_{\mathbf{xy}} \leq [\![\beta]\!]_{\mathbf{xy}}$$

and the result follows. □

(3.25.7) *Substitution.*

$$\frac{\Gamma \models_I \alpha}{\Gamma(x/\tau) \models_I \alpha(x/\tau)}$$

provided τ is free for x in Γ and α.

Proof. We may assume that x is free in Γ or α; otherwise the result is trivial. Suppose then that $[\![\Gamma]\!]_{x\mathbf{y}} \leq [\![\alpha]\!]_{x\mathbf{y}}$. By (2.15) and (2.18),

$$\begin{aligned}
&[\![\Gamma]\!]_{x\mathbf{y}} \circ [\![\langle \tau, y_1, \ldots, y_n\rangle]\!]_{\mathbf{zy}} \\
&\qquad \leq [\![\alpha]\!]_{x\mathbf{y}} \circ [\![\langle \tau, y_1, \ldots, y_n\rangle]\!]_{\mathbf{zy}}
\end{aligned}$$

where $z_1, \ldots, z_n$ are the additional free variables of τ. It now follows from (3.20) that

$$[\![\Gamma(x/\tau)]\!]_{\mathbf{zy}} \leq [\![\alpha(x/\tau)]\!]_{\mathbf{zy}},$$

yielding the required result. $\square$

(3.25.8) *Extensionality.*

$$\frac{\Gamma \models_I x \in \sigma \Leftrightarrow x \in \tau}{\Gamma \models_I \sigma = \tau}$$

provided x is not free in Γ, σ, or τ.

Proof. Suppose that $\Gamma \models_I x \in \sigma \Leftrightarrow x \in \tau$. Then by (3.23.3)

$$[\![x \in \sigma]\!]_{x\mathbf{z}} \circ \overline{[\![\Gamma]\!]_{x\mathbf{z}}} = [\![x \in \tau]\!]_{x\mathbf{z}} \circ \overline{[\![\Gamma]\!]_{x\mathbf{z}}}. \tag{1}$$

By (3.24.3), there is an isomorphism i making the diagram

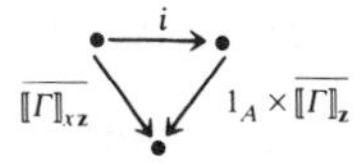

commute, where x is of type $\mathbf{A}$. That result, and (3.24.1), then give

$$\begin{aligned}
[\![x \in \sigma]\!]_{x\mathbf{z}} \circ \overline{[\![\Gamma]\!]_{x\mathbf{z}}} &= e_A \circ [\![\langle x, \sigma \rangle]\!]_{x\mathbf{z}} \circ \overline{[\![\Gamma]\!]_{x\mathbf{z}}} \\
&= e_A \circ \langle [\![x]\!]_{x\mathbf{z}}, [\![\sigma]\!]_{x\mathbf{z}} \rangle \circ \overline{[\![\Gamma]\!]_{x\mathbf{z}}} \\
&= e_A \circ \langle \pi_1, [\![\sigma]\!]_{\mathbf{z}} \circ \pi_2 \circ \mathrm{can}^{-1} \rangle \circ \mathrm{can} \circ (1_A \times \overline{[\![\Gamma]\!]_{\mathbf{z}}}) \circ i \\
&= e_A \circ (1_A \times [\![\sigma]\!]_{\mathbf{z}}) \circ (1_A \times \overline{[\![\Gamma]\!]_{\mathbf{z}}}) \circ i \\
&= e_A \circ (1_A \times [\![\sigma]\!]_{\mathbf{z}} \circ \overline{[\![\Gamma]\!]_{\mathbf{z}}}) \circ i.
\end{aligned}$$

Therefore, by (1)

$$\begin{aligned}
&e_A \circ (1_A \times [\![\sigma]\!]_{\mathbf{z}} \circ \overline{[\![\Gamma]\!]_{\mathbf{z}}}) \circ i \\
&\quad = e_A \circ (1_A \times [\![\tau]\!]_{\mathbf{z}} \circ \overline{[\![\Gamma]\!]_{\mathbf{z}}}) \circ i
\end{aligned}$$

so that

$$e_A \circ (1_A \times [\![\sigma]\!]_{\mathbf{z}} \circ \overline{[\![\Gamma]\!]_{\mathbf{z}}}) = e_A \circ (1_A \times [\![\tau]\!]_{\mathbf{z}} \circ \overline{[\![\Gamma]\!]_{\mathbf{z}}}).$$

Hence

$$[\![\sigma]\!]_{\mathbf{z}} \circ \overline{[\![\Gamma]\!]_{\mathbf{z}}} = [\![\tau]\!]_{\mathbf{z}} \circ \overline{[\![\Gamma]\!]_{\mathbf{z}}}$$

and the result follows from (3.23.3). $\square$

(3.25.9) *Equivalence.*

$$\frac{\alpha, \Gamma \models_I \beta \quad \beta, \Gamma \models_I \alpha}{\Gamma \models_I \alpha \Leftrightarrow \beta}.$$

Proof. If $\alpha, \Gamma \models_I \beta$ and $\beta, \Gamma \models_I \alpha$, then

$$[\![\alpha]\!]_{\mathbf{x}} \wedge [\![\Gamma]\!]_{\mathbf{x}} \leq [\![\beta]\!]_{\mathbf{x}} \quad \text{and} \quad [\![\beta]\!]_{\mathbf{x}} \wedge [\![\Gamma]\!]_{\mathbf{x}} \leq [\![\alpha]\!]_{\mathbf{x}}.$$

Hence

$$[\![\alpha]\!]_{\mathbf{x}} \wedge [\![\Gamma]\!]_{\mathbf{x}} = [\![\beta]\!]_{\mathbf{x}} \wedge [\![\Gamma]\!]_{\mathbf{x}}. \tag{1}$$

Therefore

$$\begin{aligned}
[\![\alpha]\!]_{\mathbf{x}} \circ \overline{[\![\Gamma]\!]_{\mathbf{x}}} &= ([\![\alpha]\!]_{\mathbf{x}} \circ \overline{[\![\Gamma]\!]_{\mathbf{x}}}) \wedge T \\
&= ([\![\alpha]\!]_{\mathbf{x}} \circ \overline{[\![\Gamma]\!]_{\mathbf{x}}}) \wedge ([\![\Gamma]\!]_{\mathbf{x}} \circ \overline{[\![\Gamma]\!]_{\mathbf{x}}}) \\
&= ([\![\alpha]\!]_{\mathbf{x}} \wedge [\![\Gamma]\!]_{\mathbf{x}}) \circ \overline{[\![\Gamma]\!]_{\mathbf{x}}} \qquad ((2.18)) \\
&= ([\![\beta]\!]_{\mathbf{x}} \wedge [\![\Gamma]\!]_{\mathbf{x}}) \circ \overline{[\![\Gamma]\!]_{\mathbf{x}}} \qquad (\text{by } (1)) \\
&= ([\![\beta]\!]_{\mathbf{x}} \circ \overline{[\![\Gamma]\!]_{\mathbf{x}}}) \wedge ([\![\Gamma]\!]_{\mathbf{x}} \circ \overline{[\![\Gamma]\!]_{\mathbf{x}}}) \qquad ((2.18)) \\
&= ([\![\beta]\!]_{\mathbf{x}} \circ \overline{[\![\Gamma]\!]_{\mathbf{x}}}) \wedge T \\
&= [\![\beta]\!]_{\mathbf{x}} \circ \overline{[\![\Gamma]\!]_{\mathbf{x}}}.
\end{aligned}$$

The result now follows from (3.23.3). □

This completes the proof of the soundness theorem. As a consequence, we have the

3.26 COROLLARY. Any pure local set theory is consistent.

Proof. Let $\mathscr{L}$ be a local language and L the corresponding pure local set theory. We set up an interpretation I of $\mathscr{L}$ in the topos **Finset** of finite sets as follows. First, $\mathbf{1}_I$ is $1 = \{0\}$, $\mathbf{\Omega}_I$ is $2 = \{0, 1\}$, and for any ground type $\mathbf{A}$ of $\mathscr{L}$, $\mathbf{A}_I$ is an arbitrary non-empty finite set. I is then extended to arbitrary types in the evident recursive way: $(\mathbf{PA})_I$ is $P(\mathbf{A}_I)$ and $(\mathbf{A}_1 \times \cdots \times \mathbf{A}_n)_I$ is $(\mathbf{A}_1)_I \times \cdots \times (\mathbf{A}_n)_I$. Finally, I is extended to function symbols by allowing, for any function symbol $\mathbf{f}$ of signature $\mathbf{A} \to \mathbf{B}$, $\mathbf{f}_I$ to be any map from $\mathbf{A}_I$ to $\mathbf{B}_I$.

Now if $\mathscr{L}$ were inconsistent, then $\vdash \mathit{false}$, so that $\vdash \alpha$, and hence, by the soundness theorem, $\models_I \alpha$ for any formula α. Let u and v be variables of type $\mathbf{P1}$. Then $\models_I u = v$, so by (3.23.3) we have $[\![u]\!]_{I,uv} = [\![v]\!]_{I,uv}$, i.e. the two projections $P1 \times P1 \underset{\pi_2}{\overset{\pi_1}{\rightrightarrows}} P1$ are the same in **Finset**. But this is clearly false since $P1 = 2$ has two elements. □

REMARK ON THE RESTRICTION ON FREE VARIABLES IN THE CUT RULE. This restriction cannot be lifted if the soundness theorem is to hold, because of the possibility of 'empty' interpretations of types. For example, suppose that I is an interpretation in **Set** of a local language $\mathscr{L}$ such that $\mathbf{A}_I = \varnothing$ for some type $\mathbf{A}$. If x is a variable of type $\mathbf{A}$, it is easily checked that, for any formula α,

$$[\![\alpha]\!]_{x\mathbf{y}} = 0 \to 2$$

so that, in particular,

$$[\![x = x]\!]_{x\mathbf{y}} \leq [\![\alpha]\!]_{x\mathbf{y}},$$

i.e.

$$x = x \models_I \alpha.$$

But since $\models_I x = x$, if the cut rule without any restriction on free variables were sound, it would follow that $\models_I \alpha$ for any formula α. Then we could argue as in the proof of (3.26) that the projections $2 \times 2 \rightrightarrows 2$ in **Set** are the same, an absurdity. (In fact, it is easy to see that this situation cannot arise if $\mathbf{A}_I \neq \varnothing$ or, more generally, whenever I is an interpretation in a topos **E** for which $\mathbf{A}_I$ has an **E**-element. This is the basis of the approach to the problem adopted by Fourman (1977).)

We note, however, that we can drop the restriction on free variables in the cut rule *provided we keep track of them.* To this end, let us write, for any sequent $\Gamma : \alpha$, and any interpretation I,

$$\Gamma \models_{I,\mathbf{x}} \alpha \qquad \text{for} \qquad [\![\Gamma]\!]_{\mathbf{x}} \leq [\![\beta]\!]_{\mathbf{x}}.$$

The proof of (3.25.6) then shows that the 'inference'

$$\frac{\alpha, \Gamma \models_{I,\mathbf{x}} \beta \quad \Gamma \models_{I,\mathbf{x}} \alpha}{\Gamma \models_{I,\mathbf{x}} \beta}$$

is valid. Thus, if we define a new provability relation $\Gamma \vdash_{\mathbf{x}} \alpha$ by

$$\Gamma \vdash_{\mathbf{x}} \alpha \qquad \text{iff} \qquad \Gamma \models_{I,\mathbf{x}} \alpha \quad \text{for all interpretations } I,$$

$\vdash_{\mathbf{x}}$ satisfies the 'unrestricted' cut rule

$$\frac{\alpha, \Gamma \vdash_{\mathbf{x}} \beta \quad \Gamma \vdash_{\mathbf{x}} \alpha}{\Gamma \vdash_{\mathbf{x}} \beta}$$

This is, in effect, the approach adopted by Lambek and Scott (1986).

The completeness theorem

We now establish the converse to the soundness theorem—the *completeness theorem.*

Given a local set theory S in a local language $\mathscr{L}$, we define the *canonical interpretation*—denoted by $C(S)$—of $\mathscr{L}$ in $\mathbf{C}(S)$ by:

$\mathbf{A}_{C(S)} = U_{\mathbf{A}}$ for each type symbol $\mathbf{A}$

$\mathbf{f}_{C(S)} = U_{\mathbf{A}} \xrightarrow{x \mapsto \mathbf{f}(x)} U_{\mathbf{B}}$ for each function symbol $\mathbf{f}$ of signature $\mathbf{A} \to \mathbf{B}$.

For each term τ, write $[\![\tau]\!]_{\mathbf{x}}$ for $[\![\tau]\!]_{C(S),\mathbf{x}}$. Then we have

3.27 PROPOSITION.

$$[\![\tau]\!]_{\mathbf{x}} = (\mathbf{x} \mapsto \tau).$$

Proof. By induction on the formation of τ. The only troublesome induction step arises when τ is $\{x : \alpha\}$. In this case

$$\begin{aligned}
[\![\{x : \alpha\}]\!]_{\mathbf{z}} &= ([\![\alpha(x/u)]\!]_{u\mathbf{z}} \circ \text{can})\hat{} \\
&= [(\langle u, z_1, \ldots, z_n\rangle \mapsto \alpha(x/u)) \circ (\langle x, y\rangle \mapsto \langle x, (y)_1, \ldots, (y)_n\rangle)]\hat{} \\
&= [\langle x, y\rangle \mapsto \alpha(z_1/(y)_1) \cdots (z_n/(y)_n)]\hat{} \\
&= y \mapsto \{x : \alpha(z_1/(y)_1) \cdots (z_n/(y)_n)\} \quad \text{(by (3.16.4) and the definition of the power transpose in } \mathbf{C}(S) \text{ (3.16.7))} \\
&= \mathbf{z} \mapsto \{x : \alpha\}.
\end{aligned}$$

This completes the induction step. □

3.28 COROLLARY

$$\Gamma \vDash_{C(S)} \alpha \quad \text{iff} \quad \Gamma \vdash_S \alpha.$$

Proof. We have, using (3.27),

$$\begin{aligned}
\vDash_{C(S)} \alpha \quad &\text{iff} \quad [\![\alpha]\!]_{\mathbf{x}} = T \\
&\text{iff} \quad (\mathbf{x} \mapsto \alpha) = (\mathbf{x} \mapsto \mathit{true}) \\
&\text{iff} \quad \vdash_S \alpha = \mathit{true} \\
&\text{iff} \quad \vdash_S \alpha.
\end{aligned}$$

In the general case, then,

$$\begin{aligned}
\Gamma \vdash_S \alpha \quad &\text{iff} \quad \bigwedge \Gamma \vdash_S \alpha \\
&\text{iff} \quad \vdash_S \bigwedge \Gamma \Rightarrow \alpha \\
&\text{iff} \quad \vDash_{C(S)} \bigwedge \Gamma \Rightarrow \alpha \\
&\text{iff} \quad \bigwedge \Gamma \vDash_{C(S)} \alpha \\
&\text{iff} \quad \Gamma \vDash_{C(S)} \alpha,
\end{aligned}$$

the last two equivalences being justified by the soundness theorem. □

(3.28) tells us that $C(S)$ may be regarded as a *canonical model of S*. As a consequence, we obtain the promised

3.29 COMPLETENESS THEOREM

(i) $\Gamma \vDash \alpha$ implies $\Gamma \vdash \alpha$;

(ii) $\dfrac{\Gamma_1 \vDash \alpha_1, \ldots, \Gamma_n \vDash \alpha_n}{\Delta \vDash \beta}$ implies $\dfrac{\Gamma_1 : \alpha_1, \ldots, \Gamma_n : \alpha_n}{\Delta : \beta}$;

(iii) $\Gamma \vDash_S \alpha$ implies $\Gamma \vdash_S \alpha$.

Proof. (i) If $\Gamma \vDash \alpha$, then in particular $\Gamma \vDash_{C(L)} \alpha$ and so $\Gamma \vdash \alpha$ by (3.28).

(ii) Let S be the set of sequents $\{\Gamma_1 : \alpha_1, \ldots, \Gamma_n : \alpha_n\}$. Then clearly

$$\Delta \vdash_S \beta \quad \text{iff} \quad \frac{\Gamma_1 : \alpha_1, \ldots, \Gamma_n : \alpha_n}{\Delta : \beta} \tag{1}$$

and

$$\Gamma_1 \vdash_S \alpha_1, \ldots, \Gamma_n \vdash_S \alpha_n. \tag{2}$$

Now suppose

$$\frac{\Gamma_1 \vDash \alpha_1, \ldots, \Gamma_n \vDash \alpha_n}{\Delta \vDash \beta} \tag{3}$$

By (2) and (3.28), we have

$$\Gamma_1 \vDash_{C(S)} \alpha_1, \ldots, \Gamma_n \vDash_{C(S)} \alpha_n.$$

Hence, by (3), $\Delta \vDash_{C(S)} \beta$, so that, by (3.28), $\Delta \vdash_S \beta$. Now (1) implies that

$$\frac{\Gamma_1 : \alpha_1, \ldots, \Gamma_n : \alpha_n}{\Delta : \beta}$$

as required.

(iii) By (3.28), $C(S)$ is a model of S. Therefore, again using (3.28),

$$\Gamma \vDash_S \alpha \quad \text{implies} \quad \Gamma \vDash_{C(S)} \alpha \quad \text{implies} \quad \Gamma \vdash_S \alpha. \quad \Box$$

The equivalence theorem

We are going to show that every topos is equivalent to a linguistic topos. To do this we start with a topos **E** and construct a theory Th(**E**) which we will show determines **E** up to equivalence.

Let **E** be a topos with specified terminal object $1_{\mathbf{E}}$, subobject classifier $\Omega_{\mathbf{E}}$, products and power objects. We define the *local language* $\mathscr{L}(\mathbf{E})$ *determined by* **E** (also called the *internal language* of **E**) as follows.

The *ground type symbols* of $\mathscr{L}(\mathbf{E})$ match the objects of **E** other than $1_{\mathbf{E}}$, $\Omega_{\mathbf{E}}$. That is, for each **E**-object A (other than $1_{\mathbf{E}}, \Omega_{\mathbf{E}}$), $\mathscr{L}(\mathbf{E})$ has a corresponding ground type symbol **A**.

In order to specify the function symbols of $\mathscr{L}(\mathbf{E})$, we associate with each type symbol $\mathbf{A}$ the $\mathbf{E}$-object $\mathbf{A}_E$ recursively as follows:

$$\mathbf{A}_E = A \qquad \text{for any ground type symbol } \mathbf{A}$$
$$(\mathbf{A}\times\mathbf{B})_E = \mathbf{A}_E \times \mathbf{B}_E$$
$$(\mathbf{PA})_E = P(\mathbf{A}_E).$$

Now the *function symbols* of $\mathscr{L}(\mathbf{E})$ are triples $(f, \mathbf{A}, \mathbf{B})$ where $\mathbf{A}$, $\mathbf{B}$ are type symbols of $\mathscr{L}(\mathbf{E})$ and $f : \mathbf{A}_E \to \mathbf{B}_E$ in $\mathbf{E}$. The *signature* of $(f, \mathbf{A}, \mathbf{B})$ is $\mathbf{A} \to \mathbf{B}$.

We shall often write just $\mathbf{f}$ for $(f, \mathbf{A}, \mathbf{B})$ when there is little likelihood of confusion. In this case $\mathbf{f}$ will be called the function symbol *associated* with the arrow f.

The *natural interpretation*—denoted by E—of $\mathscr{L}(\mathbf{E})$ in $\mathbf{E}$ is now determined by the assignments

$$\mathbf{A}_E = A \qquad \text{for any ground type symbol } \mathbf{A}$$
$$\mathbf{f}_E = f \qquad \text{for any function symbol } \mathbf{f}.$$

The local set theory Th($\mathbf{E}$)—the (*associated local set*) *theory* of $\mathbf{E}$—is now defined to be the theory in $\mathscr{L}(\mathbf{E})$ whose axioms are all sequents $\Gamma : \alpha$ such that $\Gamma \models_E \alpha$ under the natural interpretation of $\mathscr{L}(\mathbf{E})$ in $\mathbf{E}$.

The soundness theorem then immediately yields:

3.30 PROPOSITION

$$\Gamma \vdash_{\mathrm{Th}(\mathbf{E})} \alpha \quad \text{iff} \quad \Gamma \models_E \alpha. \quad \square$$

Let τ be a term of $\mathscr{L}(\mathbf{E})$ of type $\mathbf{B}$ with free variables among $x_1, \ldots, x_n$ of types $\mathbf{A}_1, \ldots, \mathbf{A}_n$ respectively. Then $[\![\tau]\!]_{E,\mathbf{x}}$ (which we shall abbreviate to $[\![\tau]\!]_{\mathbf{x}}$ as usual) is an $\mathbf{E}$-arrow; we write $\boldsymbol{\lambda}\mathbf{x}.\tau$ for the function symbol $([\![\tau]\!]_{\mathbf{x}}, \mathbf{A}_1 \times \cdots \times \mathbf{A}_n, \mathbf{B})$ in $\mathscr{L}(\mathbf{E})$. Note that $\boldsymbol{\lambda}\mathbf{x}.\tau$ then has signature $\mathbf{A}_1 \times \cdots \times \mathbf{A}_n \to \mathbf{B}$, and that

$$(\boldsymbol{\lambda}\mathbf{x}.\tau)_E = [\![\tau]\!]_{\mathbf{x}}.$$

As a consequence, we have the following result, which asserts that, in Th($\mathbf{E}$), any term of $\mathscr{L}(\mathbf{E})$ is 'representable' by a function symbol.

3.31 PROPOSITION

$$\vdash_{\mathrm{Th}(\mathbf{E})} \tau = \boldsymbol{\lambda}\mathbf{x}.\tau(\langle x_1, \ldots, x_n\rangle).$$

Proof. We have

$$[\![\tau]\!]_{\mathbf{x}} = [\![\tau]\!]_{\mathbf{x}} \circ [\![\langle x_1, \ldots, x_n\rangle]\!]_{\mathbf{x}}$$
$$= [\![\boldsymbol{\lambda}\mathbf{x}.\tau(\langle x_1, \ldots, x_n\rangle)]\!]_{\mathbf{x}},$$

so that, by (3.23.3),

$$\models_E \tau = \boldsymbol{\lambda}\mathbf{x}.\tau(\langle x_1, \ldots, x_n\rangle).$$

The result now follows from (3.30). $\square$

Our next proposition shows that function symbols in Th(**E**) behave *extensionally*.

3.32 PROPOSITION

$$\vdash_{\mathrm{Th}(\mathbf{E})} \mathbf{f}(x) = \mathbf{g}(x) \quad \text{iff} \quad f = g.$$

Proof. We have

$$\begin{aligned} \vdash_{\mathrm{Th}(\mathbf{E})} \mathbf{f}(x) = \mathbf{g}(x) \quad & \text{iff} \quad \models_E \mathbf{f}(x) = \mathbf{g}(x) \\ & \text{iff} \quad [\![\mathbf{f}(x) = \mathbf{g}(x)]\!]_x = T \\ & \text{iff} \quad \mathrm{eq} \circ \langle f, g \rangle = T \\ & \text{iff} \quad f = g \quad \text{(by (2.19(iii)).} \quad \square \end{aligned}$$

Next, we characterize *monics* in **E**.

3.33 PROPOSITION

$$f \text{ is monic iff} \quad \mathbf{f}(x) = \mathbf{f}(y) \vdash_{\mathrm{Th}(\mathbf{E})} x = y.$$

Proof. First suppose that f is monic. Then from

$$\mathbf{f}(x) = \mathbf{f}(y) \vdash_{\mathrm{Th}(\mathbf{E})} \mathbf{f}(x) = \mathbf{f}(y),$$

we deduce using (3.31) and (3.30)

$$\mathbf{f}(x) = \mathbf{f}(y) \models_E \mathbf{f}(\lambda xy.x(\langle x,y \rangle)) = \mathbf{f}(\lambda xy.y(\langle x,y \rangle))$$

So by (3.23.3),

$$f \circ [\![x]\!]_{xy} \circ \overline{[\![\mathbf{f}(x) = \mathbf{f}(y)]\!]_{xy}} = f \circ [\![y]\!]_{xy} \circ \overline{[\![\mathbf{f}(x) = \mathbf{f}(y)]\!]_{xy}}.$$

Since f is monic, it may be cancelled on both sides of this equation to give

$$[\![x]\!]_{xy} \circ \overline{[\![\mathbf{f}(x) = \mathbf{f}(y)]\!]_{xy}} = [\![y]\!]_{xy} \circ \overline{[\![\mathbf{f}(x) = \mathbf{f}(y)]\!]_{xy}}.$$

and so by (3.23.3) again

$$\mathbf{f}(x) = \mathbf{f}(y) \models_E x = y$$

whence by (3.30), $\mathbf{f}(x) = \mathbf{f}(y) \vdash_{\mathrm{Th}(\mathbf{E})} x = y$.

Conversely, suppose $f \circ g = f \circ h$. Then, by (3.32),

$$\models_E \mathbf{f}(\mathbf{g}(x)) = \mathbf{f}(\mathbf{h}(x)). \tag{*}$$

So if

$$\mathbf{f}(x) = \mathbf{f}(y) \vdash_{\mathrm{Th}(\mathbf{E})} x = y,$$

it follows from (3.30) that

$$\mathbf{f}(x) = \mathbf{f}(y) \models_E x = y$$

and hence, using (*)

$$\vDash_E \mathbf{g}(x) = \mathbf{h}(x).$$

So by (3.30) and (3.32), $g = h$, and f is monic. □

Our next objective is to show that Th(**E**) satisfies eliminability of descriptions for *all* formulae. We require two lemmas.

3.34 LEMMA. The following are equivalent:

(i) $\vdash_{\mathrm{Th}(\mathbf{E})} \alpha(y_1/\tau_1, \ldots, y_n/\tau_n)$
(ii) $[\![\alpha]\!]_{\mathbf{y}} \circ f = [\![\langle \tau_1, \ldots, \tau_n \rangle]\!]_{\mathbf{x}}$ for some arrow f, where $y_1, \ldots, y_n$ are the free variables of α and τ_i is free for y_i in α for each i.

Proof. We have

$\vdash_{\mathrm{Th}(\mathbf{E})} \alpha(y_1/\tau_1, \ldots, y_n/\tau_n)$ iff $\vDash_E \alpha(y_1/\tau_1, \ldots, y_n/\tau_n)$
iff $[\![\alpha]\!]_{\mathbf{y}} \circ [\![\langle \tau_1, \ldots, \tau_n \rangle]\!]_{\mathbf{x}} = T$ (by (3.20))
iff there is an f such that
$\overline{[\![\alpha]\!]_{\mathbf{y}}} \circ f = [\![\langle \tau_1, \ldots, \tau_n \rangle]\!]_{\mathbf{x}}$ (by (2.19(i)) □

3.35 LEMMA. If m is monic in **E**, then

$$\vdash_{\mathrm{Th}(\mathbf{E})} \boldsymbol{\chi}(\mathbf{m})(x) \Leftrightarrow \exists y . x = \mathbf{m}(y).$$

Proof. From $\vdash_{\mathrm{Th}(\mathbf{E})} \exists y(\mathbf{m}(z) = \mathbf{m}(y))$ we deduce, using (3.34), that there is an arrow f such that

$$\begin{aligned} \overline{[\![\exists y . x = \mathbf{m}(y)]\!]_x} \circ f &= [\![\mathbf{m}(z)]\!]_z \\ &= m \\ &\sim \overline{\chi(m)} \\ &= \overline{[\![\boldsymbol{\chi}(\mathbf{m})(x)]\!]_x}. \end{aligned}$$

It follows that $[\![\boldsymbol{\chi}(\mathbf{m})(x)]\!]_x \leq [\![\exists y . x = \mathbf{m}(y)]\!]_x$ and so $\boldsymbol{\chi}(\mathbf{m})(x) \vDash_E \exists y . x = \mathbf{m}(y)$, whence

$$\boldsymbol{\chi}(\mathbf{m})(x) \vdash_{\mathrm{Th}(\mathbf{E})} \exists y . x = \mathbf{m}(y).$$

Conversely, from $\chi(m) \circ m = T$, we deduce, using (3.23.1), that $\vDash_E \boldsymbol{\chi}(\mathbf{m})(\mathbf{m}(y))$, whence $\vdash_{\mathrm{Th}(\mathbf{E})} \boldsymbol{\chi}(\mathbf{m})(\mathbf{m}(y))$. Therefore $x = \mathbf{m}(y) \vdash_{\mathrm{Th}(\mathbf{E})} \boldsymbol{\chi}(\mathbf{m})(x)$, whence

$$\exists y . x = \mathbf{m}(y) \vdash_{\mathrm{Th}(\mathbf{E})} \boldsymbol{\chi}(\mathbf{m})(x).$$

The result follows. □

3.36 THEOREM (Eliminability of descriptions in Th(**E**)). If $\vdash_{\mathrm{Th}(\mathbf{E})} \exists! y \alpha$, then there is a unique **E**-arrow f such that

$$\vdash_{\mathrm{Th}(\mathbf{E})} \alpha(y/f(\langle x_1, \ldots, x_n \rangle))$$

where $x_1, \ldots, x_n, y$ are the free variables of α.

Proof. Suppose that $\vdash_{\mathrm{Th}(\mathbf{E})} \exists! y\alpha$. Then $\vdash_{\mathrm{Th}(\mathbf{E})} \exists z.\{y:\alpha\} = \{z\}$, whence, by (3.31),

$$\vdash_{\mathrm{Th}(\mathbf{E})} \exists z[\{y:\alpha\} = \boldsymbol{\lambda} z.\{z\}(z)]. \tag{1}$$

Now by (3.33), $[\![\{z\}]\!]_z$ is monic, and so by (3.35) and (1), we get, writing m for $[\![\{z\}]\!]_z$ and noting that $\mathbf{m}$ is $\boldsymbol{\lambda} z.\{z\}$,

$$\vdash_{\mathrm{Th}(\mathbf{E})} \boldsymbol{\chi}(\mathbf{m})(\{y:\alpha\}).$$

Accordingly by (3.34) there is an arrow h such that $\overline{\chi(m)} \circ h = [\![\{y:\alpha\}]\!]_{\mathbf{x}}$ and since $\overline{\chi(m)} \sim m$ it follows that there is an arrow f such that

$$m \circ f = [\![\{y:\alpha\}]\!]_{\mathbf{x}}.$$

But

$$m \circ f = [\![\{\mathbf{f}(\langle x_1, \ldots, x_n \rangle)\}]\!]_{\mathbf{x}}$$

and so

$$[\![\{y:\alpha\}]\!]_{\mathbf{x}} = [\![\{\mathbf{f}(\langle x_1, \ldots, x_n \rangle)\}]\!]_{\mathbf{x}}$$

whence

$$\models_E \{y:\alpha\} = \{\mathbf{f}(\langle x_1, \ldots, x_n \rangle)\}$$

so that

$$\vdash_{\mathrm{Th}(\mathbf{E})} \{y:\alpha\} = \{\mathbf{f}(\langle x_1, \ldots, x_n \rangle)\}.$$

Therefore

$$\vdash_{\mathrm{Th}(\mathbf{E})} y = \mathbf{f}(\langle x_1, \ldots, x_n \rangle) \Leftrightarrow \alpha$$

and so

$$\vdash_{\mathrm{Th}(\mathbf{E})} \alpha(y/\mathbf{f}(\langle x_1, \ldots, x_n \rangle)).$$

If g satisfies $\vdash_{\mathrm{Th}(\mathbf{E})} \alpha(y/\mathbf{g}(\langle x_1, \ldots, x_n \rangle))$, then

$$\vdash_{\mathrm{Th}(\mathbf{E})} \mathbf{g}(\langle x_1, \ldots, x_n \rangle) = \mathbf{f}(\langle x_1, \ldots, x_n \rangle)$$

and hence, by (3.32), $f = g$. □

We are finally in a position to prove the

3.37 EQUIVALENCE THEOREM. For any topos $\mathbf{E}$,

$$\mathbf{E} \simeq \mathbf{C}(\mathrm{Th}(\mathbf{E})).$$

Proof. Define a map $F: \mathbf{E} \to \mathbf{C}(\mathrm{Th}(\mathbf{E}))$ by

$$F(A) = U_{\mathbf{A}} \qquad \text{for } \mathbf{E}\text{-objects } A,$$

and if $A \xrightarrow{f} B$ is an arrow of $\mathbf{E}$,

$$F(f) = (x \mapsto \mathbf{f}(x)) : U_{\mathbf{A}} \to U_{\mathbf{B}}.$$

Then F is a functor, since

$$\begin{aligned} F(g \circ f) &= \{\langle x,z \rangle : \mathbf{g}(\mathbf{f}(x)) = z\} \\ &= \{\langle x,z \rangle : \exists y(\mathbf{f}(x) = y \wedge \mathbf{g}(y) = z)\} \\ &= Fg \circ Ff, \end{aligned}$$

and

$$\begin{aligned} F(1_A) &= \{\langle x,y \rangle : x = y\} \\ &= 1_{FA}. \end{aligned}$$

We claim that F is an equivalence of categories. To do this we construct a quasi-inverse (1.15) $G : \mathbf{C}(\mathrm{Th}(\mathbf{E})) \to \mathbf{E}$ for F.

Given $X = \{x : \alpha\}$ of type $\mathbf{PA}$ in $C(\mathrm{Th}(E))$, define

$$GX = \mathrm{dom}\overline{[\![\alpha]\!]_x} \qquad \text{in } \mathbf{E}.$$

If

$$X = \{x : \alpha\} \xrightarrow{f} \{y : \beta\} = Y$$

in $\mathbf{C}(\mathrm{Th}(\mathbf{E}))$, we define $Gf : GX \to GY$ as follows. Writing i for the $\mathbf{E}$-arrow $\overline{[\![\alpha]\!]_x}$ we have

$$\overline{[\![\alpha]\!]_x} \circ 1_A = i = [\![\mathbf{i}(u)]\!]_u,$$

so that, by (3.34),

$$\vdash_{\mathrm{Th}(\mathbf{E})} \alpha(x/\mathbf{i}(u)), \tag{1}$$

where u is a variable of type $\mathbf{GX}$. Hence

$$\vdash_{\mathrm{Th}(\mathbf{E})} \exists! y \langle \mathbf{i}(u), y \rangle \in f.$$

Write j for $\overline{[\![\beta]\!]_y}$. Since j is monic, it follows from (3.33) that

$$\vdash_{\mathrm{Th}(\mathbf{E})} \exists! v \langle \mathbf{i}(u), \mathbf{j}(v) \rangle \in f,$$

where v is a variable of type $\mathbf{GY}$. So by (3.36) there is a unique arrow $g : GX \to GY$ such that

$$\vdash_{\mathrm{Th}(\mathbf{E})} \langle \mathbf{i}(u), \mathbf{j}(\mathbf{g}(u)) \rangle \in f. \tag{2}$$

We define $G(f) = g$.

Now from the definition of GX, the diagram

$$\begin{array}{ccc} GX & \longrightarrow & 1 \\ {\scriptstyle i}\downarrow & & \downarrow{\scriptstyle \top} \\ A & \xrightarrow{[\![x \in X]\!]_x} & \Omega \end{array}$$

is a pullback, and from (2), for any $f : X \to Y$ in $\mathbf{C}(\mathrm{Th}(\mathbf{E}))$, Gf is the unique arrow $GX \to GY$ such that

$$\vdash_{\mathrm{Th}(\mathbf{E})} \langle \mathbf{i}(u), (\mathbf{j} \circ \mathbf{Gf})(u) \rangle \in f. \tag{3}$$

Now if $g: Y \to Z$ in $\mathbf{C}(\mathrm{Th}(\mathbf{E}))$, let k be obtained from Z as was i from X. Then it follows easily from (3) that

$$\vdash_{\mathrm{Th}(\mathbf{E})} \langle \mathbf{i}(u), (\mathbf{k} \circ \mathbf{Gg} \circ \mathbf{Gf})(u) \rangle \in g \circ f,$$

and so $G(g \circ f) = Gg \circ Gf$. Moreover,

$$\vdash_{\mathrm{Th}(\mathbf{E})} \langle \mathbf{i}(u), \mathbf{i}(u) \rangle \in 1_X,$$

so that $G1_X = 1_{GX}$. Accordingly G is a functor $\mathbf{C}(\mathrm{Th}(\mathbf{E})) \to \mathbf{E}$.

It follows from (1) that $\vdash_{\mathrm{Th}(\mathbf{E})} \mathbf{i}(u) \in X$, so we have an arrow

$$FGX = U_{GX} \xrightarrow{u \mapsto \mathbf{i}(u)} X$$

in $\mathbf{C}(\mathrm{Th}(\mathbf{E}))$. Write η_X for this arrow. Then (3) asserts that the $\mathbf{C}(\mathrm{Th}(\mathbf{E}))$-diagram

$$\begin{array}{ccc} FGX & \xrightarrow{FGf} & FGY \\ \downarrow{\scriptstyle \eta_X} & & \downarrow{\scriptstyle \eta_Y} \\ X & \xrightarrow{f} & Y \end{array}$$

commutes, so that $\eta = (\eta_X)$ is a natural transformation $FG \to 1_{\mathbf{C}(\mathrm{Th}(\mathbf{E}))}$. We claim that η is a natural isomorphism. To prove this we show that the Th(**E**)-set

$$h = \{ \langle x, u \rangle : \mathbf{i}(u) = x \}$$

is the inverse of η_X.

Since i is monic, and $\boldsymbol{\chi}(\mathbf{i})$ is $\boldsymbol{\lambda} x . \alpha$, it follows from (3.35) that

$$\vdash_{\mathrm{Th}(\mathbf{E})} \boldsymbol{\lambda} x . \alpha(x) \Leftrightarrow \exists ! u (\mathbf{i}(u) = x).$$

But by (3.31),

$$\vdash_{\mathrm{Th}(\mathbf{E})} \alpha \Leftrightarrow \boldsymbol{\lambda} x . \alpha(x).$$

Hence

$$\vdash_{\mathrm{Th}(\mathbf{E})} \alpha \Leftrightarrow \exists ! u (\mathbf{i}(u) = x),$$

so that

$$\alpha \vdash_{\mathrm{Th}(\mathbf{E})} \exists ! u (\mathbf{i}(u) = x).$$

Therefore

$$\alpha \vdash_{\mathrm{Th}(\mathbf{E})} \exists ! u (\langle x, u \rangle \in h),$$

so that

$$h : X \to U_X = FGX$$

in $\mathbf{C}(\mathrm{Th}(\mathbf{E}))$. Obviously η_X and h are mutually inverse, so each η_X is an isomorphism and η a natural isomorphism as claimed.

Finally, it follows easily from the definition of G that $GF = 1_{\mathbf{E}}$.

Accordingly G is a quasi-inverse for F and the theorem is proved. □

The equivalence $F:\mathbf{E} \to \mathbf{C}(\mathrm{Th}(\mathbf{E}))$ constructed in (3.37) is called the *canonical inclusion* of $\mathbf{E}$ in $\mathbf{C}(\mathrm{Th}(\mathbf{E}))$, and its quasi-inverse $G:\mathbf{C}(\mathrm{Th}(\mathbf{E})) \to \mathbf{E}$ the *canonical equivalence* between $\mathbf{C}(\mathrm{Th}(\mathbf{E}))$ and $\mathbf{E}$; it is often written $\mathbf{C}(\mathrm{Th}(\mathbf{E})) \xrightarrow{\text{can}} \mathbf{E}$. For an object X of $\mathbf{C}(\mathrm{Th}(\mathbf{E}))$ and an arrow f of $\mathbf{C}(\mathrm{Th}(\mathbf{E}))$ we shall write

$$[X]_{\mathbf{E}} \text{ for } GX, \qquad [f]_{\mathbf{E}} \text{ for } Gf,$$

and call them the *canonical interpretations* of X and f in $\mathbf{E}$. If follows from the proof of (3.37) that, for any object A of $\mathbf{E}$, and any arrow $A \xrightarrow{f} B$,

$$[U_{\mathbf{A}}]_{\mathbf{E}} = A, \qquad [\mathbf{f}(x)]_{\mathbf{E}} = f.$$

To conclude this section we make the following important definitions. A local set theory S in a local language $\mathscr{L}$ is said to be

- *well-termed* if, whenever $\vdash_S \exists! x\alpha$, there is a term τ whose free variables are those of α with x deleted, such that $\vdash_S \alpha(x/\tau)$;
- *well-typed* if for any S-set X there is a type symbol $\mathbf{A}$ such that $U_{\mathbf{A}} \cong X$ in $\mathbf{C}(S)$.

Thus well-termedness means the eliminability of descriptions for arbitrary terms.

Then 3.36 and (the proof of) 3.37 yield the

3.38 COROLLARY. For any topos $\mathbf{E}$, $\mathrm{Th}(\mathbf{E})$ is well-termed and well-typed. □

We also note the

3.39 PROPOSITION. (i) For any local set theory S, there is a bijective correspondence between $\mathbf{C}(S)$-arrows $1 \to \Omega$ and sentences of the language $\mathscr{L}$ of S, where we identify two sentences α, β whenever $\vdash_S \alpha \Leftrightarrow \beta$.

(ii) If in addition S is well-termed, then for any S-set of type $\mathbf{PA}$ there is a bijective correspondence between $\mathbf{C}(S)$-arrows $1 \to X$ and closed terms τ of type $\mathbf{A}$ such that $\vdash_S \tau \in X$, where we identify two such terms σ, τ whenever $\vdash_S \sigma = \tau$.

Proof. (i) Given a sentence α, associate the $\mathbf{C}(S)$-arrow $1 \xrightarrow{* \mapsto \alpha} \Omega$ with α. Conversely, if $1 \xrightarrow{f} \Omega$ in $\mathbf{C}(S)$, then $\vdash_S \exists! \omega \langle *, \omega \rangle \in f$ and so, by (3.8), there is a sentence α (viz., $\langle *, \mathit{true} \rangle \in f$) such that $\vdash_S \langle *, \alpha \rangle \in f$; we associate α with f. These correspondences are obviously mutually inverse.

(ii) Similar to (i), invoking well-termedness in lieu of (3.8) □

REMARK. Given a local set theory S, let $\mathbf{C}^*(S)$ be the subcategory of $\mathbf{C}(S)$ whose objects are all S-sets of the form $U_{\mathbf{A}}$ and whose arrows are all S-maps of the form $x \mapsto \tau$ with τ a term. $\mathbf{C}^*(S)$ is called the *category of S-types and terms*. Then it is easy to see that S is well-termed iff the insertion functor $\mathbf{C}^*(S) \hookrightarrow \mathbf{C}(S)$ is full and well-typed iff $\mathbf{C}^*(S) \hookrightarrow \mathbf{C}(S)$ is dense. Thus S is well-termed and well-typed iff $\mathbf{C}^*(S) \hookrightarrow \mathbf{C}(S)$ is an equivalence of categories.

Translations and logical functors

Let $\mathscr{L}$ and $\mathscr{L}'$ be local languages. A *translation* of $\mathscr{L}$ into $\mathscr{L}'$, written $K: \mathscr{L} \to \mathscr{L}'$, is a map K which assigns to each type symbol $\mathbf{A}$ of $\mathscr{L}$ a type symbol $K(\mathbf{A})$ of $\mathscr{L}'$ and to each function symbol $\mathbf{f}$ of $\mathscr{L}$ of signature $\mathbf{A} \to \mathbf{B}$ a function symbol $K(\mathbf{f})$ of signature $K(\mathbf{A} \to K(\mathbf{B})$ in such a way that

$$K(\mathbf{1}) = \mathbf{1}$$

$$K(\mathbf{\Omega}) = \mathbf{\Omega}$$

$$K(\mathbf{A}_1 \times \cdots \times \mathbf{A}_n) = K(\mathbf{A}_1) \times \cdots \times K(\mathbf{A}_n)$$

$$K(\mathbf{PA}) = \mathbf{P}K(\mathbf{A}).$$

Any translation $K: \mathscr{L} \to \mathscr{L}'$ may be extended to the terms of $\mathscr{L}$ in the obvious recursive way†, so that, for each term τ of $\mathscr{L}$ of type $\mathbf{A}$, $K(\tau)$ is a term of $\mathscr{L}'$ of type $K(\mathbf{A})$.

Now let S, S' be local set theories in $\mathscr{L}$, $\mathscr{L}'$ respectively. A translation $K: \mathscr{L} \to \mathscr{L}'$ is called a *translation of S into S'*, and we write $K: S \to S'$, if, for any sequent $\Gamma : \alpha$ of $\mathscr{L}$,

$$\Gamma \vdash_S \alpha \quad \text{implies} \quad K(\Gamma) \vdash_{S'} K(\alpha)$$

(where, if $\Gamma = \{\alpha_1, \ldots, \alpha_n\}$, $K(\Gamma) = \{K(\alpha_1), \ldots, K(\alpha_n)\}$).

If we have, under the same conditions,

$$\Gamma \vdash_S \alpha \quad \text{iff} \quad K(\Gamma) \vdash_{S'} K(\alpha)$$

then K is called a *conservative translation* of S into S'. There is an obvious correspondence between models of S in a topos $\mathbf{E}$ and translations $S \to \mathrm{Th}(\mathbf{E})$ (in fact the two concepts amount to the same thing). In particular, the *identity translation* $\mathrm{Th}(\mathbf{E}) \to \mathrm{Th}(\mathbf{E})$ corresponds to the *natural interpretation* of $\mathrm{Th}(\mathbf{E})$ in $\mathbf{E}$.

Let $\mathbf{E}$ and $\mathbf{E}'$ be toposes with specified terminal object, products, subobject classifier, power objects and evaluation arrows: a functor

† That is, by defining $K(*) = *, \ldots, K(\mathbf{f}(\tau)) = K\mathbf{f}(K\tau), \ldots, K(\sigma \in \tau) = K\sigma \in K\tau$, etc.

$F:\mathbf{E}\to\mathbf{E}'$ which preserves all these is called a *logical functor*. It is easy to see that the canonical inclusion $\mathbf{E}\to\mathbf{C}(\mathrm{Th}(\mathbf{E}))$ is a logical functor.

If $K:S\to S'$ is a translation, then for terms σ, τ of $\mathscr{L}$

$$\vdash_S \sigma=\tau \qquad \text{implies} \qquad \vdash_{S'} K(\sigma)=K(\tau)$$

from which we infer that K induces a map $\mathbf{C}(K)$ from the class of S-sets to the class of S'-sets, given by

$$\mathbf{C}(K)([\sigma]_S)=[K(\sigma)]_{S'}.$$

It is readily checked that $\mathbf{C}(K)$ is a logical functor $\mathbf{C}(S)\to\mathbf{C}(S')$. Write **Loc** for the category of (small†) local set theories and translations, and **Top** for the category of (small) toposes and logical functors. It is now easily seen that $\mathbf{C}$ is a functor $\mathbf{Loc}\xrightarrow{\mathbf{C}}\mathbf{Top}$.

Conversely, any logical functor $F:\mathbf{E}\to\mathbf{E}'$ induces a translation $\mathrm{Th}(F):\mathrm{Th}(\mathbf{E})\to\mathrm{Th}(\mathbf{E}')$ in the obvious way, and it is not hard to see that Th is then a functor $\mathbf{Top}\xrightarrow{\mathrm{Th}}\mathbf{Loc}$. Now it would be pleasant to be able to assert that there exists an adjunction $\mathbf{C}\dashv\mathrm{Th}$. For this to be the case, there must be a natural transformation $\eta:1_{\mathbf{Loc}}\to\mathrm{Th}\circ\mathbf{C}$ such that, for any theory S, any topos $\mathbf{E}$ and any translation $S\xrightarrow{K}\mathrm{Th}(\mathbf{E})$, there is a unique logical functor $\mathbf{C}(S)\xrightarrow{F}\mathbf{E}$ such that the diagram

$$\begin{array}{ccc} S & \xrightarrow{\eta_S} & \mathrm{Th}(\mathbf{C}(S)) \\ & {}_{K}\searrow & \downarrow{\scriptstyle \mathrm{Th}(F)} \\ & & \mathrm{Th}(\mathbf{E}) \end{array} \qquad (*)$$

commutes. Of course, η would then be the unit of the adjunction.

Now there *is* a natural candidate for η. Namely, given a local set theory S in a local language $\mathscr{L}$, define $\eta_S:\mathscr{L}\to\mathscr{L}(\mathbf{C}(S))$ by

$$\eta_S(\mathbf{A})=U_{\mathbf{A}}, \qquad \eta_S(\mathbf{f})=(x\mapsto\mathbf{f}(x))$$

for type symbols $\mathbf{A}$, function symbols $\mathbf{f}$. It is readily seen that η_S is a translation of $\mathscr{L}$ into $\mathscr{L}(\mathbf{C}(S))$. Furthermore, *$\eta_S$ is a conservative translation of S into* $\mathrm{Th}(\mathbf{C}(S))$. For it is easily shown by induction on the formation of terms that, for any term τ of $\mathscr{L}$,

$$[\![\tau]\!]_{C(S),\mathbf{x}}=[\![\eta_S(\tau)]\!]_{C(S),K(\mathbf{x})} \qquad (**)$$

where $[\![\cdot]\!]_{C(S),\mathbf{x}}$ on the l.h.s. of (**) denotes the canonical interpretation of $\mathscr{L}$ in $\mathbf{C}(S)$) and on the r.h.s. it denotes the natural interpretation of

† A local set theory is *small* if the collection of symbols of its language forms a set.

$\mathscr{L}(\mathbf{C}(S))$ in $\mathbf{C}(S)$. It follows that, for any sequent $\Gamma:\alpha$ of $\mathscr{L}$,

$$\begin{array}{llll} \Gamma \vdash_S \alpha & \text{iff} & \Gamma \vDash_{C(S)} \alpha & \text{(by (3.28))} \\ & \text{iff} & K(\Gamma) \vDash_{C(S)} K(\alpha) & \text{(by (**))} \\ & \text{iff} & K(\Gamma) \vdash_{\mathrm{Th}(C(S))} K(\alpha) & \text{(by (3.30))} \end{array}$$

as claimed.

Unfortunately, however, when one attempts to construct the functor $F:\mathbf{C}(S)\rightarrow\mathbf{E}$ making the diagram (*) commute, one finds that the natural candidate for F—viz., the composition $\mathbf{C}(S)\xrightarrow{\mathbf{C}(K)}\mathbf{C}(\mathrm{Th}(\mathbf{E}))\xrightarrow{\text{can}}\mathbf{E}$—is only a logical functor 'up to isomorphism', which furthermore is unique only 'up to isomorphism', so that one cannot claim to have an adjunction without frenzied hand-waving. A calmer solution to this problem has been proposed by Lambek and Scott (1986) who suggest tightening up the definition of topos by specifying the kernels of arrows to Ω, inducing a corresponding tightening of the definition of logical functor. This has the effect of removing the 'up to isomorphism' qualification on F, thereby yielding the desired adjunction. We shall not pursue this further, but refer the reader to Lambek and Scott's book.

The fact that the 'unit' arrow $S \hookrightarrow \mathrm{Th}(\mathbf{C}(S))$ is a conservative translation has the following useful consequence.

3.40 PROPOSITION. For any local set theory S, there is a well-termed and well-typed local set theory S' and a conservative translation of S into S'.

Proof. By the above observation and (3.38). □

Adjoining indeterminates

Let $\mathscr{L}$ be a local language, and $\mathbf{A}$ a type symbol of $\mathscr{L}$. A term of the form $\mathbf{f}(*)$, where $\mathbf{f}$ is a function symbol of $\mathscr{L}$ of signature $\mathbf{1}\rightarrow\mathbf{A}$, is called a *constant* of type $\mathbf{A}$. Let S be a local set theory in $\mathscr{L}$, and let α be a formula with exactly one free variable x of type $\mathbf{A}$. Let $\mathscr{L}(\mathbf{c})$ be the language obtained from $\mathscr{L}$ by adding a new function symbol $\mathbf{c}$ of signature $\mathbf{1}\rightarrow\mathbf{A}$, write c for $\mathbf{c}(*)$, and let $S(\alpha)$ be the theory in $\mathscr{L}(\mathbf{c})$ whose axioms are those of S plus all sequents of the form

$$:\beta(x/c),$$

where $\alpha \vdash_S \beta$. Note that $\vdash_{S(\alpha)} \alpha(x/c)$.

Then, in $S(\alpha)$, c behaves as a *universal term* or *indeterminate* satisfying α in the following sense.

3.41 THEOREM. Let $\mathcal{S}'$ be a theory in a local language $\mathcal{L}'$ and let K be a translation of S into S'. Then for any constant $c' = \mathbf{c}'(*)$ of $\mathcal{L}'$ of type $K(\mathbf{A})$ such that, writing x' for $K(x)$, $\vdash_{S'} K(\alpha)(x'/c')$, there is a unique translation $K' : S(\alpha) \rightarrow S'$ extending K such that $K'(c) = c'$.

The proof of this theorem depends on the following lemma, which will also turn out to be useful in other contexts.

3.42 LEMMA. For any sequent $\Gamma : \gamma$ of $\mathcal{L}$ we have

$$\Gamma(x/c) \vdash_{S(\alpha)} \gamma(x/c) \quad \text{iff} \quad \alpha, \Gamma \vdash_S \gamma.$$

Proof. If $\alpha, \Gamma \vdash_S \gamma$, then $\alpha, \Gamma \vdash_{S(\alpha)} \gamma$ so that $\alpha(x/c)$, $\Gamma(x/c) \vdash_{S(\alpha)} \gamma(x/c)$ by the substitution rule, and since $\vdash_{S(\alpha)} \alpha(x/c)$, it follows that $\Gamma(x/c) \vdash_{S(\alpha)} \gamma(x/c)$.

Conversely, suppose that $\Gamma(x/c) \vdash_{S(\alpha)} \gamma(x/c)$; let P be a proof from $S(\alpha)$ with the sequent $\Gamma(x/c) : \gamma(x/c)$ as conclusion. Let y be a variable of type $\mathbf{A}$ not occurring in P. For each sequent $\Delta : \delta$ of $\mathcal{L}(\mathbf{c})$ let $\Delta_y : \delta_y$ be the result of replacing each occurrence of c (if any) by y. Now replace each sequent $\Delta : \delta$ in P by the sequent $\alpha(x/y)$, $\Delta_y : \delta_y$, thus obtaining a new tree P'. The effect of this operation is to convert: (a) any basic axiom into a valid sequent, (b) any axiom $\Delta : \delta$ of S into the sequent $\alpha(x/y)$, $\Delta : \delta$ which is derivable from S, (c) any new axiom: $\beta(x/c)$ of $S(\alpha)$ into the sequent $\alpha(x/y) : \beta(x/y)$, which, since $\alpha \vdash_S \beta$, is also derivable from S, (d) any instance of a rule of inference into an instance of the same rule of inference, and finally (e) the conclusion $\Gamma(x/c) : \gamma(x/c)$ into the sequent $\alpha(x/y)$, $\Gamma(x/y) : \gamma(x/y)$. It therefore follows that P' yields a proof from S with conclusion $\alpha(x/y)$, $\Gamma(x/y) : \gamma(x/y)$. Hence $\alpha(x/y)$, $\Gamma(x/y) \vdash_S \gamma(x/y)$ and so $\alpha, \Gamma \vdash_S \gamma$ by the substitution rule. □

Now we can supply the

Proof of Theorem 3.41. Assume the conditions of the theorem. Extend K to a translation K' of $\mathcal{L}(\mathbf{c})$ into $\mathcal{L}'$ by defining $K'(\mathbf{c}) = \mathbf{c}'$. We claim that K' is a translation of $S(\alpha)$ into S'. Given any sequent $\Gamma : \beta$ of $\mathcal{L}(\mathbf{c})$, we may without loss of generality assume that it is of the form $\Delta(x/c) : \gamma(x/c)$ where $\Delta : \gamma$ is a sequent of $\mathcal{L}$. And then we have

$$\begin{aligned}
\Gamma \vdash_{S(\alpha)} \beta \quad & \text{iff} && \Delta(x/c) \vdash_{S(\alpha)} \gamma(x/c) \\
& \text{iff} && \alpha, \Delta \vdash_S \gamma \qquad \text{(by 3.42)} \\
& \text{implies} && K(\alpha), K(\Delta) \vdash_{S'} K(\gamma) \\
& \text{implies} && K(\alpha)(x'/c'), K(\Delta)(x'/c') \vdash_{S'} K(\gamma)(x'/c') \\
& \text{implies} && K(\Delta)(x'/c') \vdash_{S'} K(\gamma)(x'/c') \\
& && (\text{since} \vdash_{S'} K(\alpha)(x'/c'))
\end{aligned}$$

implies $\quad K'(\Delta(x/c)) \vdash_{S'} K'(\gamma(x/c))$

iff $\quad K'(\Gamma) \vdash_{S'} K'(\beta)$.

Therefore K' is a translation of $S(\alpha)$ into S'; the uniqueness condition is obvious. □

By virtue of this theorem we call $S(\alpha)$ the theory obtained from S by adjoining an *indeterminate* satisfying α. If I is an S-set and α is the formula $x \in I$, we write S_I for $S(\alpha)$ and call it the theory obtained from S by adjoining an *indeterminate I-element.* If α is the formula $x = x$, then $S(\alpha)$ is written $S(\mathbf{A})$ and called the theory obtained from S by adjoining an *indeterminate of type* $\mathbf{A}$.

Lastly, consider the theory $L_0(\mathbf{A})$ where L_0 is the pure local set theory in the pure local language $\mathscr{L}_0$ and $\mathbf{A}$ is a type symbol of $\mathscr{L}_0$. Now it is evident that L_0 is an *initial object* in the category **Loc**. It follows that, for any local set theory S in a local language $\mathscr{L}$, and any constant d of $\mathscr{L}$ of type $\mathbf{A}$ (noting that any type symbol of $\mathscr{L}_0$ may considered a type symbol of $\mathscr{L}$), there is a unique translation K of $L_0(\mathbf{A})$ into S such that $K(c) = d$. In this sense $L_0(\mathbf{A})$ is the universal theory of an indeterminate of type $\mathbf{A}$.

EXERCISES

(1) Show that the following conditions are equivalent:

(i) $\vdash_S \exists x \alpha$;

(ii) $S(\alpha)$ is a conservative extension of S, i.e. for any sequent $\Gamma : \gamma$ of $\mathscr{L}$, we have $\Gamma \vdash_{S(\alpha)} \gamma$ iff $\Gamma \vdash_S \gamma$.

(2) Show that the following conditions are equivalent:

(i) $\vdash_S \neg\exists x \alpha$

(ii) $S(\alpha)$ is inconsistent.

Introduction of function values

In set theory it is customary to introduce the function value $f(x)$ when f is a function and x is in its domain. In this final section we will sketch a proof that this device can also be legitimately employed within the context of a local set theory.

Let S be a local set theory in a local language $\mathscr{L}$ and let $X \xrightarrow{f} Y$ be an S-map with X and Y of types $\mathbf{PA}, \mathbf{PB}$ respectively. Let $\mathscr{L}^*$ be the language obtained from $\mathscr{L}$ by adding a new function symbol $\mathbf{f}^*$ of signature $\mathbf{A} \to \mathbf{B}$ and let S^* be the local set theory in $\mathscr{L}^*$ whose axioms are those of S together with the sequent

$$x \in X : \langle x, \mathbf{f}^*(x) \rangle \in f.$$

In S^*, accordingly, $\mathbf{f}^*(x)$ may be regarded as the *value of f at x*.

We claim that the *inclusion translation* $S \hookrightarrow S^*$ *is conservative*, i.e. for any sequent $\Gamma : \alpha$ of $\mathscr{L}$

$$\Gamma \vdash_S \alpha \qquad \text{iff} \qquad \Gamma \vdash_{S^*} \alpha. \tag{1}$$

This is proved by defining an interpretation of the terms of $\mathscr{L}^*$ in the terms of $\mathscr{L}$ 'having at most one element' in a manner to be described.

Define the map $\tau \mapsto [\tau]$ for terms τ of $\mathscr{L}^*$ recursively as follows:

$$\begin{aligned}
[*] &= \{*\} \\
[x] &= \{x\} \\
[\mathbf{f}^*(\tau)] &= \{y : \exists x \in [\tau].\langle x, y\rangle \in f\} \\
[\mathbf{g}(\tau)] &= \{\mathbf{g}(x) : x \in [\tau]\} \qquad \text{for any function symbol } \mathbf{g} \text{ of } \mathscr{L} \\
[\langle \tau_1, \ldots, \tau_n\rangle] &= \{\langle x_1, \ldots, x_n\rangle : x_1 \in [\tau_1] \wedge \cdots \wedge x_n \in [\tau_n]\} \\
[(\tau)_i] &= \{(x)_i : x \in [\tau]\} \\
[\{x : \alpha\}] &= \{\{x : \mathit{true} \in [\alpha]\}\} \\
[\sigma = \tau] &= \{(x = y) : x \in [\sigma] \wedge y \in [\tau]\} \\
[\sigma \in \tau] &= \{(x \in y) : x \in [\sigma] \wedge y \in [\tau]\}.
\end{aligned}$$

Clearly $[\tau]$ is a term of $\mathscr{L}$, for any term τ of $\mathscr{L}^*$.

One now shows easily by induction on the formation of τ that

$$x \in [\tau] \vdash_{S^*} x = \tau \qquad \text{for any } \mathscr{L}^*\text{-term } \tau. \tag{2}$$

(so that $[\tau]$ has *at most* the one element τ) and that

$$\vdash_S [\sigma] = \{\sigma\} \qquad \text{for any } \mathscr{L}\text{-term } \sigma. \tag{3}$$

We define the *interpretation* $\alpha_{\mathscr{L}}$ in $\mathscr{L}$ of an $\mathscr{L}^*$-formula α to be the formula $\mathit{true} \in [\alpha]$. It follows immediately from (3) that

$$\vdash_S \alpha_{\mathscr{L}} \Leftrightarrow \alpha \qquad \text{for any } \mathscr{L}\text{-formula } \alpha. \tag{4}$$

Now one shows that

$$\vdash_{S^*} \alpha_{\mathscr{L}} \Leftrightarrow \alpha \qquad \text{for any } \mathscr{L}^*\text{-formula } \alpha. \tag{5}$$

To get (5), we argue as follows. Using (2), we have

$$\alpha_{\mathscr{L}} \vdash_{S^*} \mathit{true} \in [\alpha] \vdash_{S^*} \alpha = \mathit{true} \vdash_{S^*} \alpha$$

and conversely, using (3),

$$\begin{aligned}
\alpha \vdash_{S^*} \alpha = \mathit{true} &\vdash_{S^*} \alpha = \mathit{true} \wedge [\mathit{true}] = \{\mathit{true}\} \\
&\vdash_{S^*} [\alpha] = [\mathit{true}] \wedge \mathit{true} \in [\mathit{true}] \\
&\vdash_{S^*} \mathit{true} \in [\alpha] \\
&\vdash_{S^*} \alpha_{\mathscr{L}}.
\end{aligned}$$

Next, one shows that, for any sequent $\Gamma:\alpha$ of $\mathscr{L}^*$,

$$\Gamma \vdash_{S^*} \alpha \qquad \text{iff} \qquad \Gamma_{\mathscr{L}} \vdash_S \alpha_{\mathscr{L}}, \tag{6}$$

where if $\Gamma = \{\alpha_1, \ldots, \alpha_n\}$, $\Gamma_{\mathscr{L}} = \{\alpha_{1\mathscr{L}}, \ldots, \alpha_{n\mathscr{L}}\}$. To prove this, one observes first that $\Gamma_{\mathscr{L}} \vdash_S \alpha_{\mathscr{L}}$ implies $\Gamma \vdash_{S^*} \alpha$ by (5). To prove the reverse implication, let P be a proof from S^* with $\Gamma:\alpha$ as conclusion. Replace in P each sequent $\Delta:\beta$ by the sequent $\Delta_{\mathscr{L}}:\beta_{\mathscr{L}}$. The effect of this operation is to convert (a) any basic axiom into a valid sequent, (b) any axiom of S into (something equivalent to) itself, (c) the axiom $x \in X: \langle x, \mathbf{f}^*(x)\rangle \in f$ into an S-derivable sequent, (d) any instance of a rule of inference into an instance of a valid derived rule of inference, (e) the conclusion $\Gamma:\alpha$ into $\Gamma_{\mathscr{L}}:\alpha_{\mathscr{L}}$. In this way we obtain a proof from S with conclusion $\Gamma_{\mathscr{L}}:\alpha_{\mathscr{L}}$ so that $\Gamma_{\mathscr{L}} \vdash_S \alpha_{\mathscr{L}}$ and (6) is proved.
(6) and (4) now together yield (1).

What this result means is that we can add function values to any local set theory S without materially altering it. (Obviously, we may do this for more than one function at a time.) In practice, we shall write

$$f(x) \quad \text{or} \quad fx \qquad \text{for} \qquad \mathbf{f}^*(x)$$

and assume that

$$x \in X \vdash_S \langle x, f(x)\rangle \in f.$$

This will be done, where necessary, without further comment.

4

Fundamental properties of toposes

Our primary purpose in this chapter is to put the equivalence theorem (3.37) to work in establishing fundamental properties of toposes. Briefly the procedure is the following. To show that all toposes have a certain essentially categorical property P, one shows that the argument that **Set** has P can be translated to any linguistic topos—which as a result has P—and then invokes the equivalence theorem to conclude that all toposes have P. In this way we will show, for example, that toposes are finitely complete, cocomplete, Cartesian closed, and also have the property that every arrow has a unique epic–monic factorization. Another result to be proved in a related manner is that slicing a topos preserves the topos property, and that, as a consequence, any pullback functor between slice toposes has a right adjoint. We continue with a discussion of how, given a topos **E**, certain syntactic properties of Th(**E**) are correlated with simple essentially categorical properties of **E**. The chapter ends with a presentation of a natural semantics for local set theories generalizing those of Beth and Kripke.

Some fundamental properties of toposes

We now establish some basic facts about toposes by the method outlined above. In each case we need only verify the assertion for a linguistic topos (i.e., of the form $\mathbf{C}(S)$).

4.1 PROPOSITION. Any topos has equalizers, and so is finitely complete.

Proof. By the equivalence theorem, it suffices to show that any linguistic topos $\mathbf{C}(S)$ has equalizers. To this end let $X \underset{g}{\overset{f}{\rightrightarrows}} Y$ be a diagram in $\mathbf{C}(S)$ and put

$$Z = \{x \in X : \forall y(\langle x,y\rangle \in f \Leftrightarrow \langle x,y\rangle \in g)\}.$$

Then it is easy to verify that

$$Z \xrightarrow{x \mapsto x} X \underset{g}{\overset{f}{\rightrightarrows}} Y$$

is an equalizer diagram in $\mathbf{C}(S)$. □

4.2 PROPOSITION. Any topos is finitely cocomplete.

Proof. This follows from (4.2.1)–(4.2.3) below.

(4.2.1) *Any topos has an initial object.*

Proof. Define 0 to be $\varnothing_{\mathbf{1}}$. To see that 0 is initial in $\mathbf{C}(S)$, given an S-set $X = \{y_{\mathbf{A}} : \alpha\}$, define $0 \to X$ to be $\varnothing_{\mathbf{1}\times\mathbf{A}}$. This is the unique arrow $0 \to X$, for if $0 \xrightarrow{f} X$ in $\mathbf{C}(S)$, we have

$$\langle x,y\rangle \in f \vdash_S x \in 0 \vdash_S \mathit{false}$$

and so

$$\vdash_S \langle x,y\rangle \in f \Leftrightarrow \langle x,y\rangle \in \varnothing_{\mathbf{1}\times\mathbf{A}}$$

whence $f = \varnothing_{\mathbf{1}\times\mathbf{A}}$. □

(4.2.2) *Any topos has binary coproducts.*

Proof. Let X and Y be S-sets. Then $X + Y$ is an S-set. Define injection arrows $X \xrightarrow{\sigma_1} X + Y \xleftarrow{\sigma_2} Y$ in $\mathbf{C}(S)$ by

$$\sigma_1 = (x \mapsto \langle \{x\}, \varnothing\rangle)$$
$$\sigma_2 = (y \mapsto \langle \varnothing, \{y\}\rangle)$$

Given $X \xrightarrow{f} Z \xleftarrow{g} Y$ in $\mathbf{C}(S)$, define $h : X + Y \to Z$ by

$$h = \{\langle\langle u,v\rangle, z\rangle : \langle u,v\rangle \in X + Y \wedge [\exists x(u = \{x\} \wedge \langle x,z\rangle \in f) \vee \exists y(v = \{y\} \wedge \langle y,z\rangle \in g)]\}.$$

It is now easily verified that $h \circ \sigma_1 = f$, $h \circ \sigma_2 = g$, and that, if $k \circ \sigma_1 = f$, $k \circ \sigma_2 = g$, then $k = h$. Therefore $X \xrightarrow{\sigma_1} X + Y \xleftarrow{\sigma_2} Y$ is a coproduct diagram. □

We note in this connection that, by (3.16.3), *the injections σ_1 and σ_2 are monic.*

(4.2.3) *Any topos has coequalizers.*

Proof. Given an S-set Y of type $\mathbf{PA}$, write Equiv(u) for the conjunction

$$\begin{aligned} &u \subseteq Y \times Y \wedge \forall y \in Y \langle y,y\rangle \in u \\ &\quad \wedge \forall y \in Y\, \forall z \in Y(\langle y,z\rangle \in u \Leftrightarrow \langle z,y\rangle \in u) \\ &\quad \wedge \forall x \in Y\, \forall y \in Y\, \forall z \in Y(\langle x,y\rangle \in u \wedge \langle y,z\rangle \in u \Rightarrow \langle x,z\rangle \in u). \end{aligned}$$

Equiv(u) asserts that u is an *equivalence relation* on Y. If R is an S-set such that $\vdash_S$ Equiv(R), define the *quotient* Y/R to be the S-set

$$\{\{x : \langle x,y\rangle \in R\} : y \in Y\}$$

and define $\pi : Y \to Y/R$ by

$$\pi = (y \mapsto \{x : \langle x,y\rangle \in R\}).$$

Now suppose we are given a diagram $X \overset{f}{\underset{g}{\rightrightarrows}} Y$ in $\mathbf{C}(S)$. Define

$$R = \bigcap \{u : \mathrm{Equiv}(u) \wedge \forall x\, \forall y\, \forall z(\langle x,y\rangle \in f \wedge \langle x,z\rangle \in g \Rightarrow \langle y,z\rangle \in u\}.$$

Then it is easy to verify that $\vdash_S \mathrm{Equiv}(R)$ and that

$$X \overset{f}{\underset{g}{\rightrightarrows}} Y \xrightarrow{\pi} Y/R$$

is a coequalizer diagram in $\mathbf{C}(S)$. □

This completes the proof of Proposition 4.2. □

4.3 PROPOSITION. Any topos is Cartesian closed.

Proof. It is enough to show that any linguistic topos has exponentials.

Given S-sets X,Z, we have (in Chapter 3) already defined the S-set Z^X. Let $\mathrm{ev}_{Z,X}: X \times Z^X \to Z$ be the S-map

$$\mathrm{ev}_{Z,X} = \{\langle\langle x,u\rangle, z\rangle : \langle x,z\rangle \in u\}.$$

Given an S-map $X \times Y \xrightarrow{f} Z$ define $\hat{f} : Y \to Z^X$ to be the S-map

$$\hat{f} = (y \mapsto \{\langle x,z\rangle : \langle\langle x,y\rangle, z\rangle \in f\}).$$

Clearly, $\mathrm{ev}_{Z,X} \circ (1_X \times \hat{f}) = f$, and it is easily verified that if $Y \xrightarrow{g} Z^X$ satisfies $\mathrm{ev}_{Z,X} \circ (1_X \times g) = f$, then $g = \hat{f}$. Therefore $(Z^X, \mathrm{ev}_{Z,X})$ is, in $\mathbf{C}(S)$, the exponential of Z by X. □

4.3.1 COROLLARY. A topos $\mathbf{E}$ is degenerate iff $0 \cong 1$ in $\mathbf{E}$. □

Let $A \xrightarrow{f} B$ be an arrow in a category $\mathbf{C}$. A *decomposition of* f is a pair of arrows $A \overset{k}{\twoheadrightarrow} C \overset{m}{\rightarrowtail} B$ with k epic, m monic and $m \circ k = f$. The decomposition (k, m) of f is said to be

(i) *unique* (up to isomorphism) if, for any decomposition $A \overset{k'}{\twoheadrightarrow} C' \overset{m'}{\rightarrowtail} B$ of f there is an isomorphism $i: C \cong C'$ such that the diagram

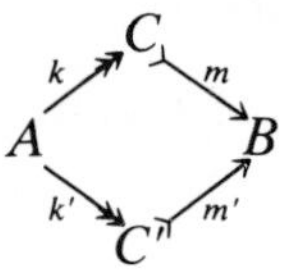

commutes;

(ii) *minimal* if, for any factorization $A \xrightarrow{l} D \overset{n}{\rightarrowtail} B$ of f with n monic, there is a (unique) $j: C \to D$ such that the diagram

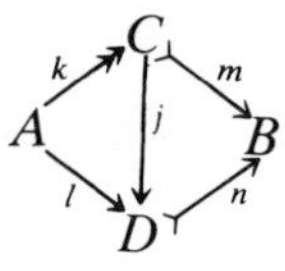

commutes (so that $m \subseteq n$).

The monic $C \overset{m}{\rightarrowtail} B$ (or just its domain C) in a unique decomposition (k,m) of f is called the *image* of f.

4.4 PROPOSITION. Any arrow in a topos has a unique minimal decomposition.

To prove (4.4) we first need

4.4.1 LEMMA. An arrow $X \xrightarrow{f} Y$ in $\mathbf{C}(S)$ is epic if and only if

$$y \in Y \vdash_S \exists x(\langle x,y\rangle \in f). \qquad (*)$$

Proof. Suppose f is epic. Let

$$g = Y \xrightarrow{y \mapsto \exists x(\langle x,y\rangle \in f)} \Omega$$

$$h = Y \xrightarrow{y \mapsto true} \Omega.$$

Then clearly $g \circ f = h \circ f$, so that, since f is assumed epic, $g = h$, and (*) easily follows.

Conversely, assume (*), and suppose we have a commutative diagram

$$X \xrightarrow{f} Y \underset{h}{\overset{g}{\rightrightarrows}} Z.$$

Then

$$y \in Y \vdash_S \exists x\, \exists u\, \exists v(\langle x,y\rangle \in f \wedge \langle y,u\rangle \in g \wedge \langle y,v\rangle \in h)$$

whence, since $g \circ f = h \circ f$,

$$y \in Y \vdash_S \exists x\, \exists u\, \exists v(\langle x,y\rangle \in f \wedge \langle y,u\rangle \in g \wedge \langle y,v\rangle \in h \wedge u = v)$$

so that

$$y \in Y \vdash_S \exists u(\langle y,u\rangle \in g \wedge \langle y,u\rangle \in h)$$

and hence $g = h$. □

Proof of Proposition 4.4. Given $X \xrightarrow{f} Y$ in $\mathbf{C}(S)$, define

$$Z = \{y : \exists x(\langle x,y\rangle \in f)\}.$$

Then clearly $\vdash_S f \in Z^X$. If we let $k = (f,X,Z)$, then $X \xrightarrow{k} Z$ and k is epic, by (4.4.1). Writing m for $Z \xrightarrow{y \mapsto y} Y$, m is monic and clearly (k,m) is a decomposition of f. To establish the uniqueness of (k,m), consider a second decomposition

$$X \xrightarrow{k'} Z' \overset{m'}{\rightarrowtail} Y$$

of f. Then by the epicity of f we have

$$z \in Z' \vdash_S \exists x(\langle x,z\rangle \in k'),$$

so, since $m' \circ k' = f$,

$$\begin{aligned}\langle z,y\rangle \in m' &\vdash_S \exists x[\langle x,z\rangle \in k' \wedge \langle z,y\rangle \in m']\\ &\vdash_S \exists x(\langle x,y\rangle \in f)\\ &\vdash_S y \in Z.\end{aligned}$$

Therefore $\vdash_S m' \in Z^{Z'}$. Writing n for the S-map (n,Z',Z) we then have $Z' \xrightarrow{n} Z$. Clearly the diagram

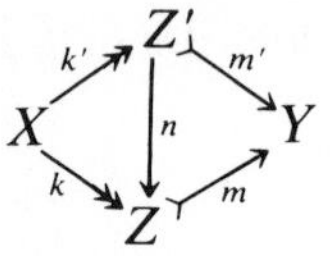

commutes. And n is an isomorphism because it is readily seen that it has inverse $Z \xrightarrow{n'} Z'$ given by

$$n' = \{\langle y,z\rangle : y \in Z \wedge \langle z,y\rangle \in n\}.$$

Finally, we want to show that (k,m) is minimal. Let

$$X \xrightarrow{l} W \overset{p}{\rightarrowtail} Y$$

be a factorization of f, with p monic. By the first part of the proof, $X \xrightarrow{l} W$ has a unique decomposition

$$X \overset{l'}{\twoheadrightarrow} V \overset{p'}{\rightarrowtail} W$$

But then

$$X \overset{l'}{\twoheadrightarrow} V \xrightarrow{n \circ p'} Y$$

is a decomposition of f and so there is an isomorphism $i: Z \to V$ such that the diagram

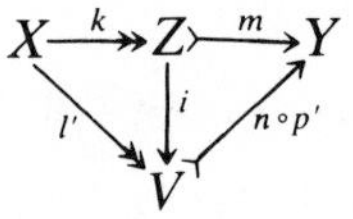

commutes. Hence

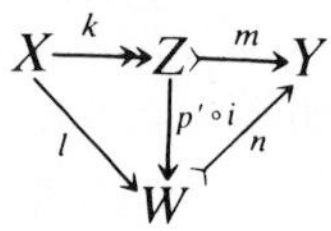

commutes and we are done. □

4.5 COROLLARY. Any topos is *balanced*, i.e. any arrow which is both epic and monic is an isomorphism.

Proof. This is a simple consequence of (4.4); the proof is left to the reader. □

4.6 LEMMA. (Characterization of pullbacks). Let

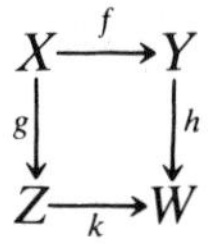

(*)

be a commutative square in $\mathbf{C}(S)$. Then (*) is a pullback iff

$$y \in Y, z \in Z \vdash_S \exists u(\langle y,u\rangle \in h \wedge \langle z,u\rangle \in k) \Rightarrow \exists! x(\langle x,y\rangle \in f \wedge \langle x,z\rangle \in g). \qquad (**)$$

Proof. Suppose (**) holds and that the diagram

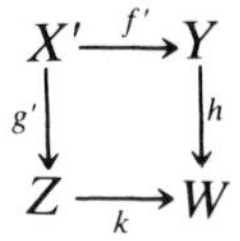

commutes. Define

$$j = \{\langle x',x\rangle : \exists y(\langle x,y\rangle \in f \wedge \langle x',y\rangle \in f') \wedge \exists z(\langle x,z\rangle \in g \wedge \langle x',z\rangle \in g')\};$$

then it is easy to verify that $X' \xrightarrow{j} X$, and that j is the unique arrow making the diagram

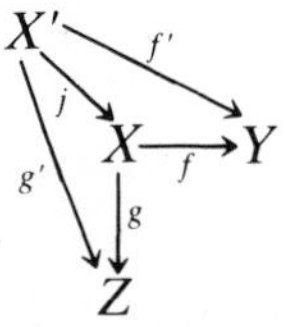

commute. Therefore (*) is a pullback diagram.

Conversely, suppose that (*) is a pullback. Let

$$V = \{\langle y,z\rangle : y \in Y \wedge z \in Z \wedge \exists u(\langle y,u\rangle \in h \wedge \langle z,u\rangle \in k)\}$$

Then, using the direction of the lemma already proved, the diagram

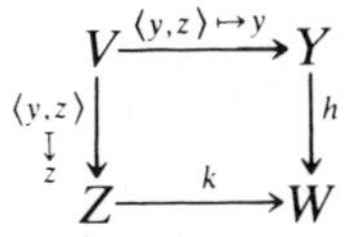

is a pullback. But then there must be an isomorphism $i : V \cong X$ such that

the diagram

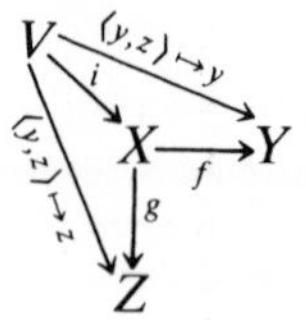

commutes, and it follows easily from this that (**) holds. □

4.7 PROPOSITION. In a topos, a pullback of an epic arrow is epic. That is, if

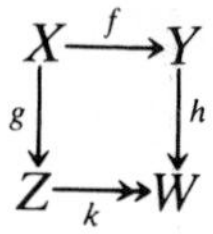

is a pullback, and k is epic, so is f.

Proof. Suppose the stated conditions hold. We argue, as usual, in $\mathbf{C}(S)$. Since k is epic, by (4.4.1) we have

$$y \in Y \vdash_S \exists z\, \exists u(\langle y,u\rangle \in h \wedge \langle z,u\rangle \in k)$$

and by (**) of Lemma 4.6, this gives

$$y \in Y \vdash_S \exists z\, \exists! x(\langle x,y\rangle \in f \wedge \langle x,z\rangle \in g)$$
$$\vdash_S \exists x(\langle x,y\rangle \in f).$$

So, by (4.4.1), f is epic. □

4.8 PROPOSITION. In a topos (binary) coproducts are disjoint. That is, the diagram

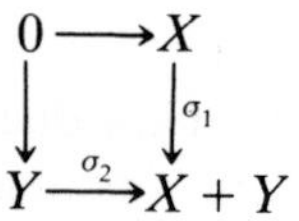

is a pullback.

Proof. In $\mathbf{C}(S)$ this follows easily from Lemma 4.6 and the definition of σ_1 and σ_2 given in the proof of (4.2.2). □

4.9 PROPOSITION. In a topos, $\Omega \cong P1$.

Proof. In $\mathbf{C}(S)$, it is easy to check that the S-map

$$U_{\mathbf{\Omega}} \xrightarrow{\omega \mapsto \{x_1 : \omega\}} U_{\mathbf{P1}}$$

is simultaneously epic and monic, and hence an isomorphism. □

The structure of Ω and Sub(A) in a topos

We know (Chapter 2) that in a topos—indeed in any category with a subobject classifier—the partially ordered set Sub(A) of subobjects of an object A is a *lower semilattice.* The techniques we have developed now enable us to show (among other things) that, *in a topos,* Sub(A) *is actually a Heyting algebra.*

Let S be a local set theory. Define the *entailment relation* $\leqslant$ on Ω to be the S-set

$$\leqslant \; = \{\langle \omega, \omega' \rangle : \omega \Rightarrow \omega'\}.$$

Given an S-set X, define the inclusion relation $\subseteq_X$ (sometimes written just $\subseteq$) to be the S-set

$$\subseteq_X = \{\langle u, v \rangle \in PX \times PX : u \subseteq v\}.$$

4.10 PROPOSITION.
(i) $\vdash_S \langle \Omega, \leqslant \rangle$ *is a Heyting algebra.*
(ii) $\vdash_S \langle PX, \subseteq_X \rangle$ *is a Heyting algebra.*

Proof. (i). First,

$$\vdash_S \leqslant \textit{is a partial ordering on } \Omega$$

follows in the usual way from the facts concerning implication proved in (3.2). Next,

$$\vdash_S \langle \Omega, \leqslant \rangle \textit{ is a lattice}$$

follows from the facts (easily derived from (3.1) and (3.6)) that

$$\vdash_S \omega \wedge \omega' \textit{ is the } \leqslant\textit{-infimum of } \{\omega, \omega'\};$$

$$\vdash_S \omega \vee \omega' \textit{ is the } \leqslant\textit{-supremum of } \{\omega, \omega'\}.$$

Clearly,

$$\vdash_S \textit{true is the } \leqslant\textit{-largest element of } \Omega;$$

$$\vdash_S \textit{false is the } \leqslant\textit{-least element of } \Omega.$$

Finally,

$$\vdash_S [\omega \leqslant (\omega' \Rightarrow \omega'')] \Leftrightarrow (\omega \wedge \omega' \leqslant \omega'')$$

follows from (3.2).

(ii) This is proved in exactly the same way as for intuitive sets. For instance, it is clear that

$$u \subseteq X,\, v \subseteq X \vdash_S u \cup v \; (u \cap v) \textit{ is the } \subseteq\textit{-supremum (infimum) of } \{u, v\} \textit{ in } PX;$$

and, if we define

$$u \rightarrow v = \bigcup \{w \subseteq X : u \cap w \subseteq v\},$$

then

$$u \subseteq X,\ v \subseteq X,\ w \subseteq X \vdash_S (u \cap w \subseteq v) \Leftrightarrow (w \subseteq u \rightarrow v). \quad \square$$

Let $\Omega(S)$ be the collection of *sentences* of the language of S, where we identify two sentences α, β whenever $\vdash_S \alpha \Leftrightarrow \beta$. Define the relation $\leqslant$ on $\Omega(S)$ by

$$\alpha \leqslant \beta \quad \text{iff} \quad \vdash_S \alpha \Rightarrow \beta.$$

If X is any S-set, let Pow(X) be the collection of all S-sets U such that $\vdash_S U \subseteq X$, and define the relation $\subseteq$ on Pow(X) by

$$U \subseteq V \quad \text{iff} \quad \vdash_S U \subseteq V.$$

Then it follows immediately from Proposition 4.10 that

4.11 COROLLARY. For any local set theory S and any S-set X, $(\Omega(S), \leqslant)$ and (Pow(X), $\subseteq$) are Heyting algebras. $\square$

Now suppose given a topos **E**. By taking $S = \mathrm{Th}(\mathbf{E})$ in Proposition 4.10 and invoking (3.30), we see that

(4.12.1) $\vDash_{\mathrm{E}} \langle \Omega, \leqslant \rangle$ *is a Heyting algebra*

and, for any **E**-object A,

(4.12.2) $\vDash_E \langle PA, \subseteq_A \rangle$ *is a Heyting algebra.*

These facts may be expressed by asserting that Ω and PA are *internal* Heyting algebras in **E**. As we now show, the corresponding *external* assertions also hold concerning the sets $\mathbf{E}(1,\Omega)$ and $\mathbf{E}(1,PA)$ of **E**-elements of Ω and PA.

4.13 COROLLARY. For any topos **E** and any **E**-object A, (Sub(A), $\subseteq$) is a Heyting algebra.

Proof. By the equivalence theorem, it suffices to prove this when **E** is linguistic. So let S be a local set theory and X an S-set. For each $U \in \mathrm{Pow}(X)$, let i_U be the S-monic

$$U \xrightarrow{x \mapsto x} X.$$

It is easy to see that the map

$$U \mapsto [i_U]$$

is an order isomorphism between (Pow(X), $\subseteq$) and (Sub(X), $\subseteq$). So the result follows from Corollary 4.11. $\square$

Since Sub$(A) \cong \mathbf{E}(1, PA)$, it follows from this that $\mathbf{E}(1, PA)$ (with the ordering inherited from the isomorphism with Sub(A)) *is a Heyting algebra.* This is the external result corresponding to (4.12.2). Moreover, since $(\mathbf{E}(A, \Omega), \leqslant) \cong (\mathrm{Sub}(A), \subseteq)$ it follows from Corollary 4.13 that $(\mathbf{E}(A, \Omega), \leqslant)$ is a Heyting algegra. In particular, taking $A = 1$, we see that *the ordered set* $(\mathbf{E}(1, \Omega), \leqslant)$ *of* **E**-*elements of* Ω *is a Heyting algebra:* this is the external version of (4.12.1).

It is easy to see that the *least* element of $\mathbf{E}(1, \Omega)$ is the characteristic arrow of the monic $0 \rightarrowtail 1$; we denote this arrow by

$$\perp : 1 \to \Omega.$$

It is also quite easy to compute the supremum of a pair of subobjects in a topos. For let $B \overset{m}{\rightarrowtail} A$, $C \overset{n}{\rightarrowtail} A$ be monics with common codomain A. Let $B + C \overset{k}{\twoheadrightarrow} D \overset{p}{\rightarrowtail} E$ be a decomposition of $B + C \xrightarrow{\binom{m}{n}} A$. Then $[p]$ is the supremum in Sub(A) of $\{[m], [n]\}$.

Recall that a partially ordered set is *complete* if every subset has an infimum and a supremum.

4.14 PROPOSITION. Let S be a local set theory. Then

$$\vdash_S \langle \Omega, \leqslant \rangle \textit{ is complete}$$

and, for any S-set X,

$$\vdash_S \langle PX, \subseteq_X \rangle \textit{ is complete.}$$

Proof. It is readily checked that

$$u \subseteq \Omega \vdash_S (\mathit{true} \in u) \textit{ is the } \leqslant\textit{-supremum of } u$$

$$u \subseteq \Omega \vdash_S (\forall \omega \in u \,.\, \omega) \textit{ is the } \leqslant\textit{-infimum of } u$$

$$v \subseteq X \vdash_S \bigcup v \textit{ is the } \subseteq_X\textit{-supremum of } v$$

$$v \subseteq X \vdash_S \bigcap v \textit{ is the } \subseteq_X\textit{-infimum of } v. \quad \square$$

By taking $S = \mathrm{Th}(\mathbf{E})$ in Proposition 4.10 and invoking (3.30) we get:

4.15 COROLLARY. For any topos **E**,

$$\models_E \langle \Omega, \leqslant \rangle \textit{ is complete,}$$

$$\models_E \langle PA, \subseteq \rangle \textit{ is complete.} \quad \square$$

That is, Ω and PA are *internally complete* in **E**. In general, we cannot infer the corresponding external result unless **E** is defined over **Set**: we take up this matter in (4.28).

REMARK. Define the arrows $\wedge, \vee, \Rightarrow : \Omega \times \Omega \to \Omega$ and $\neg : \Omega \to \Omega$ by

$$\begin{aligned} \wedge &= [\langle \omega, \omega' \rangle \mapsto \omega \wedge \omega']_{\mathbf{E}} \\ \vee &= [\langle \omega, \omega' \rangle \mapsto \omega \vee \omega']_{\mathbf{E}} \\ \Rightarrow &= [\langle \omega, \omega' \rangle \mapsto \omega \Rightarrow \omega']_{\mathbf{E}} \\ \neg &= [\omega \mapsto \neg \omega]_{\mathbf{E}} \end{aligned}$$

where $[\cdot]_{\mathbf{E}}$ denotes the canonical interpretation of objects and arrows of $\mathbf{C}(\mathrm{Th}(\mathbf{E}))$ in $\mathbf{E}$. Then the assertion that $\langle \Omega, \leqslant \rangle$ is an internal Heyting algebra translates into the commutativity of certain diagrams involving these arrows. For example, the assertion that

$$\vDash_E (\omega \wedge \omega' \leqslant \omega'') \Leftrightarrow (\omega \leqslant \omega' \Rightarrow \omega'')$$

is equivalent to the assertion that the diagram

$$\begin{array}{ccccc} & & \Omega \times (\Omega \times \Omega) \xrightarrow{1_\Omega \times \wedge} \Omega \times \Omega & & \\ & \overset{\text{can}}{\nearrow} & & \searrow^{\Rightarrow} & \\ \Omega \times \Omega \times \Omega & & & & \Omega \\ & \underset{\text{can}}{\searrow} & & \nearrow_{\Rightarrow} & \\ & & (\Omega \times \Omega) \times \Omega \xrightarrow{\Rightarrow \times 1_\Omega} \Omega \times \Omega & & \end{array}$$

commutes. The remaining conditions for a Heyting algebra are expressed similarly. The arrows $\wedge, \vee, \Rightarrow, \neg$ are called the *logical operations on Ω*.

EXERCISES

(1) Show that, in a topos $\mathbf{E}$,

(i) $\wedge : \Omega \times \Omega \to \Omega$ is the characteristic arrow of the monic $1 \xrightarrow{\langle \top, \top \rangle} \Omega \times \Omega$;

(ii) $\vee : \Omega \times \Omega \to \Omega$ is the characteristic arrow of the image of the arrow

$$\Omega + \Omega \xrightarrow{\binom{\langle \top_\Omega, 1_\Omega \rangle}{\langle 1_\Omega, \top_\Omega \rangle}} \Omega \times \Omega;$$

(iii) $\Rightarrow : \Omega \times \Omega \to \Omega$ is the characteristic arrow of the equalizer of the arrows $\Omega \times \Omega \overset{\pi}{\underset{\wedge}{\rightrightarrows}} \Omega$;

(iv) $\neg : \Omega \to \Omega$ is the characteristic arrow of $1 \xrightarrow{\perp} \Omega$.

(2) Let I be an interpretation of a language $\mathscr{L}$ in a topos $\mathbf{E}$, and let α, β be formulae in $\mathscr{L}$. Show that

$$\begin{aligned} [\![\alpha \wedge \beta]\!]_{\mathbf{x}} &= \wedge \circ [\![\langle \alpha, \beta \rangle]\!]_{\mathbf{x}} \\ [\![\alpha \vee \beta]\!]_{\mathbf{x}} &= \vee \circ [\![\langle \alpha, \beta \rangle]\!]_{\mathbf{x}}, \\ [\![\alpha \Rightarrow \beta]\!]_{\mathbf{x}} &= \Rightarrow \circ [\![\langle \alpha, \beta \rangle]\!]_{\mathbf{x}} \\ [\![\neg \alpha]\!]_{\mathbf{x}} &= \neg \circ [\![\alpha]\!]_{\mathbf{x}}. \end{aligned}$$

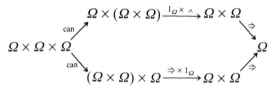

Slicing a topos

In this section we prove what is sometimes called the *fundamental theorem for toposes:* if **E** is a topos, and A any **E**-object, then $\mathbf{E}/A$ is a topos. (That is, 'no matter how thin you slice it, it's still a topos'.) This will be achieved by deriving the local set theory associated with $\mathbf{E}/A$ from the local set theory associated with **E**.

Let S be a local set theory and let I be an S-set.. We define *the category* $\mathbf{C}(S)^I$ *of I-indexed S-sets* as follows. An *object* of $\mathbf{C}(S)^I$ is an S-set of the form

$$M = \{\langle i,M_i\rangle : i \in I\}$$

with M_i a term of power type with at most the free variable i. (Thus M is an I-indexed set of S-sets.) An *arrow* $f: M \to N$ between $\mathbf{C}(S)^I$-objects $M = \{\langle i,M_i\rangle : i \in I\}$, $N = \{\langle i,N_i\rangle : i \in I\}$ is an S-set of the form

$$\{\langle i,f_i\rangle : i \in I\}$$

such that

$$\vdash_S \forall i \in I(f_i \in N_i^{M_i}).$$

(Thus $f: M \to N$ in $\mathbf{C}(S)^I$ is an I-indexed S-set of maps $M_i \to N_i$.) Composition and identity arrows in $\mathbf{C}(S)^I$ are defined in the obvious way.

4.16 LEMMA. There is an equivalence

$$\mathbf{C}(S)/I \simeq \mathbf{C}(S)^I.$$

Proof. Define the functor $F: \mathbf{C}(S)/I \to \mathbf{C}(S)^I$ by

$$F(X \xrightarrow{f} I) = \{\langle i, f^{-1}(i)\rangle : i \in I\}$$

where, as usual,

$$f^{-1}(i) = \{x : \langle x,i\rangle \in f\}$$

and, for

$$h:(X \xrightarrow{f} I) \to (Y \xrightarrow{g} I)$$

in $\mathbf{C}(S)/I$,

$$F(h) = \{\langle i,h_i\rangle : i \in I\rangle$$

with

$$h_i = \{\langle x,y\rangle : x \in f^{-1}(i) \wedge \langle x,y\rangle \in h\}.$$

Clearly F is a functor. We claim that it is an equivalence between $\mathbf{C}(S)/I$ and $\mathbf{C}(S)^I$.

F is faithful. If $X \overset{h}{\underset{h'}{\rightrightarrows}} I$ and $F(h) = F(h')$, then $\vdash_S \forall i \in I(h_i = h_i')$ and from this and the obvious fact that $\vdash_S X = \bigcup_{i \in I} f^{-1}(i)$, it follows that $h = h'$.

F is full. Given

$$k : F(X \xrightarrow{f} I) \rightarrow F(Y \xrightarrow{g} I)$$

define $h : X \rightarrow Y$ by

$$h = \{\langle x,y \rangle : \exists i \in I \,.\, \langle x,y \rangle \in k_i\}.$$

It is easy to show that $F(h) = k$.

F is dense. Given an object $M = \{\langle i,m_i \rangle : i \in I\}$ in $\mathbf{C}(S)^I$, let $X = \coprod_{i \in I} M_i$ and define $X \xrightarrow{f} I$ by

$$f = \{\langle \langle i,x \rangle, i \rangle : i \in I \wedge x \in M_i\}.$$

Now define the arrow $k : F(X \xrightarrow{f} I) \rightarrow M$ by $k = \{\langle i,k_i \rangle : i \in I\}$ with

$$k_i = \{\langle \langle i,x \rangle, x \rangle : x \in M_i\}.$$

Then it is easy to check that k has inverse $k' = \{\langle i,k_i' \rangle : i \in I\}$ with

$$k_i' = \{\langle x, \langle i,x \rangle \rangle : x \in M_i\}.$$

Therefore k is an isomorphism and F is dense. □

Recall that S_I denotes the theory obtained from S by adjoining an indeterminate I-element c. We shall need

4.17 LEMMA. Let γ be any formula not containing c, and let x be a variable free for i in γ where, if I is of type **PA**, i is a variable of type **A**. Then

$$\vdash_{S_I} \gamma(x/c) \quad \text{iff} \quad \vdash_S \forall i \in I \gamma(x/i).$$

Proof. By (3.42),

$$\begin{aligned} \vdash_{S_I} \gamma(x/c) \quad & \text{iff} \quad x \in I \vdash_S \gamma \\ & \text{iff} \quad i \in I \vdash_S \gamma(x/i) \\ & \text{iff} \quad \vdash_S i \in I \Rightarrow \gamma(x/i) \\ & \text{iff} \quad \vdash_S \forall i \in I \gamma(x/i). \quad \square \end{aligned}$$

4.18 LEMMA. There is an isomorphism of categories

$$\mathbf{C}(S_I) \cong \mathbf{C}(S)^I.$$

Proof. Define the functor $G : \mathbf{C}(S_I) \rightarrow \mathbf{C}(S)^I$ as follows. Given an S_I-set $X = \{x : \alpha(x, z/c)\}$, define the $\mathbf{C}(S)^I$-object $G(X)$ by

$$G(X) = \{\langle i,X_i \rangle : i \in I\}$$

where

$$X_i = \{x : \alpha(x, z/i)\}.$$

Given an S_I-map

$$X = \{x : \alpha(x, z/c)\} \xrightarrow{f} \{y : \beta(y, z/c)\} = Y,$$

there is a term $\tau(u)$ such that $\vdash_{S_I} f = \tau(u/c)$. Then

$$\vdash_{S_I} \tau(u/c) \in Y^X$$

so that, by (4.17),

$$\vdash_S \forall i \in I[\tau(u/i) \in Y_i^{X_i}].$$

Accordingly we define $G(f): G(X) \to G(Y)$ by

$$G(f) = \{\langle i, \tau(u/i)\rangle : i \in I\}.$$

G is well-defined since, e.g., if X and Y are S_I-sets, and $X = Y$, then $\vdash_{S_I} X = Y$, whence, by (4.17), $\vdash_S \forall i \in I.\ X_i = Y_i$ and so $G(X) = G(Y)$. Notice that reversing this argument shows that G is injective on objects.

G is obviously a functor; we claim that it is an isomorphism $\mathbf{C}(S_I) \cong \mathbf{C}(S)^I$.

G is faithful. Given $f, g : X \to Y$ in S_I, let $f = \tau(u/c)$, $g = \sigma(u/c)$. If $G(f) = G(g)$ in $\mathbf{C}(S)^I$, then

$$\vdash_S \forall i \in I[\tau(u/i) = \sigma(u/i)]$$

whence

$$\vdash_{S_I} \tau(u/c) = \sigma(u/c),$$

so that $f = g$.

G is full. Given S_I-sets X, Y and $G(X) \xrightarrow{h} G(Y)$ in $\mathbf{C}(S)^I$, then for some term τ, $h = \{\langle i, \tau(u/i)\rangle : i \in I\}$ with

$$\vdash_S \forall i \in I[\tau(u/i) \in Y_i^{X_i}].$$

Therefore

$$\vdash_{S_I} \tau(u/c) \in Y^X,$$

so that $X \xrightarrow{\tau(u/c)} Y$ in $\mathbf{C}(S_I)$. Clearly $G(\tau(u/c)) = h$.

G is bijective on objects. We already know that G is injective on objects. Given a $\mathbf{C}(S)^I$-object $M = \{\langle i, M_i\rangle : i \in I\}$, let $M_i = \tau(x/i)$ for some τ and define $N = \tau(x/c)$. Then N is an S_I-set and

$$\begin{aligned} G(N) &= \{\langle i, N_i \rangle : i \in I\} \\ &= \{\langle i, M_i \rangle : i \in I\} \\ &= M. \end{aligned}$$

So G is surjective on objects and the proof is complete. □

These two lemmas yield the promised

4.19 FUNDAMENTAL THEOREM FOR TOPOSES. For any topos $\mathbf{E}$, and any $\mathbf{E}$-object A, the slice category $\mathbf{E}/A$ is a topos.

Proof. Let $S = \mathrm{Th}(\mathbf{E})$ and let F be the canonical inclusion of $\mathbf{E}$ in $\mathbf{C}(S)$. Then, writing I for FA, we have, using Lemmas 4.16 and 4.18,

$$\mathbf{E}/A \simeq \mathbf{C}(S)/I \simeq \mathbf{C}(S)^I \simeq \mathbf{C}(S_I).$$

Since $\mathbf{C}(S_I)$ is a topos, so is $\mathbf{E}/A$. □

Recall that any $\mathbf{E}$-arrow $A \xrightarrow{f} B$ induces a pullback functor $f^* : \mathbf{E}/B \to \mathbf{E}/A$. We are going to show that, when $\mathbf{E}$ is a topos, f^* has a right (and a left) adjoint.

Consider first the special case in which $B = 1$. Now $\mathbf{E}/1 \cong \mathbf{E}$ and it is readily seen that the pullback functor $A^* : \mathbf{E} \cong \mathbf{E}/1 \to \mathbf{E}/A$ induced by $A \to 1$ has object function

$$X \mapsto (X \times A \xrightarrow{\pi_2} A).$$

4.20 LEMMA. $A^* : \mathbf{E} \to \mathbf{E}/A$ has a left adjoint $A_! : \mathbf{E}/A \to \mathbf{E}$, namely the forgetful functor with object function

$$(X \xrightarrow{f} A) \mapsto X.$$

Proof. The adjunction is trivial. □

4.21 LEMMA. Let S be a local set theory and I an S-set. Then the pullback functor $I^* : \mathbf{C}(S) \to \mathbf{C}(S)/I$ has a right adjoint $I_* : \mathbf{C}(S)/I \to \mathbf{C}(S)$ whose object function is given by

$$(Y \xrightarrow{f} I) \mapsto \prod_{i \in I} f^{-1}(i).$$

Proof. This is proved exactly as it would be in **Set**. It is easily seen that we have natural bijections

$$\frac{(X \times I \xrightarrow{\pi_2} I) \to (Y \xrightarrow{f} I)}{X \to \prod_{i \in I} f^{-1}(i)}$$

The details are left to the reader. □

This lemma and the equivalence theorem immediately yield:

4.22 COROLLARY. If $\mathbf{E}$ is a topos and A an $\mathbf{E}$-object, the pullback functor $A^*:\mathbf{E}\to\mathbf{E}/A$ has a right adjoint $A_*:\mathbf{E}/A\to E$. $\square$

REMARK. It is readily checked that under the equivalence between $\mathbf{C}(S_I)$ and $\mathbf{C}(S)/I$ the pullback functor I^* corresponds to the logical functor $\mathbf{C}(S)\to\mathbf{C}(S_I)$ induced by the inclusion translation $S\hookrightarrow S_I$. It follows that I^* preserves products and equalizers, and is accordingly *left exact.* Therefore the pair (I_*,I^*) defines a geometric morphism $\mathbf{C}(S)/I\to\mathbf{C}(S)$. So the equivalence theorem implies that, for any topos $\mathbf{E}$ and any $\mathbf{E}$-object A, the pair (A_*,A^*) is a geometric morphism $\mathbf{E}/A\to\mathbf{E}$.

We can now prove the fundamental

4.23 PULLBACK FUNCTOR THEOREM FOR TOPOSES. Let $\mathbf{E}$ be a topos and let $A\xrightarrow{f}B$ be an $\mathbf{E}$-arrow. Then the pullback functor $f^*:\mathbf{E}/B\to\mathbf{E}/A$ has a right adjoint $f_*:\mathbf{E}/A\to\mathbf{E}/B$.

Proof. First, we may regard f as an object of the slice category $\mathbf{E}/B$, which by the fundamental theorem is a topos. Now it is readily seen that the slice category $(\mathbf{E}/B)/f$ is isomorphic to $\mathbf{E}/A$. So, writing $\mathbf{E}'$ for $\mathbf{E}/B$, the pullback functor $\mathbf{E}/B\xrightarrow{f^*}\mathbf{E}/A$ may be identified with the pullback functor $\mathbf{E}'\to\mathbf{E}'/f$, which by Lemma 4.21 has a right adjoint. Hence so does f^*, and we are done. $\square$

REMARK. (i) Again it can be shown that the pullback functor $f^*:\mathbf{E}/B\xrightarrow{f^*}\mathbf{E}/A$ is left exact, so that (f_*,f^*) defines a geometric morphism $\mathbf{E}/A\to\mathbf{E}/B$.

(ii) It is trivial to show that f^* has a left adjoint $f_!:\mathbf{E}/A\to\mathbf{E}/B$ whose object function is given by

$$(X\xrightarrow{g}A)\mapsto(X\xrightarrow{f\circ g}B).$$

(iii) It is instructive to describe f^*, f_* and $f_!$ in the case where $\mathbf{E}$ is the category of sets. Identify $\mathbf{Set}/I$ with $\mathbf{Set}^I$, the category of I-indexed sets and write $\langle A_i:i\in I\rangle$ for $\{\langle i,A_i\rangle:i\in I\}$. Then, given $f:I\to J$ in $\mathbf{Set}$, the pullback functor f^* amounts to the operation of relabelling along f, i.e.

$$f^*(\langle A_j:j\in J\rangle)=\langle A_{f(i)}:i\in I\rangle.$$

The functors f_* and $f_!$ are the operations of forming products and coproducts over the fibres of f, i.e.

$$f_*(\langle A_i:i\in I\rangle)=\Big\langle \prod_{i\in f^{-1}(j)} A_i:j\in J\Big\rangle$$

$$f_!(\langle A_i:i\in I\rangle)=\Big\langle \coprod_{i\in f^{-1}(j)} A_i:j\in J\Big\rangle.$$

Coproducts in a topos

We now employ the results we have obtained to discuss the behaviour of coproducts in a topos.

4.24 PROPOSITION. In a topos, coproduct injections are monic, and coproducts are disjoint, that is, if $\{A_i : i \in I\}$ is a family of objects such that $\coprod_{i \in I} A_i$ exists, then for any $i, j \in I$ with $i \neq j$ the diagram

$$\begin{array}{ccc} 0 & \longrightarrow & A_i \\ \downarrow & & \downarrow{\scriptstyle \sigma_i} \\ A_j & \xrightarrow{\sigma_j} & \coprod A_i \end{array}$$

is a pullback.

Proof. This is a straightforward consequence of the corresponding facts—already established—for binary coproducts in a topos. □

4.25 COPRODUCT PULLBACK THEOREM. Suppose that $\{B_i \xrightarrow{f_i} B : i \in I\}$ is a coproduct diagram in a topos **E**. Let $A \xrightarrow{f} B$ and for each $i \in I$ let

$$\begin{array}{ccc} A_i & \longrightarrow & B_i \\ {\scriptstyle g_i}\downarrow & & \downarrow{\scriptstyle f_i} \\ A & \xrightarrow{f} & B \end{array}$$

be a pullback. Then $\{A_i \xrightarrow{g_i} A : i \in I\}$ is a coproduct diagram.

Proof. Consider the composition of functors

$$\mathbf{E} \xrightarrow{B^*} \mathbf{E}/B \xrightarrow{f^*} \mathbf{E}/A \xrightarrow{A_!} \mathbf{E} = \mathbf{E} \xrightarrow{F} \mathbf{E}.$$

F sends any arrow $\bullet \to B$ to its pullback $\bullet \to A$ along f. Now B^*, f^* and A all have right adjoints and therefore (1.27) preserve colimits, in particular the coproduct diagram $\{B_i \xrightarrow{f_i} B : i \in I\}$. Therefore F preserves it too, that is, $\{A_i \xrightarrow{g_i} A : i \in I\}$ is a coproduct diagram. □

4.26 COROLLARY. Suppose that for each $i \in I$ the diagram

$$\begin{array}{ccc} A_i & \longrightarrow & B_i \\ \downarrow & & \downarrow \\ A & \xrightarrow{f} & B \end{array} \qquad (*)$$

is a pullback in a topos **E**. Then so is the diagram

$$\begin{array}{ccc} \coprod A_i & \longrightarrow & \coprod B_i \\ \downarrow & & \downarrow \\ A & \xrightarrow{f} & B \end{array}$$

with the canonical arrows.

Proof. Form the pullback

$$\begin{array}{ccc} P & \longrightarrow & \coprod B_i \\ \downarrow & & \downarrow \\ A & \xrightarrow{f} & B \end{array}$$

Since (*) commutes, there is for each $i \in I$ a (unique) arrow $A_i \to P$ such that

$$\begin{array}{ccc} A_i & \longrightarrow & B_i \\ \downarrow & & \downarrow \\ P & \longrightarrow & \coprod B_i \end{array}$$

commutes. Then in the commutative diagram

$$\begin{array}{ccc} A_i & \longrightarrow & B_i \\ \downarrow & & \downarrow \\ P & \longrightarrow & \coprod B_i \\ \downarrow & & \downarrow \\ A & \longrightarrow & B \end{array}$$

the whole rectangle and the bottom square are pullbacks. Hence (1.4) the top square is a pullback so that, by (4.25), P is the coproduct of $\{A_i : i \in I\}$. The result follows. □

Recall that a topos is said to be *defined over* **Set** if it has arbitrary set-fold copowers of its terminal object 1. We use the facts about coproducts in a topos to establish

4.27 THEOREM. For any topos **E**, the following are equivalent:

(i) **E** is defined over **Set**;
(ii) there is a geometric morphism $\mathbf{E} \to \mathbf{Set}$;
(iii) for any **E**-object A, **E** has arbitrary (set-indexed) coproducts of subobjects of A;
(iv) for any **E**-object A, **E** has arbitrary set-fold copowers of A;
(v) **E** has arbitrary (set-indexed) coproducts of subobjects of 1.

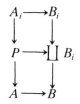

Proof. (i)⇒(ii). Assuming (i), define $\mu_* : \mathbf{E} \to \mathbf{Set}$, $\mu^* : \mathbf{Set} \to \mathbf{E}$ by $\mu_*(A) = \mathbf{E}(1, A)$ and $\mu^*(I) = \coprod_I 1$. The action of μ^* on an arrow $I \xrightarrow{f} J$ in **Set** is determined by defining $\mu^*(f)$ to be the unique arrow in **E** making the diagrams

$$\begin{array}{ccc} 1 & \xrightarrow{\sigma_i} & \coprod_I 1 \\ & {\scriptstyle \sigma_{f(i)}} \searrow & \downarrow {\scriptstyle \mu^*(f)} \\ & & \coprod_J 1 \end{array}$$

commute, where σ_i is the canonical injection.

For simplicity let us write $\hat{\imath}, \hat{I}, \hat{f}$ for σ_i, $\coprod_I 1$, $\mu^*(f)$ respectively. It is easy to see that there is a natural bijection of arrows

$$\frac{\hat{I} \to A \quad \text{in } \mathbf{E}}{I \to E(1,A) \quad \text{in } \mathbf{Set}}$$

establishing an adjunction $\mu^* \dashv \mu_*$.

To see that μ^* is left exact, it suffices to show that it preserves 1, products and equalizers.

That μ^* preserves 1 is left as an easy exercise to the reader.

To show that μ^* preserves products, we observe first that, since $\mathbf{E}$ is Cartesian closed, by (1.29),

$$\begin{aligned} \coprod_I 1 \times \coprod_J 1 \cong \coprod_J 1 \times \coprod_I 1 &\cong \coprod_I \Big(\coprod_J \times 1\Big) \\ &\cong \coprod_I \Big(\coprod_J 1\Big). \end{aligned}$$

So it suffices to show that

$$\coprod_I \Big(\coprod_J 1\Big) \cong \coprod_{I \times J} 1. \qquad (*)$$

For each $i \in I$ let f_i be the unique arrow making the diagrams

$$\begin{array}{ccc} 1 & \xrightarrow{\sigma_j} & \coprod_J 1 \\ & {\scriptstyle \tau_{ij}} \searrow & \downarrow {\scriptstyle f_i} \\ & & \coprod_{I \times J} 1 \end{array}$$

commute for all $j \in J$, where the σ_j, τ_{ij} are canonical injections, and let k be the unique arrow making the diagrams

$$\begin{array}{ccc} \coprod_J 1 & \xrightarrow{\xi_i} & \coprod_I \Big(\coprod_J 1\Big) \\ & {\scriptstyle f_i} \searrow & \downarrow {\scriptstyle k} \\ & & \coprod_{I \times J} 1 \end{array}$$

commute for all $i \in I$, where the ξ_i are canonical injections. Let l be the unique arrow making the diagrams

$$\begin{array}{ccc} 1 & \xrightarrow{\tau_{ij}} & \coprod_{I\times J} 1 \\ & \searrow^{\tau_i \circ \sigma_j} & \downarrow l \\ & & \coprod_I \left(\coprod_J 1\right) \end{array}$$

commute for all $i \in I, j \in J$. It is easy to see that the diagrams

$$\begin{array}{ccc} 1 & \xrightarrow{\tau_{ij}} & \coprod_{I\times J} 1 \\ & \searrow^{\tau_{ij}} & \downarrow k\circ l \\ & & \coprod_{I\times J} 1 \end{array} \qquad \begin{array}{ccc} \coprod_J 1 & \xrightarrow{\xi_i} & \coprod_I \left(\coprod_J 1\right) \\ & \searrow^{\xi_i} & \downarrow l\circ k \\ & & \coprod_I \left(\coprod_J 1\right) \end{array}$$

commute for all $i \in I, j \in J$. By the uniqueness condition on arrows from coproducts, $k \circ l$ and $l \circ k$ are identity arrows, so that k is an isomorphism. This proves (*).

To show that μ^* preserves equalizers, we argue as follows. Let $f, g: I \rightrightarrows J$ in **Set** and let

$$K = \{i \in I : f(i) = g(i)\} \hookrightarrow I$$

be their equalizer. Form the equalizer $C \to \hat{I}$ of $\hat{I} \underset{\hat{g}}{\overset{\hat{f}}{\rightrightarrows}} \hat{J}$ in **E**. It suffices to show that $C \cong \tilde{K}$.

Now it follows immediately from the definition of equalizer that the diagram

$$\begin{array}{ccc} C & \xrightarrow{k} & \hat{I} \\ {\scriptstyle k}\downarrow & & \downarrow{\scriptstyle \hat{f}} \\ \hat{I} & \xrightarrow{\hat{g}} & \hat{J} \end{array} \qquad (*)$$

is a pullback in **E**. For each $i \in I$ form the pullback

$$\begin{array}{ccc} U_i & \rightarrowtail & 1 \\ \downarrow & & \downarrow{\scriptstyle \hat{i}} \\ C & \xrightarrow{k} & \hat{I} \end{array} \qquad (**)$$

Combining this with (*) shows that the diagram

$$\begin{array}{ccc} U_i & \rightarrowtail & 1 \\ \downarrow & & \downarrow{\scriptstyle f(i)^\wedge} \\ 1 & \xrightarrow{g(i)^\wedge} & \hat{J} \end{array}$$

is a pullback. If $i \in K$, then $f(i) = g(i)$, so $U_i \cong 1$. If $i \notin K$, then $f(i) \neq g(i)$, so $U_i \cong 0$ by (4.24). Therefore, since (**) is a pullback and $\{1 \xrightarrow{\hat{i}} \hat{I} : i \in I\}$ a coproduct diagram, so, by the coproduct pullback theorem, is $\{U_i \to C : i \in I\}$. Hence $C \cong \coprod_{i \in I} U_i$. But since $U_i \cong 1$ when $i \in K$ and $U_i \cong 0$ when $i \notin K$, it follows easily that $\coprod_{i \in I} U_i \cong \coprod_{i \in K} 1 = \hat{K}$. Accordingly $C \cong \hat{K}$, giving (ii).

(ii) $\Rightarrow$ (i). This was proved in (2.21).

(i) $\Rightarrow$ (iii). Assume (i), and let $\{U_i \overset{u_i}{\rightarrowtail} A : i \in I\}$ be a family of (monics representing) subobjects of A. Since **E** is Cartesian closed, it follows from (1.29) that

$$A \times \coprod_I 1 \cong \coprod_I A \times 1 \cong \coprod_I A$$

and in particular $\coprod_I A$ exists. For each $i \in I$ let χ_i be the characteristic arrow of u_j; then

$$\begin{array}{ccc} U_i & \longrightarrow & 1 \\ \downarrow & & \downarrow{\scriptstyle \top} \\ A & \xrightarrow{\chi_i} & \Omega \end{array} \qquad (*)$$

is a pullback. Let

$$\chi = (\chi_i)_{i \in I} : \coprod_I A \to \Omega;$$

and form the pullback

$$\begin{array}{ccc} U & \longrightarrow & 1 \\ \downarrow & & \downarrow{\scriptstyle \top} \\ \coprod_I A & \xrightarrow{\chi} & \Omega \end{array}$$

Since (*) commutes, there is for each $i \in I$ a (unique) arrow $U_i \overset{f_i}{\rightarrowtail} U$ such that the diagram

$$\begin{array}{ccccc} U_i & \xrightarrow{f_i} & U & \longrightarrow & 1 \\ {\scriptstyle u_i}\downarrow & & {\scriptstyle \bar{\chi}}\downarrow & & \downarrow{\scriptstyle \top} \\ A & \xrightarrow{\sigma_i} & \coprod_I A & \xrightarrow{\chi} & \Omega \end{array}$$

commutes. But the whole rectangle is the pullback diagram (*) and the right-hand square is the pullback diagram (**). Hence the left-hand square

$$\begin{array}{ccc} U_i & \xrightarrow{f_i} & U \\ {\scriptstyle u_i}\downarrow & & \downarrow{\scriptstyle \bar{\chi}} \\ A & \xrightarrow{\sigma_i} & \coprod_I A \end{array}$$

is a pullback. Since $\{A \xrightarrow{\sigma_i} \coprod_I A : i \in I\}$ is a coproduct diagram, so is $\{U_i \xrightarrow{f_i} U : i \in I\}$ and hence the coproduct of the U_i exists in $\mathbf{E}$.

Finally, (iii)⇒(iv) and (v), (iv)⇒(i) and (v)⇒(i) are obvious. This completes the proof. □

As a consequence of this result, we infer the 'external' version of (4.15) for the case in which **E** is defined over **Set**.

4.28 COROLLARY. Let **E** be a topos defined over **Set**. Then, for any **E**-object A, Sub(A) is complete, and hence a locale. In particular, so is $\mathbf{E}(1,\Omega) \cong \mathrm{Sub}(1)$.

Proof. Let $\{U_i \overset{u_i}{\rightarrowtail} A : i \in I\}$ be any collection of monics into A. Then $\coprod_{i \in I} U_i$ exists; let $u : U \rightarrowtail A$ be the monic component of the minimal decomposition

$$\coprod_{i \in I} U_i \twoheadrightarrow U \overset{u}{\rightarrowtail} A$$

of $\coprod_{i \in I} U_i \xrightarrow{(\sigma_i)_{i \in I}} A$. Then it is easy to verify that the subobject $[u]$ of A is the supremum in Sub(A) of the collection of subobjects $\{[u_i] : i \in I\}$. □

Syntactic properties of local set theories versus essentially categorical properties of toposes

We next establish a correlation between certain syntactic properties of local set theories, and certain essentially categorical properties of the associated toposes. Let S be a local set theory in a local language $\mathscr{L}$. We make the following

4.29 DEFINITION.

(i) S is said to be *classical* if $\vdash_S \omega \vee \neg \omega$.

(ii) S is said to be *complete* if for any sentence α, either $\vdash_S \alpha$ or $\vdash_S \neg\alpha$.

(iii) S is said to be *term-complete* if for any type symbol $\mathbf{A}$ and any formula α with at most the free variable x of type $\mathbf{A}$, $\vdash_S \alpha(x/\tau)$ for all closed terms τ of type $\mathbf{A}$ implies $\vdash_S \forall x \alpha$.

(iv) If $\mathbf{A}$ is a type symbol of $\mathscr{L}$, an $\mathbf{A}$-*singleton* is a closed term U of type $\mathbf{PA}$ such that

$$\vdash_S \forall x \in U \, \forall y \in U . x = y.$$

$\mathbf{A}$ is said to be S-*singleton-complete* if for any formula α with at most the free variable x of type $\mathbf{A}$, $\vdash_S \forall x \in U \alpha$ for all $\mathbf{A}$-singletons U implies $\vdash_S \forall x \alpha$. S is said to be *singleton-complete* if any type symbol of $\mathscr{L}$ is S-singleton-complete.

(v) S is said to be *strongly witnessed* if under the same conditions as (iii), $\nvdash_S \neg\exists x\alpha$ implies $\vdash_S \alpha(x/\tau)$ for some closed term τ.

(vi) S is said to be *witnessed* if under the same conditions as (iii), $\vdash_S \exists x\alpha$ implies $\vdash_S \alpha(x/\tau)$ for some closed term τ.

(vii) S is said to be *singleton-witnessed* if for any type symbol $\mathbf{A}$ and any formula α with at most the free variable x of type $\mathbf{A}$, there is an $\mathbf{A}$-singleton U such that $\vdash_S \forall x \in U\alpha$ and $\vdash_S \exists x.x \in U \Leftrightarrow \exists x\alpha$.

(viii) S is said to satisfy the *axiom of choice* (AC) if, for any S-sets X, Y and any formula $\alpha(x,y)$ with at most the variables x,y free, $\vdash_S \forall x \in X\exists y \in Y\alpha(x,y)$ implies $\vdash_S \forall x \in X\alpha(x,fx)$ for some $f: X \to Y$.

We note that to say S is classical is to assert that the *law of excluded middle* holds in S. We also have the following equivalent conditions for classicality. If X is an S-set of type $\mathbf{PA}$, an S-set Y of type $\mathbf{PA}$ is called a *complement* for X if

$$X \cup Y = A \qquad \text{and} \qquad X \cap Y = \varnothing.$$

4.30 PROPOSITION. Any of the following conditions are equivalent to the classicality of S.

(i) $\vdash_S \Omega = \{true, false\}$;

(ii) $\vdash_S \neg\neg\omega \Rightarrow \omega$;

(iii) any S-set has a complement.

Proof. (i) is obviously equivalent to classicality and the equivalence of (ii) to it is a standard argument. As for (iii), if S is classical, then for any S-set X, the S-set $\{_x : \neg(x \in X)\}$ is obviously a complement for X. Conversely, suppose (iii) holds. Then $\{true\}$ has a complement U. Since $\vdash_S \{true\} \cap U = \varnothing$, we have

$$\begin{aligned} \vdash_S \omega \in U &\Rightarrow \neg(\omega = true) \\ &\Rightarrow \neg\omega \\ &\Rightarrow \omega = false. \end{aligned}$$

Hence $\vdash_S U = \{false\}$ and so

$$\vdash_S \Omega = \{true\} \cup U = \{true, false\},$$

giving (i). □

We also note the standard facts that, if S is classical, then

$$\vdash_S \exists x\alpha \Leftrightarrow \neg\forall x\neg\alpha$$

and

$$\vdash_S \forall x\alpha \Leftrightarrow \neg\exists x\neg\alpha.$$

The reader is invited to construct proofs of these assertions for himself.

We next establish some interconnections among the properties formulated in (4.29).

4.31 THEOREM. Let S be a well-termed, well-typed consistent local set theory. (In particular, S may be Th(**E**) for a non-degenerate topos **E**.) Then we have the following implications and equivalences among possible properties of S:

$$\text{(i)}\quad \left\{\begin{array}{ccccccc} & \text{term-complete} & \Rightarrow & \text{strongly witnessed} & \Rightarrow & \text{complete} & \\ \swarrow & & \searrow & & \searrow & & \searrow \\ \text{singleton-complete} & & \text{classical} & & \text{singleton-witnessed} & \Rightarrow & \text{witnessed} \end{array}\right.$$

$$\text{(ii)}\quad \left\{\begin{array}{l} \text{singleton-witnessed} + \text{classical} \Rightarrow \text{singleton-complete} \\ \qquad \text{strongly witnessed} \Leftrightarrow \text{complete} + \text{witnessed} \end{array}\right.$$

$$\text{(iii)}\quad \left\{\begin{array}{l} \text{term-complete} \Leftrightarrow \text{classical} + \text{strongly witnessed} \\ \qquad\qquad\qquad \Leftrightarrow \text{classical} + \text{singleton witnessed} + \text{complete} \\ \qquad\qquad\qquad \Leftrightarrow \text{singleton-complete} + \text{complete} \end{array}\right.$$

(iv) $AC \Rightarrow$ classical + singleton-witnessed $\Rightarrow$ singleton-complete

Proof. (i) *term-complete* $\Rightarrow$ *strongly witnessed.* Let p be an isomorphism between $U_{\mathbf{B}}$, for some type symbol $\mathbf{B}$, and $\{x : \alpha\}$. Then if $\nvdash_S \neg\exists x\alpha$, it follows that $\nvdash_S \neg\exists y.y = y$, where y is a variable of type $\mathbf{B}$, whence $\nvdash_S \forall y \neg(y = y)$. If S is term-complete, it follows that there is a closed term σ of type $\mathbf{B}$ (such that $\nvdash_S \neg(\sigma = \sigma)$, but this is trivial). Now $\vdash_S \exists! x \in X.\langle x, \sigma\rangle \in p$, so since S is well-termed there is a closed term τ of type $\mathbf{A}$ such that $\vdash_S \langle \tau, \sigma\rangle \in p$; in particular $\vdash_S \tau \in X$, whence $\vdash_S \alpha(x/\tau)$.

strongly witnessed $\Rightarrow$ *complete.* Suppose α is a sentence and $\nvdash_S \neg\alpha$. If u is a variable of type $\mathbf{1}$, it follows that $\nvdash_S \forall u \neg\alpha$, whence $\nvdash_S \neg\exists u\alpha$. If S is strongly witnessed, then there is a closed term τ of type $\mathbf{1}$ (i.e. $*$) such that $\vdash_S \alpha(u/\tau)$, i.e. $\vdash_S \alpha$.

strongly witnessed $\Rightarrow$ *singleton-witnessed.* If $\vdash_S \neg\exists x\alpha$, take $U = \{x : \alpha\}$ and if $\nvdash_S \neg\exists x\alpha$, then, using the strong witnessing property of S, take $U = \{\tau\}$ where τ is a closed term of type $\mathbf{A}$ such that $\alpha(x/\tau)$. Then in either case U is an $\mathbf{A}$-singleton satisfying the appropriate conditions for singleton-witnessing.

singleton-witnessed $\Rightarrow$ *witnessed.* If $\vdash_S \exists x\alpha$ and S is singleton-witnessed, let U be an $\mathbf{A}$-singleton such that $\vdash_S \forall x \in U.\alpha$ and $\vdash_S \exists x.x \in U \Leftrightarrow \exists x\alpha$. Then $\vdash_S \exists! x.x \in U$ and so since S is well-termed there is a closed term τ of type $\mathbf{A}$ such that $\vdash_S \tau \in U$, i.e. $\vdash_S \alpha(x/\tau)$.

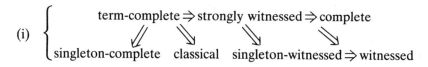

term-complete $\Rightarrow$ *singleton-complete.* If $\vdash_S \forall x \in U.\alpha$ for all **A**-singletons U, then $\vdash_S \forall x \in \{\tau\}\alpha$ for all closed terms τ of type **A**, i.e. $\vdash_S \alpha(x/\tau)$. If S is term-complete, it follows that $\vdash_S \forall x\alpha$.

term-complete $\Rightarrow$ *classical.* If S is term-complete, it is complete, so for any sentence α, $\vdash_S \alpha$ or $\vdash_S \neg\alpha$, whence $\vdash_S \alpha \vee \neg\alpha$. By term-completeness, $\vdash_S \forall\omega.(\omega \vee \neg\omega)$, i.e. classicality.

(ii) *singleton-witnessed + classical* $\Rightarrow$ *singleton-complete.* Suppose $\vdash_S \forall x \in U.\alpha$ for any **A**-singleton U. If S is singleton-witnessed, let V be an **A**-singleton satisfying

$$\vdash_S \forall x \in V\neg\alpha \tag{*}$$

and

$$\vdash_S \exists x.x \in V \Leftrightarrow \exists x\neg\alpha. \tag{**}$$

From (*) it follows that $\vdash_S \forall x \in V.\mathit{false}$, whence $\vdash_S \neg\exists x.x \in V$, so by (**) $\vdash_S \neg\exists x\neg\alpha$. If S is classical, it follows that $\vdash_S \forall x\alpha$, as required.

strongly witnessed $\Leftrightarrow$ *complete + witnessed.* One way round has already been established in (i). For the converse, suppose that $\nvdash_S \neg\exists x\alpha$. If S is complete, then $\vdash_S \exists x\alpha$, so if S is also witnessed, there is a closed term τ such that $\vdash_S \alpha(x/\tau)$, as required.

(iii) *term-complete* $\Leftrightarrow$ *classical + strongly witnessed.* One way round has already been established in (i). For the converse, suppose S is classical and strongly witnessed. If $\vdash_S \alpha(x/\tau)$ for all closed terms τ, then, since S is consistent, $\nvdash_S \neg\alpha(x/\tau)$ for all τ and strong witnessing implies $\vdash_S \neg\exists x\neg\alpha$. Since S is classical, it follows that $\vdash_S \forall x\alpha$, as required.

To establish the remaining equivalences in (iii), by (i) and (ii) it suffices to establish:

singleton-complete + complete $\Rightarrow$ *term-complete.* So suppose S is singleton-complete and complete, and $\vdash_S \alpha(x/\tau)$ for all closed terms τ of type **A**. If U is an **A**-singleton, then by completeness of S either $\vdash_S \exists x.x \in U$ or $\vdash_S \neg\exists x.x \in U$. In the former case, by the well-termedness of S, there is a closed term τ such that $\vdash_S U = \{\tau\}$ and so, since $\vdash_S \alpha(x/\tau)$, we have $\vdash_S \forall x \in U.\alpha$. If, on the other hand, $\vdash_S \neg\exists x.x \in U$, then certainly $\vdash_S \forall x \in U.\alpha$. Therefore $\vdash_S \forall x \in U.\alpha$ for *any* **A**-singleton U, and singleton-completeness yields $\vdash_S \forall x\alpha$ as required.

(iv) $AC \Rightarrow$ *classical.* Assume that S satisfies AC, and write 2 for $\{0,1\}$. Then we evidently have

$$\omega \vee \omega' \vdash_S \exists x \in 2 \quad [x = 0 \Rightarrow \omega \wedge x = 1 \Rightarrow \omega']$$

So by AC there is a map

$$f : \{\langle \omega, \omega' \rangle : \omega \vee \omega'\} \to 2$$

such that

$$\omega \vee \omega' \vdash_S f(\langle \omega, \omega' \rangle) = 0 \Rightarrow \omega \tag{1}$$

$$\omega \vee \omega' \vdash_S f(\langle \omega, \omega' \rangle) = 1 \Rightarrow \omega'. \tag{2}$$

Note that

$$\omega \vee \omega' \vdash_S f(\langle \omega, \omega' \rangle) = 0 \vee f(\langle \omega, \omega' \rangle) = 1.$$

Evaluating f at $\langle \omega, \mathit{true} \rangle$ and using (1) gives

$$\vdash_S \omega \vee f(\langle \omega, \mathit{true} \rangle) = 1,$$

while evaluating f at $\langle \mathit{true}, \omega \rangle$ and using (2) gives

$$\vdash_S f(\langle \mathit{true}, \omega \rangle) = 0 \vee \omega.$$

Hence

$$\vdash_S \omega \vee [f(\langle \omega, \mathit{true} \rangle) = 1 \wedge f(\langle \mathit{true}, \omega \rangle) = 0].$$

Writing $\phi(\omega)$ for

$$f(\langle \omega, \mathit{true} \rangle) = 1 \wedge f(\langle \mathit{true}, \omega \rangle) = 0,$$

we thus have

$$\vdash_S \omega \vee \phi(\omega) \tag{3}$$

Now

$$\omega = \mathit{true} \wedge \phi(\omega) \vdash_S 0 = 1 \vdash_S \mathit{false}$$

so

$$\omega \wedge \phi(\omega) \vdash \mathit{false}$$

whence

$$\phi(\omega) \vdash_S \omega \Rightarrow \mathit{false} \vdash_S \neg\omega.$$

Therefore, using (3),

$$\vdash_S \omega \vee \neg\omega,$$

as required. (Notice that this argument does *not* use the assumption that S is well-termed and well-typed.)

$AC \Rightarrow$ singleton-witnessed. Assuming AC, let u be a variable of type **1** and let $X = \{u : \exists x \alpha\}$, $Y = \{x : \alpha\}$. Then

$$\vdash_S \forall u \in X \exists x \in Y . \alpha,$$

so by AC there is a map $X \xrightarrow{f} Y$ such that

$$\vdash_S \forall u \in X \alpha(x/fu).$$

Let $U = \{x : \langle *, x\rangle \in f\}$. It is easy to check that U is a singleton satisfying the appropriate conditions for singleton-witnessing.

The proof of (4.31) is finally complete. □

REMARK. That classicality is a consequence of AC may also be proved by using the latter to construct a complement for any S-set. Here is a picturesque (and informal) version of the argument.

Suppose that S satisfies AC; let X be an S-set of type $\mathbf{PA}$. We want to construct a complement for X.

Writing A for $U_{\mathbf{A}}$, first form the coproduct $A + A$, which we shall think of as the union of two disjoint copies of A. Regard the elements of the first copy as being coloured black and those of the second as being coloured white. Thus each element of A has been 'split' into a 'black' copy and a 'white' copy.

Next, we identify each copy of an element of X in the first (black) copy of A with its mate in the second (white) copy; the elements thus arising we agree to colour grey, say. In this way we obtain a set Y consisting of black, white and grey elements, together with an epic map $A + A \twoheadrightarrow Y$.

Now we use AC to assign to each element $y \in Y$ an element $y' \in A + A$ in such a way that y' is sent to y by the map $A + A \twoheadrightarrow Y$ above.

The whole process—which we shall denote by P—thus transforms each element of $A + A$ into an element—possibility the same—of $A + A$.

Now, for $n = 0, 1, 2$, define

$$A_n = \{a \in A : P \text{ effects a change of colour in exactly } n \text{ copies of } a\}.$$

Then clearly $A = A_0 \cup A_1 \cup A_2$; $A_1 = X$ and $A_2 = \varnothing$. It follows that A_0 is a complement for X.

When S is Th($\mathbf{E}$), where $\mathbf{E}$ is a topos, the conditions on S formulated above are correlated with certain properties of $\mathbf{E}$, which we now determine.

A collection $\mathcal{G}$ of objects of E (more generally, of any category) is called a *generating class* or *class of generators* for $\mathbf{E}$ if for any pair of distinct arrows $A \underset{g}{\overset{f}{\rightrightarrows}} B$ of $\mathbf{E}$ there is a member G of $\mathcal{G}$ and an arrow $G \xrightarrow{h} A$ such that $f \circ h \neq g \circ h$. If $\mathcal{G}$ consists of a single object G, then G is called a *generator* for $\mathbf{E}$. A nondegenerate topos is said to be *extensional* if its terminal object 1 is a generator, and *subextensional* if the collection of subobjects of 1 is a generating class, i.e. if for any distinct arrows $A \underset{g}{\overset{f}{\rightrightarrows}} B$ of $\mathbf{E}$ there is a subobject $U \rightarrowtail 1$ and an arrow $U \xrightarrow{h} A$ such that $f \circ h \neq g \circ h$.

An object A of $\mathbf{E}$ (or of any category) is said to be *projective* if given any diagram of the form

$$\begin{array}{ccc} A & & \\ \downarrow & & \\ B & \twoheadleftarrow & C \end{array}$$

with $C \twoheadrightarrow B$ epic, there is an arrow $A \to C$ such that

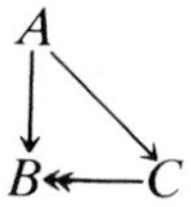

commutes.

E satisfies the *axiom of choice* (**AC**) if every object in **E** is projective. It is easy to see that this is equivalent to the following statement: for any epic $A \overset{f}{\twoheadrightarrow} B$ in **E**, there is a (necessarily monic) $B \xrightarrow{g} A$ such that $f \circ g = 1_B$.

We always assume that **Set** *satisfies* **AC**.

E is *Boolean* if the arrow

$$1+1 \xrightarrow{\binom{\top}{\bot}} \Omega$$

is an isomorphism, and *bivalent* if $\top, \bot : 1 \to \Omega$ are the *only* arrows $1 \to \Omega$ in **E**.

Our next proposition establishes a correspondence between properties of a local set theory and properties of its associated topos.

4.32 PROPOSITION. Let S be a well-termed, well-typed local set theory and let **E** be a topos. Then we have the concordance shown in Table 4.1 between properties of S (respectively Th(**E**)) and properties of **C**(S) (respectively **E**).

Proof. We prove a few of these equivalences, leaving the rest to the reader.

S term-complete $\Leftrightarrow$ **C**(S) *extensional.* Suppose S is term-complete. Since S is well-termed and well-typed, it suffices to show that **C***(S) is extensional (where **C***(S) was defined just after (3.39)). So let **A**,**B** be type symbols and suppose that $f,g : \mathbf{A} \to \mathbf{B}$ are arrows in **C***(S) such that, for any $1 \xrightarrow{h} \mathbf{A}$ in **C***(S), we have $fh = gh$. Then for some terms σ, ξ, f is

Table 4.1

S, Th(**E**)	**C**(S), **E**
consistent	nondegenerate
classical	Boolean
complete	bivalent
term-complete	extensional
singleton-complete	subextensional
strongly witnessed	for any object $A \not\cong 0$ there is an arrow $1 \to A$
witnessed	1 is projective
singleton-witnessed	every subobject of 1 is projective
AC	**AC**

$(x \mapsto \sigma)$ and g is $(x \mapsto \xi)$ and the condition just stated becomes: for any closed term τ of type $\mathbf{A}$, we have $\vdash_S \sigma(\tau) = \xi(\tau)$. Since S is term-complete, it follows that $\vdash_S \forall x(\sigma(x) = \xi(x))$, whence $f = g$. So $\mathbf{C}^*(S)$ and hence also $\mathbf{C}(S)$, is extensional.

Conversely, suppose $\mathbf{C}(S)$ is extensional. Let $\mathbf{A}$ be a type symbol and $\alpha(x)$ a formula with a free variable x of type $\mathbf{A}$. Let f be the map $(x \mapsto \alpha): X \to \Omega$. If $\vdash_S \alpha(\tau)$ for all closed terms τ of type $\mathbf{A}$, it follows that the diagram

$$1 \to A \overset{f}{\underset{T_A}{\rightrightarrows}} \Omega$$

commutes for all such τ. Since $\mathbf{C}(S)$ is extensional (and S well-termed), we deduce that $f = T_A$, or in other words $\vdash_S \forall x(\alpha(x) = \mathit{true})$, i.e. $\vdash_S \forall x \alpha$. So S is term-complete.

S singleton-witnessed $\Leftrightarrow$ every subobject of 1 is projective in $\mathbf{C}(S)$. Suppose S is singleton-witnessed. Given a diagram

$$\begin{array}{ccc} X & & \\ \downarrow g & & \\ Y & \xleftarrow{f} & V \end{array}$$

in $\mathbf{C}(S)$, with g epic and V a subobject of 1, we may assume without loss of generality that $\vdash_S V \subseteq 1$. Let X be of type $\mathbf{PA}$ and let α be the formula $x \in X \wedge \exists y \in Y. g(x) = f(y)$. Then by singleton-witnessing there is an $\mathbf{A}$-singleton U such that

$$\vdash_S \forall x \in U. \alpha, \tag{1}$$

$$\vdash_S \exists x. x \in U \Leftrightarrow \exists x \alpha. \tag{2}$$

Now define $h = V \times U$. It follows easily from (1) and (2) that $h: V \to X$ and $gh = f$. Therefore V is projective.

Conversely, suppose that the projectivity condition holds. Given a formula $\alpha(x)$, let $X = \{x : \alpha\}$ and let $V = \{u : \exists x \alpha\}$, where u is a variable of type $\mathbf{1}$. Then V is a subobject of 1 and the map $g = (x \mapsto *): X \to V$ is epic. By the projectivity condition, there is a map $f: V \to X$ (such that $gf = 1_V$). It follows easily that $U = \{x : \exists u \in V. \langle u, x \rangle \in f\}$ is an $\mathbf{A}$-singleton satisfying (1) and (2) above. So S is singleton-witnessed.

S satisfies $AC \Leftrightarrow \mathbf{C}(S)$ satisfies $\mathbf{AC}$. Given $Y \xrightarrow{g} X$ in $\mathbf{C}(S)$, let α be the formula $\langle y, x \rangle \in g$. Then $\vdash_S \forall x \in X \exists y \in Y. \alpha(x, y)$. So by AC in S there is $X \xrightarrow{f} Y$ such that $\vdash_S \forall x \in X \alpha(x, fx)$, from which it follows easily that $gf = 1_X$. So $\mathbf{AC}$ holds in $\mathbf{C}(S)$.

Conversely, suppose $\mathbf{AC}$ holds in $\mathbf{C}(S)$ and $\vdash_S \forall x \in X \exists y \in Y \alpha(x, y)$ for a given formula α. Define $Z = \{\langle x,y\rangle \in X \times Y : \alpha\}$ and $g = (\langle x,y\rangle \mapsto x) : Z \to X$, $k = (\langle x,y\rangle \mapsto y) : Z \to Y$. Then g is epic, so by **AC** there is $X \xrightarrow{h} Z$ such that $gh = 1_X$. If we now define $f = kh : X \to Y$, it is easy to see that $\vdash_S \forall x \in X \alpha(x, fx)$. So AC holds in S. □

From (4.32) and (4.31(iv)) it follows that *any topos satisfying* **AC** *is Boolean.*

Let A be an object of a topos **E**, and let $B \overset{m}{\rightarrowtail} A$ be a subobject of A. A *complement* for B is a subobject $C \overset{n}{\rightarrowtail} A$ such that $B + C \xrightarrow{\binom{m}{n}} A$ is an isomorphism. Using (4.30) and (4.32), it follows easily that **E** *is Boolean iff any subobject in* **E** *has a complement.* This should be borne in mind when considering some of the examples below.

Examples

(i) **Set** is extensional and satisfies **AC**.

(ii) *For any partially ordered set P*, $\mathbf{Set}^P$ *is subextensional.* Sketch of proof: Given two distinct arrows $\alpha, \beta : F \to G$ in $\mathbf{Set}^P$, choose $p \in P$ and $a \in F(p)$ such that $\alpha_p(a) \neq \beta_p(a)$. Define $U \in \mathbf{Set}^P$ by $U(q) = 1$ if $p \leq q$, $U(q) = \varnothing$ if $p \not\leq q$; then U is a subobject of 1 in $\mathbf{Set}^P$. Define $\eta : U \to F$ by $\eta_q(0) = F_{pq}(a)$ if $p \leq q$ and $\eta_q = \varnothing$ if $p \not\leq q$. Then $\alpha \circ \eta \neq \beta \circ \eta$.

(iii) *If M is a monoid, the topos* $\mathbf{Set}^M$ *of M-sets is Boolean iff M is a group.* For if M is a group, then the set-theoretical complement of any sub-M-set Y of an M-set X is itself a sub-M-set and therefore the complement (in $\mathbf{Set}^M$) of Y. Conversely, if M is not a group, then the set N of noninvertible elements of M (regarded as an M-set with the natural multiplication on the left by M) is a sub-M-set of M with no complement since any sub-M-set of M must—as is readily verified—intersect N.

(iv) *If M is a monoid,* $\mathbf{Set}^M$ *is bivalent.* For the terminal object in $\mathbf{Set}^M$ is the one-point set 1 with trivial M-action and evidently this has only the two subobjects 0, 1. Thus, if M is not a group, $\mathbf{Set}^M$ is bivalent but *not* Boolean (and its associated local set theory complete but not classical, so *a fortiori* not singleton-complete).

(v) *If G is a group with at least two elements, then* 1 *is not projective in* $\mathbf{Set}^G$. For $G \to 1$ in $\mathbf{Set}^G$ is certainly epic, but an arrow $1 \to G$ in $\mathbf{Set}^G$ corresponds to an element $e \in G$ such that $g.e = e$ for all $g \in G$, which cannot exist unless G has only one element. So the theory associated with $\mathbf{Set}^G$ is classical but not witnessed.

(vi) *If M is a nontrivial monoid, then* $\mathbf{Set}^M$ *is not subextensional.* For let S be Th($\mathbf{Set}^M$); then S is complete. If S were singleton-complete (i.e. $\mathbf{Set}^M$ subextensional), then by (4.31) S would be classical and witnessed; but if M is not a group, S is non-classical by (iii), and if M is a group, then S is not witnessed by (v).

(vii) Let M be the monoid with two elements 1 and e such that $e^2 = e$.

Then for any object $X \not\cong 0$ of $\mathbf{Set}^M$ there is an arrow $1 \to X$. For if X is a non-empty M-set, choose $a \in X$ and let $b = e.a$. Then it is easy to check that $0 \mapsto b$ is an arrow $1 \to X$ in $\mathbf{Set}^M$. Thus, in this case, if $S = \mathrm{Th}(\mathbf{Set}^M)$, *$S$ is strongly witnessed but not classical.*

EXERCISES

(1) Let S be a well-termed, well-typed local set theory. Show that the following are equivalent:

(i) S satisfies AC
(ii) S_X is singleton-witnessed for every S-set X.

(2) S is said to satisfy the *implicit axiom of choice* (IAC) if, for any S-sets X, Y and any formula $\alpha(x,y)$ with at most the variables x,y free, $\vdash_S \forall x \in X\, \exists y \in Y \alpha(x,y)$ implies $\vdash_S \exists f \in Y^X\, \forall x \in X\, \exists y \in Y.\ \langle x,y\rangle \in f \wedge \alpha(x,y)$.

Show that the following are equivalent:

(i) S satisfies AC
(ii) S is singleton-witnessed and satisfies IAC.

(3) *Hilbertian theories and the strong axiom of choice.* A local set theory S is called *Hilbertian* if the following condition is satisfied. For any variable x such that $\nvdash_S \neg\exists x.x = x$, and any formula α, there is a term τ of the same type as x, whose free variables are those of α apart from x, such that

$$\exists x \alpha \vdash_S \alpha(x/\tau). \tag{*}$$

(In $\alpha(x/\tau)$ we assume that the bound variables in α have been changed in such a way as to render τ free for x in α.)

The term τ in (*) is usually written $\varepsilon_x \alpha$: the 'ε' here is a version of *Hilbert's ε-symbol.* Thus (*) becomes

$$\exists x \alpha \vdash \alpha(x/\varepsilon_x \alpha).$$

(i) Suppose that S is Hilbertian. Show that S satisfies AC. Suppose in addition that S is well-termed and well-typed. Show that S is strongly witnessed, and use (4.31) to deduce that S is complete and term-complete.

A topos $\mathbf{E}$ is said to satisfy the *strong axiom of choice* (**SAC**) if, for any arrow $X \xrightarrow{f} Y$ with $X \not\cong 0$, there is $Y \xrightarrow{g} X$ with $fgf = f$.

(ii) Show that $\mathbf{E}$ satisfies **SAC** iff $\mathrm{Th}(\mathbf{E})$ is Hilbertian.

Full theories

The concept of a *full* local set theory, to be introduced presently, corresponds to the essentially categorical property of being *defined over* **Set**.

Let S be a local set theory in a local language $\mathscr{L}$. The theory S is said to be *full* if for each (intuitive) set I there is a type symbol $\hat{\mathbf{I}}$ of $\mathscr{L}$ together with a family $\{\hat{i}: i \in I\}$ of closed terms, each of type $\hat{\mathbf{I}}$, satisfying the following condition, called the *universal property of* $\hat{\mathbf{I}}$:

For any type symbol $\mathbf{A}$ and any family $\{\tau_i : i \in I\}$ of closed terms, each of type $\mathbf{A}$, there is a term $\tau(x)$ of type $\mathbf{A}$ with exactly one free variable x of type $\hat{\mathbf{I}}$ such that $\vdash_S \tau_i = \tau(\hat{i})$ for all $i \in I$ and, for any term $\sigma(x)$ of type $\mathbf{A}$ with at most the free variable x such that $\vdash_S \tau_i = \sigma(\hat{i})$ for all $i \in I$, we have $\vdash_S \tau = \sigma$. (This last condition is called the *uniqueness condition on* τ.)

We shall write $\hat{I}$ *for the S-set* $U_{\hat{\mathbf{I}}}$.

If S is well-termed and well-typed—in particular if $S = \mathrm{Th}(\mathbf{E})$ for some topos $\mathbf{E}$—then it is easy to see (using (3.39)) that $\hat{I} \cong \coprod_I 1$ in $\mathbf{C}(S)$. Hence

4.33 PROPOSITION. For any topos $\mathbf{E}$,

$$\mathrm{Th}(\mathbf{E}) \text{ is full} \Leftrightarrow \mathbf{E} \text{ is defined over } \mathbf{Set}. \quad \square$$

We shall need the following

4.34 LEMMA. Let S be a full local set theory.

(i) Let $I_1, \ldots, I_n$ be sets, and let $\alpha(x_1, \ldots, x_n)$ be a formula with free variables $x_1, \ldots, x_n$ of types $\hat{\mathbf{I}}_1, \ldots, \hat{\mathbf{I}}_n$. If $\vdash_S \alpha(\hat{i}_1, \ldots, \hat{i}_n)$ for all $i_1 \in I_1, \ldots, i_n \in I_n$, then $\vdash_S \forall x_1, \cdots \forall x_n \alpha$.

(ii) Suppose that S is well-termed. Then for any sets I, J there is a term $\rho(v)$ of type $(\mathbf{I} \times \mathbf{J})\hat{}$ with v a variable of type $\hat{\mathbf{I}} \times \hat{\mathbf{J}}$ such that $\vdash_S \rho(\langle \hat{i}, \hat{j} \rangle) = (i,j)\hat{}$ for any $i \in I$, $j \in J$ and $v \mapsto \rho(v)$ is an isomorphism $\hat{I} \times \hat{J} \cong (I \times J)\hat{}$.

Proof. (i) Assume the premise, and let $\beta(x_1)$ be the formula $\alpha(x, \hat{i}_2, \ldots, \hat{i}_n)$ for fixed $i_2 \in I_2, \ldots, i_n \in I_n$. For any $i_1 \in I_1$ we have $\vdash_S \beta(\hat{i}_1) = true$, and since $\vdash_S \beta(\hat{i}_1) = \beta(\hat{i}_1)$, it follows from the uniqueness condition that $\vdash_S \beta(x_1) = true$, whence $\vdash_S \forall x_1 \beta$ or $\vdash_S \forall x_1 \alpha(x_1, \hat{i}_2, \ldots, \hat{i}_n)$. Iterating this argument yields the conclusion.

(ii) By the universal property of $\hat{\mathbf{I}}$, for each $j \in J$ there is a term $\sigma_j(x)$ of type $(\mathbf{I} \times \mathbf{J})\hat{}$ with x a variable of type $\hat{\mathbf{I}}$ such that

$$\vdash_S (i,j)\hat{} = \sigma_j(\hat{i}) \tag{1}$$

for all $i \in I$. By the universal property of $\hat{\mathbf{J}}$, there is a term $\eta(y)$ of type $\mathbf{P}(\hat{\mathbf{I}} \times (\mathbf{I} \times \mathbf{J})\hat{})$, with y a free variable of type $\hat{\mathbf{J}}$, such that

$$\vdash_S \eta(\hat{j}) = \{\langle x, \sigma_j(x) \rangle : x \in \hat{I}\} \tag{2}$$

for all $j \in J$. Then

$$\vdash_S \eta(\hat{j}) \in ((I \times J)\hat{})^{\hat{I}}$$

for all $j \in J$, so that, by (i),

$$\vdash_S \forall y.\, \eta(y) \in ((I \times J)\hat{\ })^{\hat{\imath}}.$$

It follows that

$$\vdash_S \forall x\, \exists! u.\, \langle x,u \rangle \in \eta(y)$$

where u is of type $(\mathbf{I} \times \mathbf{J})\hat{\ }$. Since S is well-termed, there is a term $\tau(x,y)$ of type $(\mathbf{I} \times \mathbf{J})\hat{\ }$ such that

$$\vdash_S \forall x.\, \langle x,\, \tau(x,y) \rangle \in \eta(y).$$

Hence

$$\vdash_S \langle \hat{\imath},\, \tau(\hat{\imath},\hat{\jmath}) \rangle \in \eta(\hat{\jmath})$$

so that, by (2),

$$\vdash_S \tau(\hat{\imath},\hat{\jmath}) = \sigma_j(\hat{\imath}).$$

Therefore, by (1),

$$\vdash_S \tau(\hat{\imath},\hat{\jmath}) = (i,j)\hat{\ }.$$

So, if we write $\rho(v)$ for the term $\tau((v)_1, (v)_2)$ where v is of type $\hat{\mathbf{I}} \times \hat{\mathbf{J}}$, then we have

$$\vdash_S \rho(\langle \hat{\imath},\hat{\jmath} \rangle) = (i,j)\hat{\ }.$$

To see that $v \mapsto \rho(v)$ is an isomorphism $\hat{I} \times \hat{J} \to (I \times J)\hat{\ }$; we construct its inverse. By the universal property of $(\mathbf{I} \times \mathbf{J})\hat{\ }$, there is a term $\theta(u)$ of type $\hat{\mathbf{I}} \times \hat{\mathbf{J}}$ with u of type $(\mathbf{I} \times \mathbf{J})\hat{\ }$ such that

$$\vdash_S \langle \hat{\imath},\hat{\jmath} \rangle = \theta((i,j)\hat{\ }).$$

Then

$$\vdash_S \rho(\theta((i,j)\hat{\ })) = (i,j)\hat{\ }$$

and

$$\vdash_S \theta(\rho(\langle \hat{\imath},\hat{\jmath} \rangle)) = \langle \hat{\imath},\hat{\jmath} \rangle,$$

for all $i \in I$, $j \in J$, so that, by (i),

$$\vdash_S \rho(\theta(u)) = u \wedge \theta(\rho(v)) = v.$$

Therefore $u \mapsto \theta(u)$ is the inverse of $v \mapsto \rho(v)$ as claimed. □

Let S be a local set theory (not necessarily full). Recall from (4.11) the Heyting algebras $(\Omega(S), \leqslant)$ and $(\mathrm{Pow}(X), \subseteq)$, where X is any S-set. If $\mathbf{A}$ is a type symbol, we write $\mathrm{Pow}(\mathbf{A})$ for $\mathrm{Pow}(U_{\mathbf{A}})$. Note that, if S is *classical*, then clearly $\Omega(S)$ and $\mathrm{Pow}(\mathbf{A})$ are Boolean algebras.

Now we can prove what amounts to a syntactic version of (4.28).

4.35 PROPOSITION. If S is full, then

(i) $(\Omega(S), \leqslant)$ is a locale;
(ii) for any type symbol $\mathbf{A}$, $(\mathrm{Pow}(\mathbf{A}), \subseteq)$ is a locale.

Proof. Suppose $\{\alpha_i : i \in I\} \subseteq \Omega(S)$. Then by the universal property of $\hat{\mathbf{I}}$, there is a formula $\beta(x)$ with x a free variable of type $\hat{\mathbf{I}}$ such that $\vdash_S \alpha_i = \beta(\hat{i})$ for all $i \in I$. We claim that the sentence $\exists x \beta(x)$ is the supremum of $\{\alpha_i : i \in I\}$ in $\Omega(S)$. First, it is an upper bound, since

$$\alpha_i \vdash_S \beta(\hat{i}) \vdash_S \exists x \beta(x).$$

And it is the least upper bound, since if γ is a sentence such that $\alpha_i \leqslant \gamma$ for all $i \in I$, then $\vdash_S \alpha_i = \gamma \wedge \alpha_i$ for all $i \in I$, so that $\vdash_S \alpha_i = \gamma \wedge \beta(\hat{i})$ for all $i \in I$. By the uniqueness condition on β, it follows that $\vdash_S \beta(x) = \gamma \wedge \beta(x)$, so that $\vdash_S \beta(x) \Rightarrow \gamma$, whence $\vdash_S \exists x \beta(x) \Rightarrow \gamma$ and $\exists x \beta(x) \leqslant \gamma$. So $\Omega(S)$ is complete, and hence a locale.

(ii) Let $\{U_i : i \in I\} \subseteq \mathrm{Pow}(\mathbf{A})$. Then each U_i is a closed term of type $\mathbf{PA}$, so by the universal property of $\hat{\mathbf{I}}$ there is a term $\tau(x)$ such that $\vdash_S \tau(\hat{i}) = U_i$ for all $i \in I$. Define

$$U = \{y : \exists x . y \in \tau(x)\}.$$

Then U is of type $\mathbf{PA}$; we claim that U is the supremum of $\{U_i : i \in I\}$ in $\mathrm{Pow}(\mathbf{A})$. Clearly $U_i \subseteq U$ for all $i \in I$, so U is an upper bound; if V is a closed term of type $\mathbf{PA}$ such that $U_i \subseteq V$ for all $i \in I$, then $\vdash_S y \in \tau(\hat{i}) \Rightarrow y \in V$, so by (4.34)(i) we have $\vdash_S \forall x[y \in \tau(x) \Rightarrow y \in V]$, whence $\vdash_S \exists x . y \in \tau(x) \Rightarrow y \in V$ and thus $U \subseteq V$. Therefore $\mathrm{Pow}(\mathbf{A})$ is complete, and hence a locale. □

If S is full, $\Omega(S)$ is called the *locale of sentences of S*. The supremum in $\Omega(S)$ of a set $\{\alpha_i : i \in I\}$ of elements of $\Omega(S)$ will be denoted by $\bigvee_{i \in I} \alpha_i$. We note that by the proof of (4.35)(i), if $\beta(x)$ satisfies $\vdash_S \alpha_i = \beta(\hat{i})$ for all $i \in I$, then

(4.36) $$\bigvee_{i \in I} \alpha_i = \exists x \beta(x).$$

Similarly, $\mathrm{Pow}(\mathbf{A})$ is called the *locale of subsets* of $\mathbf{A}$. The supremum of a subfamily $\{U_i : i \in I\}$ of $\mathrm{Pow}(\mathbf{A})$ will be called the *union* of the subfamily, and written $\bigcup_{i \in I} U_i$ as usual.

In (4.31) we showed that the axiom of choice in S implies classicality and singleton-completeness. We finally show that the converse holds provided S is full.

4.37 PROPOSITION. Let S be a full, well-termed, well-typed local set theory. Then the following are equivalent:

(i) S satisfies AC;
(ii) S is classical and singleton-complete.

Proof. We need only prove (ii)$\Rightarrow$(i). Suppose that $\vdash_S \forall x \in X\, \exists y \in Y \alpha(x,y)$. Since S is well-typed, without loss of generality we may assume that X and Y are of the form $U_{\mathbf{A}}$ and $U_{\mathbf{B}}$. Now define

$$C = \{\langle x,y \rangle : \alpha(x,y)\}$$

and let $\mathcal{G}$ be the collection of all S-sets Z such that

$$\vdash_S Z \subseteq C \wedge \forall u \in Z\, \forall v \in Z[(u)_1 = (v)_1 \Rightarrow u = v].$$

Partially order $\mathcal{G}$ by

$$Z \subseteq Z' \quad \text{iff} \quad \vdash_S Z \subseteq Z'.$$

Since S is full, we may form unions of families of sets of the same type, and so we may apply Zorn's lemma to obtain a $\subseteq$-maximal member M of $\mathcal{G}$.

We claim that

$$\vdash_S \forall x\, \exists y \langle x,y \rangle \in M. \tag{1}$$

Write $\beta(u)$ for the formula $\forall y \in Y \neg[\langle (u)_1, y \rangle \in M]$ and let

$$D = \{u \in C : \beta(u)\}.$$

By singleton-completeness, let U be an $\mathbf{A} \times \mathbf{B}$-singleton such that

$$\vdash_S U \subseteq D \quad \text{and} \quad \vdash_S \exists u . u \in U \Leftrightarrow \exists u . u \in D.$$

Then clearly $M \cup U \in \mathcal{G}$ so, since M is maximal, we must have $U \subseteq M$, i.e. $\vdash_S U \subseteq M$. But evidently $\vdash_S U \cap M = \varnothing$, so that $\vdash_S U = \varnothing$. Therefore

$$\begin{aligned} \vdash_S \exists u . u \in D &\Rightarrow \exists u . u \in U \\ &\Rightarrow \mathit{false}. \end{aligned}$$

Hence $\vdash_S \neg \exists u . u \in D$, so since S is classical it follows from the definition of D that

$$\vdash_S \forall u \in C\, \exists y \in Y . \langle (u)_1, y \rangle \in M.$$

Together with the assumption $\vdash_S \forall x \in X\, \exists y \in Y \alpha(x,y)$, this gives (1).

It is now clear that M is a map $X \to Y$ satisfying $\vdash_S \forall x \in X \alpha(x, Mx)$. AC follows. □

As an immediate consequence, we have

4.38 PROPOSITION. The following are equivalent for a topos $\mathbf{E}$ defined over **Set**:

(i) $\mathbf{E}$ satisfies **AC**;
(ii) $\mathbf{E}$ is Boolean and subextensional. □

EXERCISE. Let $(P, \leqslant)$ be a lattice. A *filter* in P is a subset F such that (a)

$p \in F$ and $q \in F$ implies $p \wedge q \in F$, (b) $p \in F$ and $p \leqslant q$ implies $q \in F$. A filter F is *proper* if $0 \notin F$. An *ultrafilter* is a maximal proper filter.

(i) Suppose that P is a Heyting algebra. Show that a filter F in P is an ultrafilter iff it satisfies the following condition: for any $p \in P$, either $p \in F$ or $p^* \in F$ (but not both).

(ii) Let S be a local set theory, and let F be a filter in the Heyting algebra $\Omega(S)$. Let $S + F$ be the theory obtained from S by adding as axioms all sequents of the form : α, where $\alpha \in F$. Show that, if S satisfies AC, so does $S + F$.

(iii) Show that any consistent local set theory has a consistent complete extension, which satisfies AC if S does. (Use Zorn's lemma to obtain an ultrafilter in $\Omega(S)$ and then apply (i) and (ii).) State and deduce the corresponding result for toposes.

Beth–Kripke–Joyal semantics

We conclude the chapter by formulating a natural semantics for local set theories generalizing those of Beth and Kripke. Let S be a local set theory, let α be a formula of the language of S, let $u, v, \ldots$ be variables of types $\mathbf{A}, \mathbf{B}, \ldots$, let X be an S-set and let $f: X \to A$, $g: X \to B, \ldots$ be S-maps. We think of $f, g, \ldots$ as *generalized elements* of $A, B, \ldots$ *defined* at stage X. Then we write

$$X \Vdash_S \alpha[f, g, \ldots]$$

and say that X *forces* $\alpha[f, g, \ldots]$ or the generalized elements $f, g, \ldots$ *satisfy* α *at stage* X *if*

$$\vdash_S \forall x \in X \alpha(u/fx, v/gx, \ldots)$$

where x does not occur in α.

Recalling that S_X denotes the theory obtained from S by adding the indeterminate X-element c, it follows from (4.17) that

$$X \Vdash_S \alpha[f, g, \ldots] \quad \text{iff} \quad \vdash_{S_X} \alpha(u/fc, v/gc, \ldots)$$

This reduces the concept of satisfaction by generalized elements to the more familiar concept of satisfaction by elements.

It is readily seen that

$$\vdash_S \alpha \quad \text{iff} \quad X \Vdash_S \alpha[f, g, \ldots] \text{ for any } X \text{ and any } X \xrightarrow{f} A,\ X \xrightarrow{g} B, \ldots,$$

that is, provability is equivalent to being forced at all stages. Also,

$$X \Vdash_S \alpha[f] \quad \text{implies} \quad Y \Vdash_S \alpha[g \circ f] \text{ for any } Y \xrightarrow{f} X,$$

that is, satisfaction at one stage implies satisfaction at all 'previous' stages.

The rules governing the forcing relation, which are stated in the next theorem, are collectively known as *Beth–Kripke–Joyal semantics* (for local set theories).

4.39 THEOREM. Given $f: X \to A$, we have

(i) $X \Vdash_S$ *true* always;
(ii) $X \Vdash_S$ *false* iff $X = \varnothing$;
(iii) $X \Vdash_S (\alpha \wedge \beta)[f]$ iff $X \Vdash_S \alpha[f]$ and $X \Vdash_S \beta[f]$;
(iv) $X \Vdash_S (\alpha \vee \beta)[f]$ if there are S-sets U, V such that $U \cup V = X$ and $U \Vdash_S \alpha[f \mid U]$, $V \Vdash_S \beta[f \mid V]$. (Here $f \mid U$ is $\{\langle x, y\rangle \in f : x \in U\}$, the usual restriction of f to U.)
(v) $X \Vdash_S (\alpha \Rightarrow \beta)[f]$ iff, for all $Y \xrightarrow{g} X$, if $Y \Vdash_S \alpha[f \circ g]$, then $Y \Vdash_S \beta[f \circ g]$;
(vi) $X \Vdash_S \neg\alpha[f]$ iff, for all $Y \xrightarrow{g} X$, if $Y \Vdash_S \alpha[f \circ g]$, then $Y = \varnothing$;
(vii) $X \Vdash_S \forall v \alpha[f]$ iff, for all $Y \xrightarrow{g} X$ and $Y \xrightarrow{h} B$, we have $Y \Vdash_S \alpha[f \circ g, h]$;
(viii) $X \Vdash_S \exists v \alpha[f]$ iff there is an epic $Y \overset{g}{\twoheadrightarrow} X$ and $Y \xrightarrow{h} B$ such that $Y \Vdash_S \alpha[f \circ g, h]$.

Proof. We prove (v) and (viii), leaving the rest to the reader.

(v) Suppose that $X \Vdash_S (\alpha \Rightarrow \beta)[f]$ and $Y \xrightarrow{g} X$. Then

$$\vdash_S \forall x \in X[\alpha(u/fx) \Rightarrow \beta(u/fx)],$$

whence

$$\vdash_S \forall y \in Y[\alpha(u/f(gy)) \Rightarrow \beta(u/f(gy))].$$

It follows that, if $Y \Vdash_S \alpha[f \circ g]$, then $\vdash_S \forall y \in Y \alpha(u/f(gy))$, whence $\vdash_S \forall y \in Y \beta(u/f(gy))$, so that $Y \Vdash_S \beta[f \circ g]$. Conversely, if this implication holds for any $Y \xrightarrow{g} X$, define $U = \{x \in X : \alpha(u/fx)\}$ and $i = (x \mapsto x) : U \to X$. Then clearly $U \Vdash_S \alpha[f \circ i]$, whence $U \Vdash_S \beta[f \circ i]$ i.e.

$$\vdash_S \forall x \in X[\alpha(u/fx) \Rightarrow \beta(u/fx)]$$

or $X \Vdash_S (\alpha \Rightarrow \beta)[f]$.

(viii) Suppose that $X \Vdash_S \exists v \alpha[f]$. Then

$$\vdash_S \forall x \in X\, \exists v \alpha(u/fx, v). \tag{*}$$

Let

$$Y = \{\langle x, v\rangle : x \in X \wedge \alpha(u/fx, v)\}$$

and

$$g = (\langle x, v\rangle \mapsto x) : Y \to X$$
$$h = (\langle x, v\rangle \mapsto v) : Y \to B$$

Then (*) implies that g is epic; moreover

$$\vdash_S \forall y \in Y \alpha(u/f(gy)),$$

i.e.

$$Y \Vdash_S \alpha[f \circ g, h].$$

Conversely, suppose that $Y \Vdash_S \alpha[f \circ g, h]$ for some epic $Y \xrightarrow{g} X$ and $Y \xrightarrow{h} B$. Then

$$\vdash_S \forall y \in Y \alpha(u/f(gy), v/hy).$$

But since g is epic, we have

$$\vdash_S \forall x \in X\, \exists y \in Y. x = gy.$$

Amalgamating these last two assertions yields

$$\vdash_S \forall x \in X\, \exists y \in Y \alpha(u/fx, v/hy),$$

whence $\vdash_S \forall x \in X\, \exists v \alpha(u/fx, v)$, i.e. $X \Vdash_S \exists v \alpha[f]$. □

We note that the satisfaction, at a given stage X, of any formula apart from a conjunction involves the satisfaction of the constituents of the formula at stages different from or 'previous to' X. (In this connection, however, see the exercises below.)

EXERCISES

(1) An S-set X is said to be *indecomposable* if for any S-sets U, V, $U \cup V = X$ implies $U = X$ or $V = X$. Show that the following conditions are equivalent:

(i) X is indecomposable;
(ii) for any formulae α, β and any $X \xrightarrow{f} A$

$$X \Vdash_S (\alpha \vee \beta)[f] \quad \text{implies} \quad X \Vdash_S \alpha[f] \quad \text{or} \quad X \Vdash_S \beta[f].$$

(2) An S-set X is said to be *choice* if, for any S-set Y and any formula $\alpha(x,y)$, $\vdash_S \forall x \in X\, \exists y \in Y \alpha(x,y)$ implies that there is an S-map $X \xrightarrow{f} Y$ such that $\vdash_S \forall x \in X \alpha(x, fx)$. (Thus the axiom of choice holds in S exactly when every S-set is choice.) Show that the following conditions are equivalent:

(i) X is choice;
(ii) X is projective in $\mathbf{C}(S)$;
(iii) for any formula $\alpha(u,v)$ and any S-map $X \xrightarrow{f} A$,

$X \Vdash_S \exists v \alpha[f]$ implies $X \Vdash_S \alpha[f,g]$ for some S-map $X \xrightarrow{g} B$.

(3) Let $\mathscr{G}$ be a class of generators for $\mathbf{C}(S)$. Show that, for any

formula α,

$$\vdash_S \alpha \quad \text{iff} \quad X \Vdash_S \alpha[f, g, \ldots] \text{ for all } X \in \mathscr{G} \text{ and all } X \xrightarrow{f} A,\ X \xrightarrow{g} B, \ldots .$$

Now let **E** be a topos; let $X, A, B, \ldots$ be **E**-objects; let $X \xrightarrow{f} A$, $X \xrightarrow{g} B, \ldots$ be **E**-arrows and let α be a formula of $\mathscr{L}(\mathbf{E})$. We write

$$X \Vdash_{\mathbf{E}} \alpha[f, g, \ldots]$$

for $U_X \Vdash_{\mathrm{Th}(\mathbf{E})} \alpha[f, g, \ldots]$ and say that *X forces* $\alpha[f, g, \ldots]$ or $f, g, \ldots$ *satisfy* α *at stage X*.

From (4.39) we immediately deduce the following:

4.40 THEOREM. Given $X \xrightarrow{f} A$, we have

(i) $X \Vdash_{\mathbf{E}} \mathit{true}$ always;
(ii) $X \Vdash_{\mathbf{E}} \mathit{false}$ iff $X \cong 0$;
(iii) $X \Vdash_{\mathbf{E}} (\alpha \wedge \beta)[f]$ iff $X \Vdash_{\mathbf{E}} \alpha[f]$ and $X \Vdash_{\mathbf{E}} \beta[f]$;
(iv) $X \Vdash_{\mathbf{E}} (\alpha \vee \beta)[f]$ iff there are monics $U \overset{m}{\rightarrowtail} X$, $V \overset{n}{\rightarrowtail} X$ such that $U + V \xrightarrow{\binom{m}{n}} X$ is epic and $U \Vdash_{\mathbf{E}} \alpha[f \circ m]$, $V \Vdash_{\mathbf{E}} \beta[f \circ n]$;
(v) $X \Vdash_{\mathbf{E}} (\alpha \Rightarrow \beta)[f]$ iff, for all $Y \xrightarrow{g} X$, if $Y \Vdash_{\mathbf{E}} \alpha[f \circ g]$, then $Y \Vdash_{\mathbf{E}} \beta[f \circ g]$;
(vi) $X \Vdash_{\mathbf{E}} \neg\alpha[f]$ iff, for all $Y \xrightarrow{g} X$, if $Y \Vdash_{\mathbf{E}} \alpha[f \circ g]$, then $Y \cong 0$;
(vii) $X \Vdash_{\mathbf{E}} \forall v \alpha[f]$ iff, for all $Y \xrightarrow{g} X$ and $Y \xrightarrow{h} B$, we have $Y \Vdash_{\mathbf{E}} \alpha[f \circ g, h]$;
(viii) $X \Vdash_{\mathbf{E}} \exists v \alpha[f]$ iff there is an epic $Y \overset{g}{\twoheadrightarrow} X$ and $Y \xrightarrow{h} B$ such that $Y \Vdash_{\mathbf{E}} \alpha[f \circ g, h]$. □

EXERCISE. Formulate the previous group of exercises in terms of toposes.

We finally consider the special case in which **E** is the topos of presheaves on a small category.

Let **C** be a small category. We shall write **E(C)** for the topos $\mathbf{Set}^{\mathbf{C}^{\mathrm{op}}}$ of presheaves on **C**. Recall from Chapter 2 that the truth value object Ω in **E(C)** has, for any **C**-object A,

$$\Omega(A) = \text{set of all cosieves on } A$$

(where a *cosieve* on A is a set of arrows with codomain A closed under composition on the right) and for any **C**-arrow $B \xrightarrow{f} A$, $S \in \Omega(A)$, defining

$$f^*S = \{g \in \mathrm{Arr}(\mathbf{C}) : \mathrm{cod}(g) = B \ \&\ f \circ g \in S\},$$

we have

$$(\Omega f)S = f^*S.$$

The arrow $1 \xrightarrow{\top} \Omega$ is given by

$$\top_A(0) = \max^A = \{g \in \mathrm{Arr}(\mathbf{C}) : \mathrm{cod}(g) = A\}.$$

Here $\max^A$ is the *maximal cosieve* on A.

Let F be a presheaf on $\mathbf{C}$, and let $\mathbf{F}$ be the corresponding type symbol in the language $\mathscr{L}(\mathbf{E(C)})$. Let α be a formula with at most the one free variable u, of type $\mathbf{F}$: we shall call such a formula an $\mathbf{F}$-*formula*. Then the natural interpretation $[\![\alpha]\!]_u$—which we shall abbreviate simply to $[\![\alpha]\!]$—of α in $\mathbf{E(C)}$ is an $\mathbf{E(C)}$-arrow (i.e. a natural transformation) $F \to \Omega$. If α contains an additional free variable v of type $\mathbf{G}$, where G is a presheaf on $\mathbf{C}$, then $[\![\alpha]\!] =_{\mathrm{df}} [\![\alpha]\!]_{uv}$ is an arrow $F \times G \to \Omega$, and similarly in the presence of still more free variables.

4.41 PROPOSITION. For any $\mathbf{C}$-object A, any $a, a' \in FA$ and any $\mathbf{F}$-formulas α, β, we have

$$[\![u = u']\!]_A(\langle a, a'\rangle) = \{f \in \max^A : (Ff)a = (Ff)a'\};$$
$$[\![\alpha \wedge \beta]\!]_A(a) = [\![\alpha]\!]_A(a) \cap [\![\beta]\!]_A(a);$$
$$[\![a \vee \beta]\!]_A(a) = [\![\alpha]\!]_A(a) \cup [\![\beta]\!]_A(a);$$
$$[\![\alpha \Rightarrow \beta]\!]_A(a) = \{f \in \max^A : f^*([\![\alpha]\!]_A(a)) \subseteq f^*([\![\beta]\!]_A(a)\};$$
$$[\![\neg\alpha]\!]_A(a) = \{f \in \max^A : f^*([\![\alpha]\!]_A(a)) = \varnothing\}.$$

If α contains an additional free variable v of type $\mathbf{G}$,

$$[\![\exists v\alpha]\!]_A(a) = \{B \xrightarrow{f} A : \text{for some } b \in GB,\ [\![\alpha]\!]_B(\langle (Ff)a, b\rangle) = \max^B\};$$

$$[\![\forall v\alpha]\!]_A(a)$$
$$= \{B \xrightarrow{f} A : \text{for all } C \xrightarrow{g} B, \text{ all } b \in GC,\ [\![\alpha]\!]_C(\langle F(f \circ g)(a), b\rangle) = \max^C\}.$$

Proof. The easiest way to verify these assertions is to employ the logical rules that characterize each logical operator. For example, one verifies that the cosieve

$$S = \{f \in \max^A : f^*([\![\alpha]\!]_A(a)) \subseteq f^*([\![\beta]\!]_A(a))\}$$

is the largest cosieve on A such that $S \cap [\![\alpha]\!]_A(a) \subseteq [\![\beta]\!]_A(a)$ and therefore $S = [\![\alpha \Rightarrow \beta]\!]_A(a)$. The remaining cases are treated similarly. □

As an immediate consequence of (4.41) we have

4.42 COROLLARY. For $\mathbf{F}$-formulae α, β

$$\alpha \vDash_{E(C)} \beta \quad \text{iff} \quad [\![\alpha]\!]_A(a) \subseteq [\![\beta]\!]_A(a) \quad \text{for all} \quad A \in \mathrm{Ob}(\mathbf{C}), a \in FA,$$
$$\vDash_{E(C)} \alpha \quad \text{iff} \quad [\![\alpha]\!]_A(a) = \max^A \quad \text{for all} \quad A \in \mathrm{Ob}(\mathbf{C}), a \in FA. \quad □$$

For $\mathbf{C}$-objects A and $a \in FA$ we write

$$A \Vdash^{\mathbf{C}} \alpha[a]$$

for $[\![\alpha]\!]_A(a) = \max^A$ and say that A forces $\alpha[a]$ or *a satisfies α at stage A*. This notion of forcing is related to the one introduced at the beginning of this section in the following way. By the Yoneda lemma, each $a \in FA$ corresponds to a natural transformation $H^A \xrightarrow{a^*} F$, and it is not hard to verify that

(4.43) $$A \Vdash^{\mathbf{C}} \alpha[a] \quad \text{iff} \quad H^A \Vdash_{\mathbf{E(C)}} \alpha[a^*].$$

Using (4.41), we obtain the following list of rules governing the relation $\Vdash^{\mathbf{C}}$.

4.44 PROPOSITION. For any **C**-object A, any $a, a' \in FA$ and any **F**-formula α,

$$[\![\alpha]\!]_A(a) = \{B \xrightarrow{f} A : B \Vdash^{\mathbf{C}} \alpha[(Ff)a]\};$$
$$A \Vdash^{\mathbf{C}} (u = u')[\langle a, a' \rangle] \quad \text{iff} \quad a = a';$$
$$A \Vdash^{\mathbf{C}} (\alpha \wedge \beta)[a] \quad \text{iff} \quad A \Vdash^{\mathbf{C}} \alpha[a] \quad \text{and} \quad A \Vdash^{\mathbf{C}} \beta[a];$$
$$A \Vdash^{\mathbf{C}} (\alpha \vee \beta)[a] \quad \text{iff} \quad A \Vdash^{\mathbf{C}} \alpha[a] \quad \text{or} \quad A \Vdash^{\mathbf{C}} \beta[a];$$
$$A \Vdash^{\mathbf{C}} (\alpha \Rightarrow \beta)[a] \quad \text{iff for all} \quad B \xrightarrow{f} A,$$
$$B \Vdash^{\mathbf{C}} \alpha[(Ff)a] \quad \text{implies} \quad B \Vdash^{\mathbf{C}} \beta[(Ff)a];$$
$$A \Vdash^{\mathbf{C}} \neg\alpha[a] \quad \text{iff for no} \quad B \xrightarrow{f} A \quad \text{does} \quad B \Vdash^{\mathbf{C}} \alpha[(Ff)a].$$

If α contains an additional free variable v of type **G**, then

$$A \Vdash^{\mathbf{C}} \exists v \alpha[a] \quad \text{iff for some} \quad b \in GA, A \Vdash^{\mathbf{C}} \alpha[\langle a, b \rangle];$$
$$A \Vdash^{\mathbf{C}} \forall v \alpha[a] \quad \text{iff for all} \quad B \xrightarrow{f} A \quad \text{and all} \quad b \in GB,$$
$$B \Vdash^{\mathbf{C}} \alpha[\langle (Ff)a, b \rangle]. \quad \square$$

We note that the satisfaction of $\alpha \vee \beta$ and $\exists v \alpha$ in the sense of $\Vdash^{\mathbf{C}}$ at a given stage is determined by the satisfaction of α and β at the *same* stage. (This could also have been deduced from (4.43) and the readily established fact—which we leave to the reader as an exercise—that H^A is indecomposable and projective in **E(C)**.)

If P is a preordered or partially ordered set, the rules governing $\Vdash^{\mathbf{P}}$ are collectively known as *Kripke semantics*. They constitute a generalization to higher-order logic of the rules originally devised by Kripke for first-order intuitionistic logic. We leave to the reader the explicit formulation of the rules for $\Vdash^{\mathbf{P}}$ (which are immediate from (4.44)): for example,

$$p \Vdash^{\mathbf{P}} (\alpha \Rightarrow \beta)[a] \quad \text{iff} \quad \text{for all } q \leqslant p$$
$$q \Vdash^{\mathbf{P}} \alpha[a] \quad \text{implies} \quad q \Vdash^{\mathbf{P}} \beta[a]$$

and

$$\models_{E(P)} \alpha \quad \text{iff} \quad \text{for all } p \in P, \quad p \Vdash^{\mathbf{P}} \alpha.$$

E(P) is sometimes called a *Kripke model*.

EXERCISES

(1) Let P be a partially ordered set.

(i) Show that the following conditions are equivalent:

(a) $\models_{E(P)} \forall\omega(\omega \vee \neg\omega)$

(b) P is *discrete*, i.e. for all $p, q \in P$, $p \leqslant q$ implies $p = q$.

(ii) Show that the following conditions are equivalent:

(a) $\models_{E(P)} \forall\omega\, \forall\omega'[(\omega \Rightarrow \omega') \vee (\omega' \Rightarrow \omega)]$;

(b) for any $p \in P$, $\{q \in P : q \leqslant p\}$ is linearly ordered by $\leqslant$.

(2) Let N^* be the linearly ordered set of negative integers, and let α be the formula $\forall\omega(\omega \vee \neg\omega)$. Show that $\not\models_{E(N^*)} \neg\neg\alpha$ and deduce that $\not\vdash_{L_0} \neg\neg\alpha$.

5

From logic to sheaves

In this chapter we introduce the interderivable concepts of *truth set, modality* and *universal closure operation* in a local set theory S. Given a modality in S, we state what it means for an S-set to be a *sheaf* with respect to that modality, and proceed to show that the full subcategory of sheaves is a topos, and indeed a reflective subcategory of $\mathbf{C}(S)$. Upon interpreting the concept of a modality in a presheaf topos $\mathbf{Set}^{\mathbf{C}^{\mathrm{op}}}$, we obtain the concept of a *covering system* or *Grothendieck topology* on $\mathbf{C}$. The concept of sheaf with respect to a modality is then found to correspond to the concept, familiar from algebraic geometry, of sheaf with respect to a Grothendieck topology.

Truth sets, modalities, and universal closure operations

Given a theory S in a local language $\mathscr{L}$, the S-set $Tr = \{true\}$ represents the 'true' or 'assertable' formulae in S, since we evidently have

$$\vdash_S \alpha \in Tr \Leftrightarrow \alpha.$$

Tr clearly has the following properties:

$$\vdash_S true \in Tr$$

$$\alpha \vdash_S \beta \quad \text{implies} \quad \alpha \in Tr \vdash_S \beta \in Tr$$

$$(\alpha \in Tr) \in Tr \vdash_S \alpha \in Tr.$$

More generally, an S-set T such that $\vdash_S T \subseteq \Omega$ is called a *truth set* in S if it satisfies the same conditions, viz.

(tr_1) $\vdash_S true \in T$

(tr_2) $\alpha \vdash_S \beta$ implies $\alpha \in T \vdash_S \beta \in T$

(tr_3) $(\alpha \in T) \in T \vdash_S \alpha \in T.$

It is helpful to think of T as a set of *possibly true* or *almost true* formulae.

5.1 PROPOSITION. Let T be a truth set in S. Then

(i) $\alpha \in T,\ \beta \in T \vdash_S \alpha \wedge \beta \in T,$

(ii) $\alpha \in T,\ \alpha \Rightarrow \beta \in T \vdash_S \beta \in T,$

(iii) $\alpha \Rightarrow \beta \vdash_S \alpha \in T \Rightarrow \beta \in T.$

Proof. (i) Let us write α^* for $\alpha \in T$. Then from $\alpha \vdash_S \beta = \alpha \wedge \beta$ we deduce $\alpha \vdash \beta^* = (\alpha \wedge \beta)^*$, whence $\alpha \vdash_S \beta^* \Rightarrow (\alpha \wedge \beta)^*$, so that

$$\alpha, \beta^* \vdash_S (\alpha \wedge \beta)^*. \tag{1}$$

Replacing α by α^* in (1) gives

$$\alpha^*, \beta^* \vdash_S (\alpha^* \wedge \beta)^*, \tag{2}$$

and interchanging α and β in (1) yields

$$\alpha^*, \beta \vdash_S (\alpha \wedge \beta)^*. \tag{3}$$

(3) gives $\alpha^* \wedge \beta \vdash (\alpha \wedge \beta)^*$, so by ($\mathrm{tr}_2$) $(\alpha^* \wedge \beta)^* \vdash_S (\alpha \wedge \beta)^{**}$, whence $(\alpha^* \wedge \beta)^* \vdash_S (\alpha \wedge \beta)^*$ by (tr_3). This, together with (2), gives

$$\alpha^*, \beta^* \vdash_S (\alpha \wedge \beta)^*$$

as required.

(ii) follows easily from (i).

(iii) The following is a proof tree in S:

$$\dfrac{\alpha \in T,\ \alpha \Rightarrow \beta \in T : \beta \in T \qquad \dfrac{\dfrac{}{\alpha \Rightarrow \beta : (\alpha \Rightarrow \beta) = \mathit{true}} \qquad \dfrac{}{\mathit{true} \in T}}{\alpha \Rightarrow \beta : (\alpha \Rightarrow \beta) \in T}}{\alpha \Rightarrow \beta,\ \alpha \in T : \beta \in T}$$

Hence $\alpha \Rightarrow \beta,\ \alpha \in T \vdash_S \beta \in T$, whence (iii). □

Condition (iii) of this proposition asserts that T is a *filter* in Ω.

Let μ be a formula with exactly one free variable of type $\mathbf{\Omega}$. The formula μ is called a *modality* or *possibility operator* in S if it satisfies the following conditions:

(mod_1) $\quad \alpha \vdash_S \mu(\alpha)$

(mod_2) $\quad \alpha \vdash_S \beta \quad$ implies $\quad \mu(\alpha) \vdash_S \mu(\beta)$

(mod_3) $\quad \mu(\mu(\alpha)) \vdash_S \mu(\alpha)$,

for any formulae α, β. *We often write α^* for $\mu(\alpha)$.*

The proof of the next proposition is easy and is left to the reader.

5.2 PROPOSITION. Truth sets and modalities are equivalent in the following sense. For any truth set T, the formula $\mu_T = \omega \in T$ is a modality (called the modality *induced by* T) and, conversely, for any modality μ the set $T_\mu = \{\omega : \mu(\omega)\}$ is a truth set. Moreover,

$$\vdash_S T_{\mu_T} = T, \qquad \vdash_S \mu_{T_\mu} = \mu. \quad \square$$

5.3 COROLLARY. For any modality μ,

(i) $\vdash_S \mu(\alpha \wedge \beta) = \mu(\alpha) \wedge \mu(\beta)$,

(ii) $\alpha \Rightarrow \beta \vdash_S \mu(\alpha) \Rightarrow \mu(\beta)$,

(iii) $\mu(\alpha \Rightarrow \beta) \vdash_S \mu(\alpha) \Rightarrow \mu(\beta)$,
(iv) $\exists x \mu(\alpha) \vdash_S \mu(\exists x \alpha)$, whence

$$\mu(\exists x \mu(\alpha)) \vdash_S \mu(\exists x \alpha),$$

(v) $\mu(\forall x \alpha) \vdash_S \forall x \mu(\alpha)$,
where in (iv) and (v) x is free in α.

Proof. (i) follows from 5.1(i) and 5.2, and (ii) and (iii) are easy consequences of (i). As for (iv), we have $\alpha \vdash_S \exists x \alpha$, so $\mu(\alpha) \vdash_S \mu(\exists x \alpha)$, whence $\exists x \mu(\alpha) \vdash_S \mu(\exists x \alpha)$. (v) is proved similarly. □

A *modalized* (*local set*) *theory* in $\mathscr{L}$ is a pair (S, μ), where S is a local set theory in $\mathscr{L}$ and μ is a modality in S.

Truth sets and modalities are equivalent to a further notion, that of *universal closure operation*, which has a decidedly topological flavour.

Given a modality μ in S, write α^* for $\mu(\alpha)$ and for each S-set X define

$$\bar{X} = \{x : (x \in X)^*\}.$$

$\bar{X}$ is called the μ-*closure* of X.

More generally, for any term τ of power type we may define its μ-closure $\bar{\tau}$ by

$$\bar{\tau} = \{x : (x \in \tau)^*\}.$$

$\bar{\tau}$ is then a term of the same type as τ.

5.4 PROPOSITION. μ-closure has the following properties:

(i) $\vdash_S X \subseteq \bar{X}$,
(ii) $\vdash_S \bar{\bar{X}} = \bar{X}$,
(iii) $\vdash_S X \subseteq Y$ implies $\vdash_S \bar{X} \subseteq \bar{Y}$,
(iv) if $f : X \to Y$ and $\vdash_S U \subseteq Y$, then $\vdash_S f^{-1}[\bar{U}] = X \cap \overline{f^{-1}[U]}$, where we write $f^{-1}[U]$ for the term $\{x \in X : f(x) \in U\}$.

Proof. (i) We have

$$x \in X \vdash_S (x \in X)^* \vdash_S x \in \bar{X},$$

whence (i).

(ii) We have $\vdash_S \bar{X} \subseteq \bar{\bar{X}}$ from (i). Also

$$x \in \bar{\bar{X}} \vdash_S (x \in \bar{X})^* \vdash_S (x \in X)^{**} \vdash_S (x \in X)^* \vdash_S x \in \bar{X},$$

whence $\vdash_S \bar{\bar{X}} \subseteq X$ and (ii) follows.

(iii) If $\vdash_S X \subseteq Y$, then $x \in X \vdash_S x \in Y$, whence $(x \in X)^* \vdash_S (x \in Y)^*$, so that $x \in \bar{X} \vdash_S x \in \bar{Y}$, and (iii) follows.

(iv) If $f: X \to Y$ and $\vdash_S U \subseteq Y$, then†

$$x \in X \vdash_S x \in \overline{f^{-1}[U]} \Leftrightarrow (x \in f^{-1}[U])^*$$
$$\Leftrightarrow (f(x) \in U)^*$$
$$\Leftrightarrow f(x) \in \bar{U}$$
$$\Leftrightarrow x \in f^{-1}[\bar{U}],$$

whence (iv). □

Motivated by this proposition, we now define a *universal closure operation* in a local set theory S to be an assignment, to each S-set X, of an S-set $\bar{X}$ of the same type as X, satisfying

(clos$_1$) $\vdash_S X \subseteq \bar{X}$
(clos$_2$) $\vdash_S \bar{\bar{X}} = \bar{X}$
(clos$_3$) $\vdash_S X \subseteq Y$ implies $\vdash_S \bar{X} \subseteq \bar{Y}$
(clos$_4$) if $f: X \to Y$ and $\vdash_S U \subseteq Y$, then $\vdash_S f^{-1}[\bar{U}] = X \cap \overline{f^{-1}[U]}$.

Proposition 5.4 shows that each modality μ gives rise to a universal closure operation $\bar{\cdot}$ defined by

$$\bar{X} = \{x : \mu(x \in X)\}.$$

$\bar{\cdot}$ is called the universal closure operation *induced by* μ. We now show, conversely, that each universal closure operation induces a corresponding modality. To do this we require a

5.5 LEMMA. Let $\bar{\cdot}$ *be a universal closure operation in S.*

(i) If $f: X \to Y$ and $\vdash_S U \subseteq X$, then

$$\vdash_S f[\bar{U} \cap X] \subseteq \overline{f[U]}$$

where we write $f[U]$ for the term $\{f(x) : x \in U\}$.
(ii) $\vdash_S \overline{X \times Y} = \bar{X} \times \bar{Y}$ for any S-sets X, Y.
(iii) $\vdash_S \overline{X \cap Y} = \bar{X} \cap \bar{Y}$ for any S-sets X, Y of the same type.

Proof. (i) Assuming $f: X \to Y$ and $\vdash_S U \subseteq X$, we have

$$\vdash_S \bar{U} \cap X \subseteq \overline{f^{-1}[f[U]]} \cap X = f^{-1}[\overline{f[U]}]$$

so that

$$\vdash_S f[\bar{U} \cap X] \subseteq f[f^{-1}[\overline{f[U]}]] \subseteq \overline{f[U]}.$$

† Here and in the sequel we shall feel free to use evident informal notations for formulae in a local language. For example, $\Gamma \vdash_S \tau_1 = \tau_2 = \cdots = \tau_n$ will mean

$$\Gamma \vdash_S (\tau_1 = \tau_2) \wedge \cdots \wedge (\tau_{n-1} = \tau_n),$$

from which we automatically infer $\Gamma \vdash_S \tau_1 = \tau_n$.

(ii) Let X, Y be of types **PA**,**PB** respectively and let π_1, π_2 be the projections $A \times B \to A$, $A \times B \to B$ respectively†.
Then by (i)

$$\vdash_S \pi_1[\overline{X \times Y}] \subseteq \overline{\pi_1[X \times Y]} = \bar{X}$$

and similarly

$$\vdash_S \pi_2[\overline{X \times Y}] \subseteq \bar{Y}.$$

Hence

$$\vdash_S \overline{X \times Y} \subseteq \pi_1[\overline{X \times Y}] \times \pi_2[\overline{X \times Y}] \subseteq \bar{X} \times \bar{Y}. \tag{1}$$

For the reverse inclusion, we argue as follows. Let σ be the projection $A \times Y \to A$. Then

$$\begin{aligned}\vdash_S \bar{X} \times Y = \sigma^{-1}[\bar{X}] &= \overline{\sigma^{-1}[X]} \cap (A \times Y)\\ &\subseteq \overline{\sigma_1^{-1}[X]} = \overline{X \times Y}\end{aligned}$$

so that

$$\vdash_S \bar{X} \times Y \subseteq \overline{X \times Y}. \tag{2}$$

Similarly,

$$\vdash_S X \times \bar{Y} \subseteq \overline{X \times Y}. \tag{3}$$

Hence, replacing X by $\bar{X}$ in (3),

$$\vdash_S \bar{X} \times \bar{Y} \subseteq \overline{\bar{X} \times Y}. \tag{4}$$

But, by (2)

$$\vdash_S \overline{\bar{X} \times Y} \subseteq \overline{\overline{X \times Y}} = \overline{X \times Y}$$

This, together with (4), gives

$$\vdash \bar{X} \times \bar{Y} \subseteq \overline{X \times Y}$$

and so, using (1), (ii) follows.

(iii) We first show that, for any set X,

$$\vdash_S x \in \bar{X} \Leftrightarrow (x \in X) \in \overline{\{true\}}. \tag{5}$$

For let $\chi : A \to \Omega$ be the characteristic arrow of X (where X is of type **PA**). Then

$$\begin{aligned}\vdash_S \{x : (x \in X) \in \overline{\{true\}}\} &= \chi^{-1}[\overline{\{true\}}]\\ &= \overline{\chi^{-1}[\{true\}]}\\ &= \bar{X},\end{aligned}$$

and (5) follows.

† Here and in the sequel we shall write A for U_A as usual.

Now we have

$$\begin{aligned}\vdash_S x \in \bar{X} \cap \bar{Y} &\Leftrightarrow x \in \bar{X} \wedge x \in \bar{Y}\\ &\Leftrightarrow \langle x,x\rangle \in \bar{X} \times \bar{Y} = \overline{X \times Y} \qquad \text{(by (ii))}\\ &\Leftrightarrow (\langle x,x\rangle \in X \times Y) \in \overline{\{true\}} \qquad \text{(by (5))}\\ &\Leftrightarrow (x \in X \wedge x \in Y) \in \overline{\{true\}}\\ &\Leftrightarrow (x \in X \cap Y) \in \overline{\{true\}}\\ &\Leftrightarrow x \in \overline{X \cap Y} \qquad \text{(by (5))}, \quad \square\end{aligned}$$

giving (iii).

5.6 PROPOSITION. Let $\bar{\bullet}$ be a universal closure operation in a theory S. Then the formula

$$\mu(\omega) = \omega \in \overline{\{true\}}$$

is a modality in S, called the modality *induced by* $\bar{\bullet}$.

Proof. Write α^* for $\mu(\alpha)$ and define $j: \Omega \to \Omega$ by

$$j = \Omega \xrightarrow{\omega \mapsto \omega^*} \Omega.$$

Then

$$\vdash_S j^{-1}[\{true\}] = \overline{\{true\}}. \tag{1}$$

For

$$\begin{aligned}\vdash_S \omega \in j^{-1}[\{true\}] &\Leftrightarrow j(\omega) = true\\ &\Leftrightarrow \omega^* = true\\ &\Leftrightarrow \omega^*\\ &\Leftrightarrow \omega \in \overline{\{true\}}.\end{aligned}$$

Now we can verify that μ is a modality.

(i) $\alpha \vdash_S \alpha^*$. For

$$\alpha \vdash_S \alpha = true \vdash_S \alpha \in \overline{\{true\}} \vdash_S \alpha^*.$$

(ii) $\alpha \vdash_S \beta$ *implies* $\alpha^* \vdash_S \beta^*$. Let $\&: \Omega \times \Omega \to \Omega$ be the map

$$\Omega \times \Omega \xrightarrow{\langle \omega, \omega' \rangle \mapsto \omega \wedge \omega'} \Omega.$$

Now assume $\alpha \vdash_S \beta$. Then $\vdash_S \alpha = \alpha \wedge \beta$. Hence

$$\begin{aligned}\alpha^* \vdash_S \alpha \in \overline{\{true\}} \vdash_S \alpha \wedge \beta \in \overline{\{true\}}&\\ \vdash_S \langle \alpha, \beta \rangle \in \&^{-1}[\overline{\{true\}}]&\\ = \overline{\&^{-1}[\{true\}]}&\\ = \overline{\{true\} \times \{true\}}&\\ = \overline{\{true\}} \times \overline{\{true\}}& \qquad \text{(by 5.4(ii))}\\ \vdash_S \beta \in \overline{\{true\}} \vdash_S \beta^*.&\end{aligned}$$

Finally, using (5) of the proof of (5.5), we have

(iii) $\alpha^{**} \vdash_S (\alpha \in \overline{\{true\}}) \in \overline{\{true\}} \vdash_S \alpha \in \overline{\overline{\{true\}}} = \overline{\{true\}} \vdash_S \alpha^*$.

Thus μ is a modality. □

Using (5) in the proof of (5.5), it is now easy to show that the procedure of inducing a universal closure operation by a modality establishes a bijective correspondence between modalities and universal closure operations.

In (5.5) we showed that universal closure distributes over *intersection*. This is in sharp contrast with the Kuratowski closure operations familiar from topology which distribute over *union*. Nevertheless, the formal similarity in other respects between the two types of closure operation entitles us to extend some of the usual topological terminology and ideas to the context of universal closure operations. To emphasize this, we shall call a local set theory with a universal closure operation a *topologized* (*local set*) *theory*.

In sum, then, we have demonstrated the equivalence between truth sets, modalities, and universal closure operations.

We pause to consider some

Examples

(i) We have two *trivial* modalities **true** and **fix** given by

$$\mathbf{true}(\omega) = true \qquad \text{and} \qquad \mathbf{fix}(\omega) = \omega.$$

(ii) Given a sentence γ, define μ_γ by

$$\mu_\gamma(\omega) = \gamma \Rightarrow \omega.$$

It is easily verified that μ_γ is a modality; the corresponding truth set is given by

$$\{\omega : \gamma \Rightarrow \omega\},$$

the set of consequences of γ.

(iii) Given a sentence γ, define μ^γ by

$$\mu^\gamma(\omega) = \gamma \vee \omega.$$

Then μ^γ is a modality; the truth set is given by

$$\{\omega : \gamma \vee \omega\}.$$

(iv) It is easy to show that the formula $\neg\neg\omega$ is a modality; we denote it, naturally, by $\neg\neg$ and call it the *double negation modality*.

We now turn to the problem of *generating* truth sets.

Let u be a variable of type $\mathbf{P\Omega}$ and let Truth(u) be the formula

$$\mathit{true} \in u \wedge \forall\omega\,\forall\omega'[(\omega \Rightarrow \omega') \Rightarrow (\omega \in u \Rightarrow \omega' \in u)] \\ \wedge \forall\omega[(\omega \in u) \in u \Rightarrow \omega \in u].$$

Using (5.1), it is easy to see that, if $\vdash_S T \subseteq \Omega$, then T is a truth set in S if and only if $\vdash_S \mathrm{Truth}(T)$.

A *family of truth sets* is an S-set A of type $\mathbf{PP\Omega}$ such that

$$u \in A \vdash_S \mathrm{Truth}(u).$$

5.7 LEMMA. The intersection of a family of truth sets is a truth set.

Proof. Let A be a family of truth sets and let $T = \bigcap A$. Then T is a truth set:

(i) $\vdash_S \mathit{true} \in T$. This follows from $\vdash_S \forall u \in A\ (\mathit{true} \in u)$.
(ii) If $\alpha \vdash_S \beta$, then $\vdash_S \alpha \Rightarrow \beta$, so

$$\begin{aligned} \alpha \in T &\vdash_S \forall u \in A.\alpha \in u \\ &\vdash_S \forall u \in A\ (\alpha \in u \wedge \alpha \Rightarrow \beta) \\ &\vdash_S \forall u \in A.\beta \in u \\ &\vdash_S \beta \in T. \end{aligned}$$

(iii) We have

$$\begin{aligned} (\alpha \in T) \in T,\ u \in A &\vdash_S (\alpha \in T) \in u \\ &\vdash_S (\forall v \in A.\alpha \in v) \in u \\ &\vdash_S (\forall v \in A.\alpha \in v) \in u \wedge (\forall v \in A.\alpha \in v) \Rightarrow \alpha \in u \\ &\vdash_S (\alpha \in u) \in u \\ &\vdash_S \alpha \in u. \end{aligned}$$

Hence

$$(\alpha \in T) \in T \vdash_S \forall u \in A.\alpha \in u \vdash_S \alpha \in T. \quad \square$$

Given a set X such that $\vdash_S X \subseteq \Omega$, we may apply (5.7) to the family of truth sets containing X, i.e. to

$$A(X) = \{u : \mathrm{Truth}(u) \wedge X \subseteq u\}.$$

The set $\bigcap A(X)$ is then the *least* truth set containing X; we denote it by X^t and call it the truth set *generated* by X.

The connection with modalities is as follows. Let us call a modality μ *stricter* than a modality ν (and ν *laxer* than μ), and write $\mu \leqslant \nu$, if $\mu \vdash_S \nu$. This is obviously equivalent to the assertion that $\vdash_S \mathrm{Tr}_\mu \subseteq \mathrm{Tr}_\nu$. Given X such that $\vdash_S X \subseteq \Omega$, we write μ_X for the modality induced by the truth set

X^t. Then μ_X is evidently the strictest modality μ for which

$$\vdash_S \forall\omega \in X.\mu(\omega).$$

μ_X is called the modality *generated* by X.

Examples

(i) Given two modalities μ and ν, it is easy to see that $\mu \wedge \nu$ is the modality induced by the truth set $Tr_\mu \cap Tr_\nu$. So $\mu \wedge \nu$—the *meet* of μ and ν—is the laxest modality stricter than both μ and ν. Similarly, we can obtain the *join* $\mu + \nu$ of μ and ν as the modality generated by $Tr_\mu \cup Tr_\nu$ (Caution: $\mu + \nu$ is *not* the same as the disjunction $\mu \vee \nu$.) The modality $\mu + \nu$ is then the strictest modality laxer than both μ and ν. Thus the collection of modalities in a given theory forms a *lattice* under $\leqslant$ with least element **fix** and greatest element **true**.

(ii) *The forcing modality*. Given two sentences α and β, the modality $\mu^\alpha \wedge \mu_\beta$ is readily seen to be the strictest modality μ such that $\mu(\alpha) \vdash_S \mu(\beta)$, in other words, the strictest modality that 'forces' $\alpha \vdash \beta$ to be true.

EXERCISES

Given modalities μ, ν write $\mu \circ \nu$ for $\mu(\nu(\omega))$.

(1) Show that

(i) $\vdash_S \mu^\alpha + \mu^\beta = \mu^{\alpha\vee\beta}$	(ii) $\vdash_S \mu_\alpha + \mu_\beta = \mu_{\alpha\wedge\beta}$
(iii) $\vdash_S \mu^\alpha \wedge \mu^\beta = \mu^{\alpha\wedge\beta}$	(iv) $\vdash_S \mu_\alpha \wedge \mu_\beta = \mu_{\alpha\vee\beta}$
(v) $\vdash_S \mu^\alpha \wedge \mu_\alpha = \mathbf{fix}$	(vi) $\vdash_S \mu^\alpha + \mu_\alpha = \mathbf{true}$
(vii) $\vdash_S \mu^\alpha + \mu = \mu \circ \mu_\alpha$ for any modality μ	(viii) $\vdash_S \mu_\alpha + \mu = \mu_\alpha \circ \mu$ for any modality μ.

(2) Show that if μ, ν are modalities such that $\vdash_S \mu \circ \nu = \nu \circ \mu$, then $\mu \circ \nu$ is a modality and that $\vdash_S \mu \circ \nu = \mu + \nu$.

(3) Show that the following conditions on a modality μ in a local set theory S are equivalent:

(i) there is a sentence γ such that $\vdash_S \mu = \mu_\gamma$;
(ii) we have $\vdash_S \mu(\forall\omega(\mu(\omega) \Rightarrow \omega))$;
(iii) for any formulae α, β, we have

$$\mu(\alpha) \Rightarrow \mu(\beta) \vdash_S \mu(\alpha \Rightarrow \beta) \quad \text{and} \quad \forall x\mu(\alpha) \vdash_S \mu(\forall x\alpha).$$

(4) Show that the following conditions on a local set theory S are equivalent:

(i) S is classical;
(ii) for any modality μ there are sentences α, β such that $\vdash_S \mu = \mu_\alpha = \mu^\beta$.

Deduce that, if S is classical, the map $\alpha \mapsto \mu_\alpha$ is an isomorphism of the Boolean algebra $\Omega(S)$ with the lattice of modalities in S.

Sheaves

Throughout this section we suppose given a modalized theory (S,μ). We write α^* for $\mu(\alpha)$ and $\bar{\tau}$ for the μ-closure of a term τ of power type. Throughout, the term 'set' will be employed as an abbreviation of the term 'S-set'.

We make the following definitions by analogy with the standard notions of general topology. Let X, Y be sets of the same type. X is said to be (μ)-*closed* in Y if

$$\vdash_S X = \bar{X} \cap Y,$$

and (μ)-*dense* in Y if

$$\vdash_S X \subseteq Y \wedge Y \subseteq \bar{X}.$$

Clearly X is closed in Y iff

$$\vdash_S x \in X \Leftrightarrow (x \in X)^* \wedge x \in Y.$$

Notice also that if $\vdash_S X \subseteq Y$ then X is dense in $\bar{X} \cap Y$, and if $\vdash_S \bar{X} \subseteq Y$ then $\bar{X}$ is closed in Y.

A monic arrow $X \overset{m}{\rightarrowtail} Y$ is called *closed* or *dense*, and X a *closed* or *dense subobject* of Y if the image of X

$$m[X] = \{y \in Y : \exists x \in X . y = m(x)\}$$

is closed or dense in Y.

EXERCISE. Show that $\neg\neg$ is the laxest modality μ for which $0 \rightarrowtail 1$ is μ-closed.

A set X is said to be *separated* if the diagonal $\Delta_X = \{\langle x,x \rangle : x \in X\}$ is closed in $X \times X$. This corresponds to the *Hausdorff* condition on a topological space.

5.8 PROPOSITION. The following conditions on a set X are equivalent.

(i) X is separated;

(ii) $x \in X,\ x' \in X \vdash_S (x = x')^* \Leftrightarrow (x = x')$;

(iii) $x \in X \vdash_S \{x\} = \overline{\{x\}} \cap X$ (i.e., 'points of X are closed'—this corresponds to the T_1-condition on a topological space);

(iv) If $U \overset{m}{\rightarrowtail} V$ is a dense monic and $f,g : V \to X$ are such that $f \circ m = g \circ m$, then $f = g$.

Proof. (i)$\Rightarrow$(ii). Suppose X is separated. Then

$$\begin{aligned} x \in X,\ x' \in X,\ (x = x')^* &\vdash_S (\langle x,x'\rangle \in \Delta_X)^* \\ &\vdash_S \langle x,x'\rangle \in \overline{\Delta_X} \\ &\vdash_S \langle x,x'\rangle \in \Delta_X \\ &\vdash_S x = x'. \end{aligned}$$

(ii)$\Leftrightarrow$(iii) is easy, using the observation that

$$\vdash_S x = x' \Leftrightarrow x \in \{x'\}.$$

(ii)$\Rightarrow$(iv). Suppose that (ii) and the premises of (iv) hold. Then

$$\begin{aligned} y \in V &\vdash_S y \in \overline{m[U]} \\ &\vdash_S (y \in m[U])^* \\ &\vdash_S (\exists x \in X. y = m(x))^* \\ &\vdash_S (\exists x \in X. f(y) = f(m(x)) = g(m(x)) = g(y))^* \\ &\vdash_S (f(y) = g(y))^* \\ &\vdash_S f(y) = g(y), \end{aligned}$$

whence $f = g$.

(iv)$\Rightarrow$(i). Assume (iv) and let σ_1, σ_2 be the restrictions to $\overline{\Delta_X} \cap (X \times X)$ of the projections $X \times X \to X$. These coincide on Δ_X and hence on $\overline{\Delta_X} \cap (X \times X)$ by (iv). It immediately follows that $\vdash_S \overline{\Delta_X} \cap X \times X \subseteq \Delta_X$, so that Δ_X is closed in $X \times X$, i.e. X is separated. □

5.9 PROPOSITION

(i) Any subobject of a separated object is separated.
(ii) If X and Y are separated, so is $X \times Y$.
(iii) If Y is separated and $f,g: X \to Y$, then $\{x \in X : f(x) = g(x)\}$ is closed in X.

Proof. (i) Suppose that Y is separated and $m: X \rightarrowtail Y$ monic. Then we have

$$\begin{aligned} x \in X,\ x' \in X &\vdash_S x = x' \Rightarrow m(x) = m(x') \\ &\vdash_S (x = x')^* \Rightarrow (m(x) = m(x'))^* \\ &\vdash_S (x = x')^* \Rightarrow (m(x) = m(x')) \\ &\vdash_S (x = x')^* \Rightarrow x = x', \end{aligned}$$

so X is separated.

(ii) Suppose that X and Y are separated. Then

$$\begin{aligned}\langle x,y\rangle \in X\times Y, \langle x',y'\rangle \in X\times Y,\ (\langle x,y\rangle) = \langle x',y'\rangle)^* \\ \vdash_S (x=x' \wedge y=y')^* \\ \vdash_S (x=x')^* \wedge (y=y')^* \\ \vdash_S x=x' \wedge y=y' \\ \vdash_S \langle x,y\rangle = \langle x',y'\rangle,\end{aligned}$$

so $X\times Y$ is separated.

(iii) Assuming the premises, let $U=\{x\in X: f(x)=g(x)\}$. Then we have

$$x\in \bar{U}\cap X \vdash_S (f(x)=g(x))^* \vdash_S f(x)=g(x) \vdash_S x\in U,$$

so $\vdash_S \bar{U}\cap X \subseteq U$ and U is closed in X.

Now we define

$$\Omega^* = \{\omega : \omega = \omega^*\}.$$

Under the isomorphism $\Omega \cong P1$ in $\mathbf{C}(S)$ (cf. (4.9)) it is easy to check that Ω^* corresponds to the set of *closed* subsets of 1.

5.10 THEOREM. Ω^* is separated.

Proof. We have

$$\begin{aligned}\omega\in\Omega^*,\ \omega'\in\Omega^*,\ (\omega=\omega')^* \\ \vdash_S \omega=\omega^* \wedge \omega'=\omega'^* \wedge (\omega=\omega')^* \\ \vdash_S \omega=\omega^* \wedge \omega'=\omega'^* \wedge \omega^*=\omega'^* \\ \vdash_S \omega=\omega'.\quad \square\end{aligned}$$

5.11 PROPOSITION. Ω^* classifies closed subobjects. That is, given a closed monic $X \overset{m}{\rightarrowtail} Y$ there is a unique $\chi: Y\to\Omega^*$ such that $m[X]=\chi^{-1}[\{true\}]$ and conversely, given $Y\xrightarrow{h}\Omega^*$, $X=h^{-1}[\{true\}]$ is closed in Y.

Proof. Suppose that $X \overset{m}{\rightarrowtail} Y$ is a closed monic. Define $\chi: Y\to\Omega$ by

$$\chi = (y \mapsto y\in m[X]).$$

Then

$$\begin{aligned}y\in Y \vdash_S \chi(y)^* = (y\in m[X])^* &\Leftrightarrow y\in \overline{m[X]} \\ &\Leftrightarrow y\in m[X] \\ &= \chi(y).\end{aligned}$$

Hence $\chi: Y\to\Omega^*$. Also

$$\begin{aligned}y\in Y \vdash_S y\in\chi^{-1}[\{true\}] &\Leftrightarrow \chi(y)=true \\ &\Leftrightarrow y\in m[X],\end{aligned}$$

so that $m[X]=\chi^{-1}[\{true\}]$. If $\xi:Y\to\Omega^*$ satisfies $m[X]=\xi^{-1}[\{true\}]$, then

$$y\in Y\vdash_S \xi(y)=true\Leftrightarrow y\in m[X]\Leftrightarrow \chi(y)=true,$$

so that $y\in Y\vdash_S \xi(y)=(y)$ and so $\xi=\chi$.

Conversely, let $h:Y\to\Omega^*$ and put $X=h^{-1}[\{true\}]$. Then

$$\begin{aligned} x\in\bar{X}\cap Y \vdash_S x\in \overline{h^{-1}[\{true\}]} \\ \vdash_S (x\in h^{-1}[\{true\}])^* \\ \vdash_S (h(x)=true)^* \\ \vdash_S h(x)^*\vdash_S h(x)\vdash_S h(x)=true \\ \vdash_S x\in h^{-1}[\{true\}]\vdash_S x\in X \end{aligned}$$

and so X is closed. □

Now for any set X define

$$P^*(X)=\{u\subseteq X:\bar{u}\cap X=u\}.$$

Then $P^*(X)$ is the set of closed subsets of X.

5.12 PROPOSITION. $P^*(X)$ is separated.

Proof. We have

$$\begin{aligned} &u\in P^*(x),\ v\in P^*(X),\ (u=v)^* \\ &\quad\vdash_S [\forall x\in X\ (x\in u\Leftrightarrow x\in v)]^*\wedge u=\bar{u}\cap X\wedge v=\bar{v}\cap X \\ &\quad\vdash_S \forall x\in X\ (x\in u\Leftrightarrow x\in v)^*\wedge u=\bar{u}\cap X\wedge v=\bar{v}\cap X \\ &\quad\vdash_S \forall x\in X\ [(x\in u)^*\Leftrightarrow (x\in v)^*]\wedge u=\bar{u}\cap X\wedge v=\bar{v}\cap X \\ &\quad\vdash_S \forall x\in X\ [x\in\bar{u}\Leftrightarrow x\in\bar{v}]\wedge u=\bar{u}\cap X\wedge v=\bar{v}\cap X \\ &\quad\vdash_S \bar{u}\cap X=\bar{v}\cap X\wedge u=\bar{u}\cap X\wedge v=\bar{v}\cap X \\ &\quad\vdash_S u=v.\quad \square \end{aligned}$$

We now make some crucial definitions. A set F is said to be *absolutely closed* if it is separated and whenever it is a subobject of a separated set it is a *closed* subobject: that is, any monic $F\rightarrowtail X$ to a separated set X is closed. F is said to be a (μ)-*sheaf* if any map f into F has a *unique* extension to any set of which dom(f) is a dense subobject, i.e., given $X\xrightarrow{f}F$ and a dense monic $X\overset{m}{\rightarrowtail}Y$, there is a unique map $Y\xrightarrow{g}F$ such that $f=g\circ m$.

Note the analogy between absolutely closed sets and compact Hausdorff spaces: the latter are well-known to be closed subsets of any containing Hausdorff space.

We are ultimately going to show that the *category of μ-sheaves is a* *topos*. The first step in this process is the following theorem, which, *inter alia,* asserts that sheafhood and absolute closedness amount to the same thing.

5.13 THEOREM. The following are equivalent for any (S-)set F.

(i) F is a sheaf.
(ii) F is absolutely closed.
(iii) For any formula $\alpha(x)$, we have

$$(\exists!\, x \in F.\alpha)^* \vdash_S \exists! x \in F.\alpha^*$$

('possible unique existence in F implies actual unique existence in F').

(iv) F is separated and for any formula $\alpha(x)$,

$$(\exists! x \in F.\alpha)^* \vdash_S \exists x \in F.\alpha^*.$$

Proof. (i)$\Rightarrow$(ii). By (5.8), any sheaf is separated. Let F be a sheaf and let $F \overset{m}{\rightarrowtail} X$ be a monic into a separated set X. Let $G = m[F]$; then $\bar{G} \cap X$, as a subset of X, is separated. Since F is a sheaf, and m a dense monic into $\bar{G} \cap X$, there is a map $f : \bar{G} \cap X \to F$ such that $f \circ m = 1_F$. Then

$$\begin{aligned} y \in \bar{G} \cap X \vdash_S &(y \in G)^* \\ \vdash_S &(\exists x \in F.m(x) = y)^* \\ \vdash_S &[\exists x \in F(m(x) = y \wedge f(y) = f(m(x)) = x)]^* \\ \vdash_S &(y = m(f(y)))^* \\ \vdash_S &y \in G. \end{aligned}$$

Therefore $\vdash_S G = \bar{G} \cap X$, G is closed, and (ii) follows.

(ii)$\Rightarrow$(iii). Assuming (ii), let

$$\begin{aligned} X &= \{u \in P(F) : \exists! x.x \in u\} \\ &= \{u \in P(F) : \exists x \in F.u = \{x\}\}. \end{aligned}$$

Since F is separated,

$$\begin{aligned} u \in X \vdash_S &\exists x \in F[u = \{x\} = \overline{\{x\}} \cap F] \\ \vdash_S &u = \bar{u} \cap F. \end{aligned} \tag{1}$$

It follows that

$$u \in X \vdash_S u \in P^*(F)$$

so that $\vdash_S X \subseteq P^*(F)$. Since $P^*(F)$ is separated and $x \mapsto \{x\}$ is an isomorphism $F \to X$, X is closed in $P^*(F)$.

Now, for any formula α, put $U = \{x : \alpha\}$. Then we have, using (1),

$$\exists! x \in F.\alpha \vdash_S U \in X \vdash_S \bar{U} \cap F \in X$$
$$\vdash_S \exists! x \in F. x \in U \vdash_S \exists! x \in F.\alpha^*.$$

Hence, using the fact that X is closed in $P^*(F)$,

$$(\exists! x \in F.\alpha)^* \vdash_S (\exists! x \in F.\alpha^*)^* \vdash_S (\exists! x \in F. x \in \bar{U})^*$$
$$\vdash_S (\bar{U} \cap F \in X)^* \vdash_S \bar{U} \cap F \in \bar{X} = X \cap P^*(F) \vdash_S \exists! x \in F.\alpha^*.$$

This gives (iii).

(iii)$\Leftrightarrow$(iv). Assume (iii). To obtain (iv) it suffices to show that F is separated. To do this we verify (5.8)(iii). Let α be the formula $x \in \{x'\}$. We have $x' \in F \vdash_S \exists! x \in F.\alpha$, so $x' \in F \vdash_S (\exists! x \in F.\alpha)^*$, whence $x' \in F \vdash_S \exists! x \in F.\alpha^*$. That is,

$$x' \in F \vdash_S \exists! x \in F(x \in \{x'\})^*,$$

or

$$x' \in F \vdash_S \exists! x \in F. x \in \overline{\{x'\}}$$

so that $x' \in F \vdash_S \{x'\} = \{x'\} \cap F$. This is (5.8)(iii).

Conversely, assume (iv). Then to obtain (iii) it suffices to show that

$$(\exists! x \in F.\alpha)^*, \exists x \in F.\alpha^* \vdash_S \exists! x \in F.\alpha^*.$$

And to establish this we need only observe that

$$x \in F,\ y \in F,\ \alpha^*(x),\ \alpha^*(y),\ (\exists! x \in F.\alpha)^*$$
$$\vdash_S [x \in F \wedge y \in F \wedge \alpha(x) \wedge \alpha(y) \wedge \exists! x \in F.\alpha]^*$$
$$\vdash_S (x = y)^* = \vdash_S x = y,$$

since F is separated.

(iii)$\Rightarrow$(i). Assuming (iii), we know from (iii)$\Rightarrow$(iv) that F is separated. Now let $X \overset{m}{\rightarrowtail} Y$ be a dense monic and $X \xrightarrow{f} F$. Define

$$g = \{\langle y, w\rangle \in Y \times F : [\exists! x \in X(y = m(x) \wedge w = f(x))]^*\}$$

We claim that $g : Y \to F$. For

$$y \in Y \vdash_S y \in \overline{m[X]} \vdash_S (y \in m[X])^*$$
$$\vdash_S [\exists! x \in X. y = m(x)]^* \qquad \text{(since } m \text{ is monic)}$$
$$\vdash_S [\exists! w \in F\ \exists! x \in X(y = m(x) \wedge w = f(x))]^*$$
$$\vdash_S \exists! w \in F[\exists! x \in X(y = m(x) \wedge w = f(x))]^*$$
$$\vdash_S \exists! w \in F. \langle y, w\rangle \in g.$$

And $f = g \circ m$, for

$$
\begin{aligned}
z \in X, w \in F, \langle m(z), w \rangle \in g &\\
\vdash_S [\exists! x \in X(m(z) = m(x) \wedge w = f(x))]^* &\\
\vdash_S [\exists! x \in X\ (z = x \wedge w = f(x))]^* &\\
\vdash_S (w = f(z))^* &\\
\vdash_S w = f(z) \qquad \text{(since } F \text{ is separated).} &
\end{aligned}
$$

Finally, g is unique since F is separated. This establishes (iv) and completes the proof. □

5.14 COROLLARY. A closed subobject of a sheaf is a sheaf.

Proof. Let F be a sheaf and $X \overset{m}{\rightarrowtail} F$ a closed monic. Let $n : m[X] \to X$ be the inverse of m. Then for any formula $\alpha(x)$ not containing y free we have

$$
\begin{aligned}
\exists! x \in X. \alpha(x) &\vdash_S [\exists! y \in F(y \in m[X] \wedge \alpha(n(y))]^* \\
&\vdash_S \exists! y \in F[(y \in m[x])^* \wedge \alpha^*(n(y))] \\
&\vdash_S \exists! y \in F[y \in \overline{m[X]} \wedge \alpha^*(n(y))] \\
&\vdash_S \exists! y[y \in \overline{m[X]} \cap F \wedge \alpha^*(n(y))] \\
&\vdash_S \exists! y[y \in m[X] \wedge \alpha^*(n(y))] \\
&\vdash_S \exists! y[\exists! x \in X. y = m(x) \wedge \alpha^*(n(y))] \\
&\vdash_S \exists! y\ \exists! x \in X. \alpha^*(x) \\
&\vdash_S \exists! x \in X. \alpha^*(x).
\end{aligned}
$$

So by (5.13)(iii), X is a sheaf. □

5.15 COROLLARY. If X and Y are sheaves, so is $X \times Y$.

Proof. Assuming that X and Y are sheaves, we verify (5.13)(iii) for $X \times Y$. We have

$$
\begin{aligned}
(\exists! u \in X \times Y. \alpha)^* &\vdash_S (\exists! x \in X\ \exists! y \in Y. \alpha(\langle x, y \rangle))^* \\
&\vdash_S \exists! x \in X\ \exists! y \in Y. \alpha^*(\langle x, y \rangle) \\
&\vdash_S \exists! u \in X \times Y. \alpha^*. \quad \square
\end{aligned}
$$

5.16 COROLLARY. Ω^* is a sheaf.

Proof. We verify (5.13)(iv) for Ω^*. Given a formula α, let β be the formula $(\omega = \omega^*) \wedge \alpha$ and let γ be $\beta(\omega/\mathit{true})$. Then, using (3.8),

$$
\exists! \omega \in \Omega^*. \alpha \vdash_S \exists! \omega \beta \vdash_S \beta(\omega/\gamma) \vdash_S \gamma^* = \gamma \wedge \alpha(\omega/\gamma) \vdash_S \alpha(\omega/\gamma^*).
$$

Hence

$$(\exists! \omega \in \Omega^*.\alpha)^* \vdash_S \alpha^*(\omega/\gamma^*) \vdash_S \exists \omega \in \Omega^*.\alpha^*.$$

Since Ω^* is separated, the result follows. □

5.17 COROLLARY. For any set X, $P^*(X)$ is a sheaf.

Proof. We have already shown (5.12) that $P^*(X)$ is separated. For any formula α let

$$V = \{x \in X : [\exists u \in P^*(X)(\alpha \wedge x \in u)]^*\}.$$

Clearly $\vdash_S V \in P^*(X)$. Also, writing β for the formula $u \in P^*(X) \wedge \alpha$, we have

$$\begin{aligned} \beta(u/v), x \in v \vdash_S & \exists u(\beta \wedge x \in u) \\ & \vdash_S [\exists u(\beta \wedge x \in u)]^* \\ & \vdash_S x \in V, \end{aligned}$$

so that

$$\beta(u/v) \vdash_S v \subseteq V. \tag{1}$$

Moreover,

$$\begin{aligned} & \exists! u\beta, \beta(u/v), x \in V \\ & \quad \vdash_S \exists! u\beta \wedge [\exists u(\beta \wedge x \in u)]^* \wedge \beta(u/v) \\ & \quad \vdash_S (\exists! u\beta)^* \wedge [\exists u(\beta \wedge x \in u)]^* \wedge \beta^*(u/v) \wedge \beta(u/v) \\ & \quad \vdash_S [\exists! u\beta \wedge \exists u(\beta \wedge x \in u) \wedge \beta(u/v)]^* \wedge \beta(u/v) \\ & \quad \vdash_S (x \in v)^* \wedge v \in P^*(x) \\ & \quad \vdash_S x \in \bar{v} \wedge \bar{v} \cap X = v \\ & \quad \vdash_S x \in v \end{aligned}$$

This, together with (1), gives

$$\exists! u\beta, \beta(u/v) \vdash_S v = V,$$

so that

$$\exists! u\beta \vdash_S \beta(u/V).$$

Therefore

$$(\exists! u\beta)^* \vdash_S \beta^*(u/V),$$

whence

$$\begin{aligned} (\exists! u \in P^*(X).\alpha)^* \vdash_S & \beta^*(u/V) \\ & \vdash_S \exists u \in P^*(x).\beta^*. \\ & \vdash_S \exists u \in P^*(x).\alpha^*. \end{aligned}$$

Thus $P^*(X)$ satisfies (5.13)(iv) and is accordingly a sheaf. □

Let $\mathbf{Sh}_\mu(S)$ or simply $\mathbf{Sh}(S)$ be the full subcategory of $\mathbf{C}(S)$ whose objects are the μ-sheaves. We can now finally prove the central

5.18 THEOREM. $\mathbf{Sh}(S)$ is a topos.

Proof. 1 is easily seen to be a sheaf so it is the terminal object in $\mathbf{Sh}(S)$.

$\mathbf{Sh}(S)$ has binary products by (5.15).

Since Ω^* is a sheaf by (5.16) and by (5.11) classifies closed subobjects, the fact that sheaves are absolutely closed (5.13) enables one to show that Ω^* is a subobject classifier in $\mathbf{Sh}(S)$ in exactly the same way as Ω was shown to be one in $\mathbf{C}(S)$ (3.16.6).

Finally, for any sheaf X, $P^*(X)$ is a sheaf by (5.17); we show that it is a power object for X in $\mathbf{Sh}(S)$. For define $X \times P^*(X) \xrightarrow{e_X} \Omega$ by

$$e_X = (\langle x,u\rangle \mapsto x \in u).$$

Then

$$x \in X, u \in P^*(x), (x \in u)^* \vdash_S x \in \bar{u} \cap X \vdash_S x \in u,$$

so $X \times P^*(X) \xrightarrow{e_X} \Omega^*$. Given $X \times Y \xrightarrow{f} \Omega^*$, define $Y \xrightarrow{\hat{f}} PX$ by

$$\hat{f} = (y \mapsto \{x \in X : f(\langle x,y\rangle)\}).$$

Note that

$$\begin{aligned} y \in Y \vdash_S X \cap \overline{\hat{f}(y)} &= \{x \in X : f(\langle x,y\rangle)^*\} \\ &= \{x \in X : f(\langle x,y\rangle)\} \\ &= \hat{f}(y), \end{aligned}$$

so that $y \in Y \vdash_S \hat{f}(y) \in P^*(X)$. Hence $Y \xrightarrow{\hat{f}} P^*(X)$. Then

$$\begin{aligned} x \in X, y \in Y \vdash_S e_X((1 \times \hat{f})(\langle x,y\rangle)) &= e_X(\langle x, \hat{f}(y)\rangle) \\ &= (x \in \hat{f}(y)) = f(\langle x,y\rangle). \end{aligned}$$

so

$$e_X \circ (1 \times \hat{f}) = f.$$

If $e_X \circ (1 \times g) = f$, then

$$x \in X, y \in Y \vdash_S f(\langle x,y\rangle) = e_X(\langle x, g(y)\rangle = (x \in g(y)),$$

so

$$x \in X, y \in Y \vdash_S x \in \hat{f}(y) \Leftrightarrow f(\langle x,y\rangle) \Leftrightarrow x \in g(y).$$

Hence

$$y \in Y \vdash_S \hat{f}(y) = g(y), \text{ whence } g = \hat{f}.$$

We have thus verified all the conditions for $P^*(X)$ to be a power object for X in $\mathbf{Sh}(S)$, and the proof that the latter is a topos is accordingly complete. □

The sheafification functor

Having shown that $\mathbf{Sh}(S)$ is a topos, we are next going to establish that it is *reflective* in $\mathbf{C}(S)$.

For each (S-)set X let

$$[x] = \overline{\{x\}} \cap X$$

and write $l_X : X \to P^*(X)$ for the map $(x \mapsto [x])$. We shall frequently write l for l_X.

We note that, since $P^*(X)$ is separated, so is the image $l[X]$ of X.

5.19 LEMMA.

$$x \in X, y \in X \vdash_S l(x) = l(y) \Leftrightarrow (x = y)^*.$$

Proof. We have

$$\begin{aligned} x \in X, y \in X \vdash_S [x] = [y] &\Leftrightarrow x \in \overline{\{y\}} \wedge y \in \overline{\{x\}} \\ &\Leftrightarrow (x = y)^* \wedge (y = x)^* \\ &\Leftrightarrow (x = y)^*. \quad \square \end{aligned}$$

A map $f : X \to Y$ is called *almost monic* if

$$x \in X, x' \in X \vdash_S f(x) = f(x') \Rightarrow (x = x')^*.$$

Clearly, if X is separated, then any almost monic map with domain X is monic. Hence

5.20 COROLLARY. l_X is almost monic. If X is separated, then l_X is monic, so X is a subobject of $P^*(X)$. □

Also,

5.21 COROLLARY. The map $X \xrightarrow{l_X} l[X]$ is universal among maps from X to separated objects. That is, for any map $X \xrightarrow{f} F$ to a separated object F, there is a unique map $l[X] \xrightarrow{g} F$ such that the diagram

$$\begin{array}{ccc} X & \xrightarrow{f} & F \\ {\scriptstyle l}\downarrow & \nearrow{\scriptstyle g} & \\ l[X] & & \end{array}$$

commutes. Moreover, g is monic iff f is almost monic.

Proof. Given $X \xrightarrow{f} F$ to be a separated F, the uniqueness condition on $l[X] \xrightarrow{g} F$ is evident since $X \xrightarrow{l} l[X]$ is epic.

As for the existence of g, define

$$g = \{\langle u,y\rangle \in l[X] \times F : \exists x \in X[u = l(x) \wedge y = f(x)].$$

Then

$$\begin{aligned} u \in l[X] &\vdash_S \exists x \in X. u = l(x) \\ &\vdash_S \exists x \in X\ (u = l(x) \wedge \exists y \in F. y = f(x)) \\ &\vdash_S \exists y \in F\ \exists x \in X\ (u = l(x) \wedge y = f(x)) \\ &\vdash_S \exists y \in F. \langle u,y\rangle \in g \end{aligned}$$

and

$$\begin{aligned} &\langle u,y\rangle \in g, \langle u,y'\rangle \in g \\ &\quad \vdash_S \exists x \in X\ \exists x \in X\ [u = l(x) \wedge y = f(x) \wedge u = l(x') \wedge y' = f(x')] \\ &\quad \vdash_S \exists x \in X\ \exists x' \in X\ [l(x) = l(x') \wedge y = f(x) \wedge y' = f(x')] \\ &\quad \vdash_S \exists x \in X\ \exists x' \in X\ [(x = x')^* \wedge y = f(x) \wedge y' = f(x')] \qquad \text{(by 5.19)} \\ &\quad \vdash_S \exists x \in X\ \exists x' \in X\ [x = x' \wedge y = f(x) \wedge y' = f(x')]^* \\ &\quad \vdash_S (y = y')^* \\ &\quad \vdash_S y = y', \end{aligned}$$

since F is separated. Thus g is a map $g : l[X] \to F$; and evidently $f = g \circ l$.

Finally, if g is monic, then f, as the composition of a monic with the almost monic l, is easily seen to be almost monic. Conversely, if f is almost monic, then

$$\begin{aligned} &u \in l[X],\ v \in l[X], g(u) = g(v) \\ &\quad \vdash_S \exists x \in X\ \exists x' \in X\ [g(u) = g(v) \wedge u = l(x) \wedge v = l(x')] \\ &\quad \vdash_S \exists x \in X\ \exists x' \in X\ [g(l(x)) = g(l(x')) \wedge u = l(x) \wedge v = l(x')] \\ &\quad \vdash_S \exists x \in X\ \exists x' \in X\ [f(x) = f(x') \wedge u = l(x) \wedge v = l(x')] \\ &\quad \vdash_S \exists x \in X\ \exists x' \in X\ [(x = x')^* \wedge u = l(x) \wedge v = l(x')] \\ &\quad \vdash_S \exists x \in X\ \exists x' \in X\ [l(x) = l(x') \wedge u = l(x) \wedge v = l(x')] \\ &\quad \vdash_S u = v, \end{aligned}$$

so that g is monic. □

It follows from (5.21) that $X \mapsto l[X]$ is the object function of a functor left adjoint to the insertion functor from the full subcategory of separated objects into $\mathbf{C}(S)$.

For each set X define

$$\tilde{X} = \overline{l[X]} \cap P^*(X).$$

Then $l_X: X \to \tilde{X}$, and $\tilde{X}$, as a closed subset of the sheaf $P^*(X)$, is itself a sheaf.

We now show that $X \mapsto \tilde{X}: \mathbf{C}(S) \to \mathbf{Sh}(S)$ is the object function of a functor left adjoint to the insertion functor $\mathbf{Sh}(S) \hookrightarrow \mathbf{C}(S)$, so that $\mathbf{Sh}(S)$ is reflective in $\mathbf{C}(S)$. To prove this we need only establish the

5.22 THEOREM. The map $X \xrightarrow{l_X} \tilde{X}$ is universal among maps from X to sheaves. That is, for any sheaf F and any $f: X \to F$ there is a unique map $f^0: \tilde{X} \to F$ such that the diagram

$$\begin{array}{ccc} X & \xrightarrow{f} & F \\ {\scriptstyle l}\downarrow & \nearrow_{f^0} & \\ \tilde{X} & & \end{array}$$

commutes.

Proof. By (5.21), there is a unique map $g: l[X] \to F$ such that the diagram

$$\begin{array}{ccc} X & \xrightarrow{f} & F \\ {\scriptstyle l}\downarrow & \nearrow_{g} & \\ l[X] & & \end{array}$$

commutes. Now the insertion map $l[X] \hookrightarrow \tilde{X}$ is by definition a dense monic and so, since F is a sheaf, there is a unique map $f^0: \tilde{X} \hookrightarrow F$ such that the diagram

$$\begin{array}{ccc} l[X] & \xrightarrow{g} & F \\ \downarrow & \nearrow_{f^0} & \\ \tilde{X} & & \end{array}$$

commutes. It is easy to see that f^0 is the required unique map. □

5.23 COROLLARY.

(i) Let F be a sheaf and $X \xrightarrow{f} F$ an almost monic map. Then

$$\tilde{X} \cong \overline{f[X]} \cap F.$$

(ii) If $\vdash_S U \subseteq X$, then

$$\tilde{U} \cong \overline{l_X[U]} \cap \tilde{X}.$$

Proof. (i) Let $G = \overline{f[X]} \cap F$. Then G, as a closed subset of the sheaf F, is itself a sheaf. By (5.21), there is a monic map $l[X] \overset{g}{\rightarrowtail} G$ such that the diagram

$$\begin{array}{ccc} X & \xrightarrow{f} & G \\ {\scriptstyle l}\downarrow & \nearrow_{g} & \\ l[X] & & \end{array}$$

commutes. Since G is a sheaf, there is a (unique) map $\tilde{X} \xrightarrow{h} G$ such that the diagram

$$\begin{array}{ccc} l[X] & \xrightarrow{g} & G \\ {\scriptstyle i}\downarrow & \nearrow_{h} & \\ \tilde{X} & & \end{array}$$

commutes, where i is the insertion map. We claim that h is an isomorphism. To prove this we construct its inverse.

Since $l[X] \overset{g}{\rightarrowtail} G$ is a dense monic and $\tilde{X}$ is a sheaf, there is a (unique) $G \xrightarrow{k} \tilde{X}$ such that the diagram

$$\begin{array}{ccc} l[X] & \xrightarrow{i} & \tilde{X} \\ {\scriptstyle g}\downarrow & \nearrow_{k} & \\ G & & \end{array}$$

commutes. It is now easy to check that the diagrams

$$\begin{array}{ccc} l[X] & \longrightarrow & \tilde{X} \\ {\scriptstyle i}\downarrow & \nearrow_{k \circ h} & \\ \tilde{X} & & \end{array}$$

and

$$\begin{array}{ccc} l[X] & \xrightarrow{g} & G \\ {\scriptstyle g}\downarrow & \nearrow_{h \circ k} & \\ G & & \end{array}$$

commute. Since i and g are dense monics and $\tilde{X}$ and G are sheaves, the maps $k \circ h$ and $h \circ k$ are the unique ones making these diagrams commute. But $1_{\tilde{X}}$ and 1_G also make them commute, so $k \circ h = 1_{\tilde{X}}$ and $h \circ k = 1_G$. Therefore k is the inverse of h and h is an isomorphism as claimed.

(ii) This follows from (i), taking $F = \tilde{X}$, $X = U$, $f = l_X$. □

If $f: X \to Y$, then according to (5.22) there is a uniquely determined map $\tilde{f}: \tilde{X} \to \tilde{Y}$ such that $\tilde{f} \circ l_X = l_Y \circ f$, namely, $\tilde{f} = (l_Y \circ f)^\circ$. It follows that $\tilde{\ }$ *determines a functor* $L: \mathbf{C}(S) \to \mathbf{Sh}(S)$ which by (5.22) is left adjoint to the insertion functor. L is called the *sheafification functor*. Note that X is a sheaf iff $X \xrightarrow{l_X} LX$ is an isomorphism.

We now prove

5.24 THEOREM. L is left exact.

Proof. It suffices to show:

L preserves the terminal object. This is obvious since 1 is a sheaf.

L preserves equalizers. Given $X \underset{g}{\overset{f}{\rightrightarrows}} Y$, let $U = \{x \in X : f(x) = g(x)\}$ be their equalizer and let $V = \{u \in \tilde{X} : \tilde{f}(u) = \tilde{g}(u)\}$ be the equalizer of $\tilde{f}$ and $\tilde{g}$. We want to show that $\tilde{U} \cong V$, and in order to do this it suffices by (5.23)(ii) to show that

$$V = \overline{l_X[U]} \cap \tilde{X}. \tag{1}$$

First, we have

$$\begin{aligned} u \in V \vdash_S & \tilde{f}(u) = \tilde{g}(u) \\ \vdash_S & [\exists x \in X\ (u = l_X(x) \wedge \tilde{f}(u) = \tilde{g}(u))]^* \\ \vdash_S & [\exists x \in X\ (u = l_X(x) \wedge \tilde{f}(l_X(x)) = \tilde{g}(l_X(x))]^* \\ \vdash_S & [\exists x \in X\ (u = l_X(x) \wedge l_Y(f(x)) = l_Y(g(x)))]^* \\ \vdash_S & [\exists x \in X\ (u = l_X(x) \wedge (f(x) = g(x))^*)]^* \\ \vdash_S & [\exists x \in \bar{U} \cap X\ (u = l_X(x))]^* \\ \vdash_S & u \in \overline{l_X[\bar{U} \cap X]} \subseteq \overline{\overline{l_X[U]}} \subseteq \overline{l_X[U]} \end{aligned}$$

so that

$$\vdash_S V \subseteq \overline{l_X[U]} \cap \tilde{X}.$$

For the reverse inclusion, we observe that, by (5.9)(iii), V is closed in X. Then

$$\begin{aligned} u \in \tilde{X}, u \in \overline{l_X[U]} \vdash_S & [\exists x \in U. u = l_X(x)]^* \\ \vdash_S & [\exists x \in X\ (u = l_X(x) \wedge f(x) = g(x))]^* \\ \vdash_S & [\exists x \in X\ (u = l_X(x) \wedge l_Y(f(x)) = l_Y(g(x)))]^* \\ \vdash_S & [\exists x \in X\ (u = l_X(x) \wedge \tilde{f}(l_X(x)) = \tilde{g}(l_X(x))]^* \\ \vdash_S & [\exists x \in X\ (u = l_X(x) \wedge \tilde{f}(u) = \tilde{g}(u))]^* \\ \vdash_S & u \in \bar{V} \cap \tilde{X} = V \end{aligned}$$

so that

$$\vdash_S \overline{l_X[U]} \cap \tilde{X} \subseteq V.$$

This proves (1).

Finally, L preserves products. Let π_1, π_2 be the projections $X \times Y \to X$, $X \times Y \to Y$, respectively and write l for $l_{X \times Y}$. We have

$$x \in X, y \in Y \vdash_S \tilde{\pi}_1(l\langle x,y \rangle) = l_X(x) \wedge \tilde{\pi}_2(l(\langle x,y \rangle)) = l_Y(y)$$

and if we define $p : (X \times Y)^{\tilde{}} \to \tilde{X} \times \tilde{Y}$ by

$$p = (u \mapsto \langle \tilde{\pi}_1(u), \tilde{\pi}_2(u) \rangle),$$

then we have

$$x \in X, y \in Y \vdash_S p(l(\langle x,y \rangle) = \langle l_X(x), l_Y(y) \rangle.$$

We claim that p is an isomorphism. First, p is monic. For (using evident abbreviations)

$$
\begin{aligned}
&u,v \in (X \times Y)\tilde{},p(u)=p(v)\\
&\quad \vdash_S [\exists xyx'y'(u=l(\langle x,y\rangle) \wedge v=l(\langle x',y'\rangle) \wedge p(u)=p(v)]^*\\
&\quad \vdash_S [\exists xyx'y'\ (u=l(\langle x,y\rangle) \wedge v=l(\langle x',y'\rangle) \wedge l_X(x)=l_X(x')\\
&\qquad\qquad\qquad\qquad\qquad\qquad\qquad \wedge l_Y(y)=l_Y(y')]^*\\
&\quad \vdash_S [\exists xyx'y'\ (u=l(\langle x,y\rangle) \wedge v=l(\langle x',y'\rangle) \wedge (x=x')^*\\
&\qquad\qquad\qquad\qquad\qquad\qquad\qquad \wedge (y=y')^*)]^*\\
&\quad \vdash_S [\exists xyx'y'\ (u=l(\langle x,y\rangle) \wedge v=l(\langle x',y'\rangle) \wedge (x=x') \wedge (y=y')]^*\\
&\quad \vdash_S (u=v)^* \vdash_S u=v.
\end{aligned}
$$

Finally, we show that p is epic. To do this we note first that since p is monic and $(X \times Y)\tilde{}$ is a sheaf, $p[(X \times Y)\tilde{}]$ is closed in $\tilde{X} \times \tilde{Y}$. Then we have

$$
\begin{aligned}
\langle u,v\rangle \in \tilde{X} \times \tilde{Y} &\vdash_S [\exists xy\ (l_X(x)=u \wedge l_Y(y)=v)]^*\\
&\vdash_S [\exists xy. \langle u,v\rangle = \langle l_X(x), l_Y(y)\rangle]^*\\
&\vdash_S [\exists xy. \langle u,v\rangle = p(l(\langle x,y\rangle))]^*\\
&\vdash_S [\exists z \in X \times Y. \langle u,v\rangle = p(l(z))]^*\\
&\vdash_S \langle u,v\rangle \in \overline{p[l[X \times Y]]} \cap (\tilde{X} \times \tilde{Y})\\
&\qquad \subseteq \overline{p[(X \times Y)\tilde{}]} \cap (\tilde{X} \times \hat{Y})\\
&\qquad \subseteq p[(X \times Y)\tilde{}].
\end{aligned}
$$

Therefore p is epic and we are finished. □

Theorem 5.24 implies that we have a geometric morphism $i:\mathbf{Sh}(S)\to \mathbf{C}(S)$, where i_* is the insertion functor and $i^* = L$.

A map $f:X\to Y$ is *almost epic* if $f[X]$ is dense in Y, and *almost iso* if it is almost epic and almost monic.

5.25 LEMMA. Let X and Y be sheaves and $f:X\to Y$. Then

(i) f is monic in $\mathbf{Sh}(S)$ iff f is monic in $\mathbf{C}(S)$.
(ii) f is epic in $\mathbf{Sh}(S)$ iff f is almost epic in $\mathbf{C}(S)$.

Proof. (i) Clearly if f is monic in $\mathbf{C}(S)$, it is also monic in $\mathbf{Sh}(S)$. Conversely, suppose f is monic in $\mathbf{C}(S)$. Define

$$Z=\{\langle x,x'\rangle \in X \times X : f(x)=f(x')\}.$$

It is readily checked that Z is closed in $X \times X$; since the latter is a sheaf, so is Z. Now let σ_1 and σ_2 be the restrictions of the projections $X \times X\to X$ to Z. Then clearly $f \circ \sigma_1 = f \circ \sigma_2$, so since f is monic in $\mathbf{Sh}(S)$,

$\sigma_1 = \sigma_2$. Therefore

$$x,x' \in X,\ f(x)=f(x') \vdash_S \langle x,x' \rangle \in Z$$
$$\vdash_S x = \sigma_1(\langle x,x' \rangle) = \sigma_2(\langle x,x' \rangle) = x',$$

so that f is monic in $\mathbf{C}(S)$.

(ii) Suppose that f is almost epic in $\mathbf{C}(S)$. Let

$$X \xrightarrow{f} Y \overset{g}{\underset{h}{\rightrightarrows}} Z$$

be a commutative diagram in $\mathbf{Sh}(S)$, and let

$$U = \{y \in Y : g(y) = h(y)\}.$$

Then U is closed in Y and $\vdash_S f[X] \subseteq U$. But since f is almost epic, $\vdash_S \overline{f[X]} \cap Y = Y$, so

$$\vdash_S Y = \overline{f[X]} \cap Y \subseteq \bar{U} \cap Y \subseteq U.$$

It follows that $g = h$ and so f is epic in $\mathbf{Sh}(S)$.

Conversely, suppose that f is epic in $\mathbf{Sh}(S)$. Then $\overline{f[X]} \cap Y$ is a sheaf and we may consider its characteristic arrow $\chi : Y \to \Omega^*$ in $\mathbf{Sh}(S)$ given by

$$\chi = (y \mapsto (\exists x \in X.\, y = f(x))^*).$$

Define

$$\tau = \{\langle y, true \rangle : y \in Y\}.$$

Then $\chi \circ f = \tau \circ f$, so $\chi = \tau$. Hence

$$y \in Y \vdash_S (\exists x \in X.\, y = f(x))^* = \tau(y) = \chi(y) = true)$$

so that $\vdash_S Y \subseteq \overline{f[X]}$ and f is almost epic. □

5.26 PROPOSITION. Let $f : X \to Y$ in $\mathbf{C}(S)$. Then

(i) f is almost monic iff $\tilde{f}$ is monic in $\mathbf{Sh}(S)$;
(ii) f is almost epic iff $\tilde{f}$ is epic in $\mathbf{Sh}(S)$;
(iii) f is almost iso iff $\tilde{f}$ is an isomorphism in $\mathbf{Sh}(S)$.

Proof. (i) Suppose that $\tilde{f}$ is monic in $\mathbf{Sh}(S)$. Then by (5.25)(i) it is also monic in $\mathbf{C}(S)$ and so

$$x,\ x' \in X,\ f(x) = f(x') \vdash_S \tilde{f}(l_X(x)) = l_Y(f(x)) = l_Y(f(x')) = \tilde{f}(l_Y(x'))$$
$$\vdash_S l_X(x) = l_X(x')$$
$$\vdash_S (x = x')^*,$$

so that f is almost monic.

Conversely, suppose that f is almost monic. Then

$$\begin{aligned}
&u, u' \in \tilde{X},\ \tilde{f}(u) = \tilde{f}(u') \\
&\quad \vdash_S [\exists x x' \in X\ (u = l_X(x) \wedge u' = l_X(x') \wedge \tilde{f}(u) = \tilde{f}(u'))]^* \\
&\quad \vdash_S [\exists x x' \in X\ (u = l_X(x) \wedge u' = l_X(x') \wedge l_Y(f(x)) = l_Y(f(x')))]^* \\
&\quad \vdash_S [\exists x x' \in X\ (u = l_X(x) \wedge u' = l_X(x') \wedge f(x) = f(x'))]^* \\
&\quad \vdash_S [\exists x x' \in X\ (u = l_X(x) \wedge u' = l_X(x') \wedge x = x')]^* \\
&\quad \vdash_S (u = u')^* \\
&\quad \vdash_S u = u'.
\end{aligned}$$

Therefore $\tilde{f}$ is monic in $\mathbf{C}(S)$ and hence, by (5.25)(i), also in $\mathbf{Sh}(S)$.

(ii) Suppose that f is almost epic. Then $Y = \overline{f[X]} \cap Y$ and we have

$$\begin{aligned}
\vdash_S l_Y[Y] &= l_Y[\overline{f[X]} \cap Y] \\
&\subseteq \overline{l_Y[f[X]]} \\
&\subseteq \overline{\tilde{f}[l_X[X]]} \\
&\subseteq \overline{\tilde{f}[\tilde{X}]}.
\end{aligned}$$

Hence

$$\vdash_S \tilde{Y} \subseteq \overline{l_Y[Y]} \subseteq \overline{\tilde{f}[\tilde{X}]},$$

so that $\tilde{f}$ is almost epic, and thus, by (5.25)(ii), epic in $\mathbf{Sh}(S)$.

For the converse we need the following fact, whose easy proof is left to the reader:

$$\vdash_S Z \subseteq Y \quad \text{implies} \quad \vdash_S l_Y^{-1}[l_Y[Z]] \subseteq \bar{Z}.$$

Now suppose that f is epic in $\mathbf{Sh}(S)$; then by (5.25)(ii) it is almost epic, i.e. $\vdash_S \tilde{Y} \subseteq \overline{\tilde{f}[\tilde{X}]}$. Hence

$$\begin{aligned}
\vdash_S Y = l_Y^{-1}[\tilde{Y}] &\subseteq l_Y^{-1}[\overline{\tilde{f}[\tilde{X}]}] \\
&\subseteq \overline{l_Y^{-1}[\tilde{f}[\tilde{X}]]}.
\end{aligned} \tag{1}$$

Now

$$\vdash_S \tilde{f}[\tilde{X}] = \tilde{f}[\overline{l_X[X]} \cap \tilde{X}] \subseteq \overline{\tilde{f}[l_X[X]]} = \overline{l_Y[f[X]]},$$

so that

$$\begin{aligned}
\vdash_S l_Y^{-1}[\tilde{f}[\tilde{X}]] &\subseteq l_Y^{-1}[\overline{l_Y[f[X]]}] \\
&\subseteq \overline{l_Y^{-1}[l_Y[f[X]]]} \\
&\subseteq \overline{f[X]}.
\end{aligned}$$

It follows from (i) that

$$\vdash_S Y \subseteq \overline{l_Y^{-1}[\tilde{f}[\tilde{X}]]} \subseteq \overline{\overline{f[X]}} = \overline{f[X]},$$

and so f is almost epic.

(iii) follows immediately from (i) and (ii). □

It follows from (5.26) that the effect of the sheafification functor is to *convert dense monics into isomorphisms*, or, loosely, to identify almost isomorphic objects. More specifically, we have the

5.27 COROLLARY. Let $f: X \to Y$ in $\mathbf{C}(S)$. Then there are unique strictest modalities μ_1, μ_2, μ_3 in S such that $\tilde{f}$ is monic in $\mathbf{Sh}_{\mu_1}(S)$, $\tilde{f}$ is epic in $\mathbf{Sh}_{\mu_2}(S)$, and $\tilde{f}$ is an isomorphism in $\mathbf{Sh}_{\mu_3}(S)$ (where in each case $\tilde{f}$ denotes the image of f under the sheafification functor to the relevant topos of sheaves).

Proof. We first note that, for any inclusion $\vdash_S X \subseteq Y$, there is a unique strictest modality μ in S such that X is μ-dense in Y: it is easy to see that μ is the modality generated by the set

$$\{\omega : \exists y \in Y. \omega = (y \in X)\}.$$

Now, given $f: X \to Y$, let

$$Z = \{\langle x, x' \rangle \in X \times X : f(x) = f(x')\}.$$

Then $\vdash_S \Delta_X \subseteq Z$; and we can take μ_1 to be the strictest modality μ in S such that Δ_X is μ-dense in Z. And we take μ_2 to be the strictest modality μ such that $f[X]$ is μ-dense in Y. Finally, we take μ_3 to be the join $\mu_1 + \mu_2$. Using (5.26), it is readily checked that μ_1, μ_2, μ_3 satisfy the required conditions. □

The modalities μ_1, μ_2, μ_3 constructed in (5.27) may be vividly described as the strictest modalities *forcing* f to be monic, epic, or an isomorphism, respectively.

Another important consequence of (5.26) is the following

5.28 COROLLARY. For any theory S, the topos $\mathbf{Sh}_{\neg\neg}(S)$ is Boolean.

Proof. By a well-known result in intuitionistic propositional logic (see, e.g. Kleene (1952), Chapter VI, Theorem 7, 51a), we have

$$\vdash \neg\neg(\omega \vee \neg\omega). \tag{1}$$

Now let $2 = 1 + 1$ in $\mathbf{C}(S)$ and define

$$h = \begin{pmatrix} \top \\ \bot \end{pmatrix} : 2 \to \Omega^{\neg\neg} = \{\neg\neg\omega : true\}.$$

Then

$$\vdash_S h[2] = \{true, false\}.$$

But, by (1),

$$\vdash_S \forall\omega \in \Omega^{\neg\neg}.\neg\neg(\omega \vee \neg\omega),$$

whence

$$\vdash_S \forall\omega \in \Omega^{\neg\neg}.\neg\neg(\omega = true \vee \omega = false),$$

so that

$$\vdash_S \forall\omega \in \Omega^{\neg\neg}.\neg\neg(\omega \in h[2]).$$

Hence h is almost iso in S, and it follows that $\tilde{h}:\tilde{2}\to(\Omega^{\neg\neg})^{\sim}$ is an isomorphism in $\mathbf{Sh}_{\neg\neg}(S)$. But

$$\tilde{2} = (1+1)^{\sim} \cong \tilde{1} + \tilde{1} = 1 + 1 = 2$$

in $\mathbf{Sh}_{\neg\neg}(S)$ and $(\Omega^{\neg\neg})^{\sim} \cong \Omega^{\neg\neg}$ since $\Omega^{\neg\neg}$ is a sheaf. Hence h is an isomorphism in $\mathbf{Sh}_{\neg\neg}(S)$, and the result follows. □

EXERCISE. Show that the following conditions are equivalent for any modality:

(i) Ω^* is dense in Ω;
(ii) $\vdash_S \forall\omega.(\omega^* \Rightarrow \omega)^*$.

Deduce that, if the modality is of the form $\neg\neg$ or μ_γ, then $L\Omega \cong \Omega^*$.

Modalized toposes

We now consider the interpretation in toposes of the concepts introduced in this chapter.

Let (S,μ) be a modalized theory and let I be a model of S in a topos $\mathbf{E}$.

5.29 PROPOSITION. Define the arrow $j:\Omega\to\Omega$ in $\mathbf{E}$ by

$$j = [\![\mu]\!]_{\omega,I}.$$

Then, in $\mathbf{E}$, the following diagrams commute:

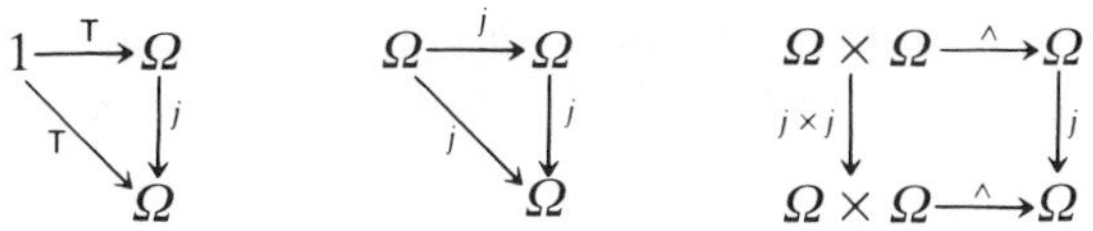

(i) (ii) (iii)

Proof. (i) Since $\vdash_S \mu(true) = true$, we have by the soundness theorem $\models_I \mu(true) = true$, which just asserts the commutativity of (i).

(ii) We have $\vdash_S \mu(\mu(\omega)) = \mu(\omega)$, so $\models_I \mu(\mu(\omega)) = \mu(\omega)$, which asserts the commutativity of (ii).

(iii) By (5.3), we have $\vdash_S \mu(\omega \wedge \omega') = \mu(\omega) \wedge \mu(\omega')$, so $\models_I \mu(\omega \wedge \omega') = \mu(\omega) \wedge \mu(\omega')$, which asserts the commutativity of (iii). □

This proposition prompts the following definition. A *modal operator* in a topos **E** is an arrow $j : \Omega \to \Omega$ in **E** such that the diagrams (i), (ii) and (iii) of 5.29 commute. The pair $(\mathbf{E}, j)$ is called a *modalized topos.*

Let $\mathscr{L}$ be a local language and μ a formula of $\mathscr{L}$ with exactly one free variable ω. A *modal interpretation* of $(\mathscr{L}, \mu)$ in a topos **E** is an interpretation I of $\mathscr{L}$ in **E** such that $[\![\mu]\!]_{\omega,I}$ is a modal operator in **E**.

Write $\mathscr{L}(\mathbf{m})$ for the language obtained from $\mathscr{L}$ by adding a new function symbol **m** of signature $\mathbf{\Omega} \to \mathbf{\Omega}$ and write m for $\mathbf{m}(\omega)$. Let M be the collection of sequents $\{\omega : \mathbf{m}(\omega), \mathbf{m}(\mathbf{m}(\omega)) : \mathbf{m}(\omega)\}$. For any sequent $\Gamma : \alpha$ of $\mathscr{L}(\mathbf{m})$, write

$$\Gamma \vdash_{\text{mod}} \alpha$$

for

$\Gamma : \alpha$ is deducible from M using the additional rule of inference

$$\frac{\gamma : \delta}{m(\gamma) : m(\delta)}.$$

We write L_{mod} for the theory in $\mathscr{L}(\mathbf{m})$ whose axioms are all sequents $\Gamma : \alpha$ such that $\Gamma \vdash_{\text{mod}} \alpha$. We note that *$(L_{\text{mod}}, m)$ is a modalized theory.* And clearly we have

$$\Gamma \vdash_{\text{mod}} \alpha \quad \text{iff} \quad \Gamma \vdash_{L_{\text{mod}}} \alpha.$$

Finally, we write

$$\Gamma \models_{\text{mod}} \alpha$$

for

$\Gamma \models_I \alpha$ for any modal interpretation I of $(\mathscr{L}, m)$.

We can now prove the

5.30 MODAL SOUNDNESS THEOREM

$$\Gamma \vdash_{\text{mod}} \alpha \quad \text{implies} \quad \Gamma \models_{\text{mod}} \alpha.$$

Proof. By the soundness theorem, it suffices to show that the axioms $\{\omega : \mathbf{m}(\omega), \mathbf{m}(\mathbf{m}(\omega)) : \mathbf{m}(\omega)\}$ and the additional inference rule

$$\frac{\alpha : \beta}{\mathbf{m}(\alpha) : \mathbf{m}(\beta)}$$

are valid under any modal interpretation of $(\mathscr{L}, m)$. So let I be a modal interpretation of $(\mathscr{L}, m)$ in a topos $\mathbf{E}$. Writing j for $[\![m]\!]_{\omega,I}$, then j is a modal operator in $\mathbf{E}$.

Now we have

(i) $\omega \models_I \mathbf{m}(\omega)$. For let

$$\begin{array}{ccc} \mathrm{dom}(\bar{j}) & \longrightarrow & 1 \\ {\scriptstyle \bar{j}}\downarrow & & \downarrow{\scriptstyle \mathsf{T}} \\ \Omega & \xrightarrow{\;j\;} & \Omega \end{array}$$

be a pullback. Then since $j \circ \mathsf{T} = \mathsf{T}$, there is a (unique) $1 \xrightarrow{f} \mathrm{dom}(\bar{j})$ such that $\bar{j} \circ f = \mathsf{T}$. Therefore $T \subseteq \bar{j}$, whence $1_\Omega = \chi(T) \leqslant \chi(\bar{j}) = j$. Hence

$$[\![\omega]\!]_\omega = 1_\Omega \leqslant j = j \circ 1_\Omega = [\![\mathbf{m}(\omega)]\!]_\omega$$

and so $\omega \models_I \mathbf{m}(\omega)$.

(ii) $\mathbf{m}(\mathbf{m}(\omega)) \models_I \mathbf{m}(\omega)$. We have

$$[\![\mathbf{m}(\mathbf{m}(\omega))]\!]_\omega = j \circ j \circ 1_\Omega = j \circ j = j = j \circ 1_\Omega = [\![\mathbf{m}(\omega)]\!]_\omega,$$

and (ii) follows.

(iii) $\alpha \models_I \beta$ implies $\mathbf{m}(\alpha) \models_I \mathbf{m}(\beta)$. We show first that

$$\models_I \mathbf{m}(\alpha) \wedge \mathbf{m}(\beta) \Leftrightarrow \mathbf{m}(\alpha \wedge \beta). \qquad (*)$$

To do this we note that the diagram

$$\begin{array}{ccccc} \bullet & \xrightarrow{\langle [\![\alpha]\!]_x,\, [\![\beta]\!]_x \rangle} & \Omega \times \Omega & \xrightarrow{\;\wedge\;} & \Omega \\ & & {\scriptstyle j \times j}\downarrow & & \downarrow{\scriptstyle j} \\ & & \Omega \times \Omega & \xrightarrow{\;\wedge\;} & \Omega \end{array}$$

commutes, so that, writing $[\![\alpha]\!]$, $[\![\beta]\!]$ for $[\![\alpha]\!]_x$, $[\![\beta]\!]_x$ etc.,

$$\begin{aligned} \wedge \circ \langle j \circ [\![\alpha]\!], j \circ [\![\beta]\!] \rangle &= \wedge \circ (j \times j) \circ \langle [\![\alpha]\!], [\![\beta]\!] \rangle \\ &= j \circ \wedge \circ \langle [\![\alpha]\!], [\![\beta]\!] \rangle \\ &= j \circ [\![\alpha \wedge \beta]\!]. \end{aligned}$$

Hence

$$\begin{aligned} [\![\mathbf{m}(\alpha) \wedge \mathbf{m}(\beta)]\!] &= \wedge \circ \langle j \circ [\![\alpha]\!], j \circ [\![\beta]\!] \rangle \\ &= j \circ [\![\alpha \wedge \beta]\!] = [\![\mathbf{m}(\alpha \wedge \beta)]\!]. \end{aligned}$$

Therefore

$$[\![\mathbf{m}(\alpha) \wedge \mathbf{m}(\beta) \Leftrightarrow \mathbf{m}(\alpha \wedge \beta)]\!] = T$$

and (*) follows.

If $\alpha \models_I \beta$, then $\models_I \alpha \Rightarrow \beta$, whence $\models_I \alpha = \alpha \wedge \beta$. So by (*) $\models_I \mathbf{m}(\alpha) = \mathbf{m}(\alpha \wedge \beta) = \mathbf{m}(\alpha) \wedge \mathbf{m}(\beta)$. Therefore $\models_I \mathbf{m}(\alpha) \Rightarrow \mathbf{m}(\beta)$, whence $\mathbf{m}(\alpha) \models_I \mathbf{m}(\beta)$. $\square$

Let (S, μ) be a modalized theory and define $j: \Omega \to \Omega$ in $\mathbf{C}(S)$ by

$$j = (\omega \mapsto \mu).$$

It is easy to check (using (5.3)(i)) that *j is a modal operator in* $\mathbf{C}(S)$; we call it the modal operator *induced by* μ.

The canonical interpretation of $\mathscr{L}$ in $\mathbf{C}(S)$ is then a modal interpretation of $(\mathscr{L}, \mu)$ in the topos $\mathbf{C}(S)$ and as a consequence we have the

5.31 MODAL COMPLETENESS THEOREM

$$\Gamma \vDash_{\text{mod}} \alpha \qquad \text{implies} \qquad \Gamma \vdash_{\text{mod}} \alpha.$$

Proof. Let j be the modal operator in $\mathbf{C}(L_{\text{mod}})$ induced by the modality m in L_{mod}. Then, if $\Gamma \vDash_{\text{mod}} \alpha$, it follows that $\Gamma \vDash_{C(L_{\text{mod}})} \alpha$, whence $\Gamma \vdash_{L_{\text{mod}}} \alpha$ by (3.28), and therefore $\Gamma \vdash_{\text{mod}} \alpha$. □

Now let $(\mathbf{E}, j)$ be a modalized topos. Then $(j, \boldsymbol{\Omega}, \boldsymbol{\Omega})$ is a function symbol—which we shall denote simply by $\mathbf{j}$—in $\mathscr{L}(\mathbf{E})$. We can now prove

5.32 PROPOSITION. The formula $\mathbf{j}(\omega)$ is a modality in Th($\mathbf{E}$).

Proof. The natural interpretation of $\mathscr{L}(\mathbf{E})$ in $\mathbf{E}$ is clearly a modal interpretation, and so, by the modal soundness theorem, the axioms (mod_1), (mod_2), (mod_3) governing modalities are valid under this interpretation. But by (3.30) we have

$$\Gamma \vdash_{\text{Th}(\mathbf{E})} \alpha \qquad \text{iff} \qquad \Gamma \vDash_E \alpha$$

and so, since the interpretation of $\mathbf{j}(\boldsymbol{\omega})$ in $\mathbf{E}$ is j, it follows that (mod_1), (mod_2), (mod_3) are true of $\mathbf{j}(\boldsymbol{\omega})$ in Th($\mathbf{E}$). Therefore $\mathbf{j}(\boldsymbol{\omega})$ is a modality in Th($\mathbf{E}$) as claimed. □

Let us define an *equivalence* between two modalized toposes $(\mathbf{E}, j)$ and $(\mathbf{E}', j')$ to be an equivalence $F: \mathbf{E} \to \mathbf{E}'$ such that $F(\Omega_{\mathbf{E}}) = \Omega_{\mathbf{E}'}$ and $F(j) = j'$. We say that $(\mathbf{E}, j)$ and $(\mathbf{E}', j')$ are *equivalent* and write $(\mathbf{E}, j) \simeq (\mathbf{E}', j')$ if there is an equivalence between them.

The proof of the equivalence theorem (3.37) now yields the

5.33 MODAL EQUIVALENCE THEOREM. Let $(\mathbf{E}, j)$ be a modalized topos and let j' be the modal operator in $\mathbf{C}(\text{Th}(\mathbf{E}))$ induced by the modality $\mathbf{j}(\omega)$ in Th($\mathbf{E}$). Then

$$(\mathbf{E}, j) \simeq (\mathbf{C}(\text{Th}(\mathbf{E}), j'). \quad \square$$

Next, we consider the interpretation in toposes of universal closure operations and sheaves.

Given a modalized topos $(\mathbf{E}, j)$, we define the *induced* universal closure operation in $\mathbf{E}$ as follows. First, we form the modalized theory $(\text{Th}(\mathbf{E}), \mathbf{j}(\omega))$. Next, we form the induced closure operation $\bar{\cdot}$ in Th($\mathbf{E}$).

We now conceive of this closure operation as acting on *subobjects* in **C**(Th(**E**)). That is, given a subobject $X \xrightarrow{m} Y$ in **C**(Th(**E**)), we define the closure $\bar{X}$ as a subobject of Y to be $\overline{m[X]}$.

Recall that $[\bullet]_{\mathbf{E}}$ denotes the canonical interpretation of objects and arrows of **C**(Th(**E**)) in **E**. We define, for a subobject $X \xrightarrow{m} A$ in **E**,

$$\bar{X} = [\overline{U_X}]_{\mathrm{E}}.$$

The operation $X \mapsto \bar{X}$ is called the *induced closure operation* in (**E**,j) and $\bar{X}$ the *j-closure* of X. From the properties of universal closure operations established earlier in this chapter it follows immediately that

$$X \rightarrowtail \bar{X}$$

$$\bar{\bar{X}} \cong \bar{X}$$

$$\overline{X \cap Y} \cong \bar{X} \cap \bar{Y}$$

$$f^{-1}[\bar{Y}] \cong \overline{f^{-1}[Y]} \qquad \text{for any } f: A \to B \quad \text{and} \quad Y \to B.$$

Using the definition of $\bar{\bullet}$ in Th(**E**), it is easy to check that, for any subobject X of an object A of **E**, with characteristic arrow χ, $\bar{X}$ is the subobject of A classified by $j \circ \chi$.

A subobject X of an object A of **E** is said to be (j-)*closed* if $X \cong \bar{X}$ and (j-) *dense* if $\bar{X} \cong A$. Clearly, if χ is the characteristic arrow of X, then X is closed iff $j \circ \chi = \chi$ and dense iff $j \circ \chi = T_A$.

An object X of **E** is said to be (j-) *separated* if the diagonal $X \xrightarrow{\delta_X} X \times X$ is a closed subobject of $X \times X$, and a (j-) *sheaf* if, given any dense $U \xrightarrow{m} V$ and any $U \xrightarrow{f} X$, there is a unique $V \xrightarrow{g} X$ such that $g \circ m = f$.

If (S, μ) is a modalized theory and j the induced modal operator in **C**(S), then for any S-set X it is easily seen that

X is μ-separated iff X is j-separated (in **C**(S))
X is a μ-sheaf iff X is a j-sheaf (in **C**(S)).

Let (**E**,j) be a modalized topos, and let j' be the modal operator in **C**(Th(**E**)) induced by the modality **j**(ω) in Th(**E**). Then clearly, for any **E**-object X,

X is a j-sheaf iff $U_{\mathbf{X}}$ is a j'-sheaf.

Therefore, if we write $\mathbf{Sh}_j(E)$ for the full subcategory of **E** whose objects are the j-sheaves, it follows that the canonical inclusion of **E** in **C**(Th(**E**)) restricts to an equivalence between $\mathbf{Sh}_j(\mathbf{E})$ and $\mathbf{Sh}_{j'}(\mathrm{Th}(\mathbf{E}))$. Since, by (5.18), the latter is a topos, we arrive at

5.34 THEOREM. For any modalized topos (**E**,j), the full subcategory $\mathbf{Sh}_j(\mathbf{E})$ of j-sheaves is a topos. □

The sheafification functor $L:\mathbf{C}(\mathrm{Th}(\mathbf{E}))\to\mathbf{Sh}_{j'}(\mathrm{Th}(\mathbf{E}))$ then induces (via the equivalence $\mathbf{E}\simeq\mathbf{C}(\mathrm{Th}(\mathbf{E}))$, a functor $L':\mathbf{E}\to\mathbf{Sh}_j(\mathbf{E})$ left adjoint to the insertion $\mathbf{Sh}_j(\mathbf{E})\hookrightarrow\mathbf{E}$. Since L is left exact (5.24), so is L'. It follows that we have a geometric morphism $i:\mathbf{Sh}_j(\mathbf{E})\to\mathbf{E}$, where i_* is the insertion functor and $i^*=L'$.

All the results concerning modalities in theories proved in this chapter transfer in a similar way to modal operators in toposes.

Modal operators and sheaves in $\mathbf{Set}^{\mathbf{C}^{\mathrm{op}}}$

We shall now investigate the special case in which the topos $\mathbf{E}$ is the topos of presheaves on a small category.

Let $\mathbf{C}$ be a small category and as usual write $\mathbf{E}(\mathbf{C})$ for the topos $\mathbf{Set}^{\mathbf{C}^{\mathrm{op}}}$ of presheaves on $\mathbf{C}$. Let F be a presheaf on $\mathbf{C}$, and let $\mathbf{F}$ be the corresponding type symbol in $\mathscr{L}(\mathbf{E}(\mathbf{C}))$. Recall that a formula of $\mathscr{L}(\mathbf{E}(\mathbf{C}))$ with at most one free variable, of type $\mathbf{F}$, is called an $\mathbf{F}$-*formula*. The natural interpretation $[\![\alpha]\!]$ of an $\mathbf{F}$-formula α in $\mathbf{E}(\mathbf{C})$ is then an $\mathbf{E}(\mathbf{C})$-arrow $F\to\Omega$.

Now suppose that we are given a *modality* μ in $\mathrm{Th}(\mathbf{E}(\mathbf{C}))$. Then its interpretation $j=[\![\mu]\!]$ is a modal operator in $\mathbf{E}(\mathbf{C})$. Let K be the subfunctor of Ω classified by j, given by

$$K(A)=\{S\in\Omega(A):j_A(S)=\max{}^A\}.$$

We determine the characteristic properties of K.

First, since K is a subfunctor of Ω, we must have, for $S\in K(A)$, $B\xrightarrow{f}A$,

$$(Kf)S=(\Omega f)S=f^*S,$$

whence

(I) $$S\in K(A),\ B\xrightarrow{f}A\Rightarrow f^*S\in K(B).$$

Next, we have, for any $\mathbf{F}$-formula α, $\alpha\vdash_{\mathrm{Th}(\mathbf{E}(\mathbf{C}))}\mu(\alpha)$, so by (4.42), for any $x\in FA$,

$$[\![\alpha]\!]_A(x)\subseteq[\![\mu_A(\alpha_A)]\!](x)=j_A([\![\alpha]\!]_A(x)).$$

Taking $\alpha=\mathit{true}$ then gives

$$\max{}^A=[\![\mathit{true}]\!]_A(0)\subseteq j_A([\![\mathit{true}]\!]_A(0))=j_A(\max{}^A).$$

Therefore

(II) $$\max{}^A\in K(A)\qquad\text{for any } \mathbf{C}\text{-object } A.$$

To continue, for any $\mathbf{F}$-formulas α,β, we have

$$\alpha\vdash_{\mathrm{Th}(\mathbf{E}(\mathbf{C}))}\beta\Rightarrow\mu(\alpha)\vdash_{\mathrm{Th}(\mathbf{E}(\mathbf{C}))}\mu(\beta),$$

so by (4.42), for all A, $x \in FA$,

$$[\![\alpha]\!]_A(x) \subseteq [\![\beta]\!]_A(x) \Rightarrow j_A([\![\alpha]\!]_A)(x) \subseteq j_A([\![\beta]\!]_A)(x). \tag{1}$$

Let $S \in \Omega(A)$ and define the subfunctor S^* of the contravariant hom-functor H^A (an $\mathbf{E}(\mathbf{C})$-object) by

$$S^*(X) = C(X,A) \cap S$$

for $X \in \mathrm{Ob}(\mathbf{C})$. Let $\alpha : H^A \to S$ be the characteristic arrow of S^*. Then it is readily checked that

$$\alpha_A(1_A) = S. \tag{2}$$

Now let $\boldsymbol{\alpha}$ be the function symbol in $\mathscr{L}(\mathbf{E}(\mathbf{C}))$ of signature $\mathbf{H}^A \to \mathbf{\Omega}$ associated with α. If u is a variable of type $\mathbf{H}^{\mathbf{A}}$, $\boldsymbol{\alpha}(u)$ is a formula, which we call the formula *associated* with S. Clearly $[\![\boldsymbol{\alpha}(u)]\!] = \alpha$.

Given $S' \in \Omega(A)$, let α', $\boldsymbol{\alpha}'(u)$ be obtained from S as were α, $\boldsymbol{\alpha}(u)$ from S. If $S \subseteq S'$, then by (2),

$$\begin{aligned}[\![\boldsymbol{\alpha}(u)]\!]_A(1_A) = \alpha_A(1_A) = S \subseteq S' &= \alpha'_A(1_A) \\ &= [\![\boldsymbol{\alpha}'(u)]\!]_A(1_A),\end{aligned}$$

so by (1),

$$j_A([\![\boldsymbol{\alpha}(u)]\!]_A(1_A)) \subseteq j_A([\![\boldsymbol{\alpha}'(u)]\!]_A(1_A)),$$

i.e. $j_A(S) \subseteq j_A(S')$. Therefore, if also $S \in K(A)$, then $j_A(S) = \max^A$, so $j_A(S') = \max^A$, whence $S' \in K(A)$. Accordingly we have proved

(III) for all $A \in \mathrm{Ob}(\mathbf{C})$, all $S, S' \in \Omega(A)$,

$$S \subseteq S' \ \&\ S \in K(A) \Rightarrow S' \in K(A).$$

Finally, we have, for any $\mathbf{F}$-formula α,

$$\mu(\mu(\alpha)) \vdash_{\mathrm{Th}(\mathbf{E}(\mathbf{C}))} \mu(\alpha),$$

whence for any $A \in \mathrm{Ob}(\mathbf{C})$, $x \in FA$,

$$j_A(j_A([\![\alpha]\!]_A(x)) \subseteq j_A([\![\alpha]\!]_A(x)). \tag{3}$$

Given $S \in \Omega(A)$, let $\alpha(u)$ be the formula associated with S. Then $[\![\alpha(u)]\!]_A(1_A) = S$, so from (3) we get

$$j_A(j_A(S)) \subseteq j_A(S),$$

whence

$$j_A(j_A(S)) = \max{}^A \Rightarrow j_A(S) = \max{}^A$$

i.e.

$$j_A(S) \in K(A) \Rightarrow S \in K(A). \tag{4}$$

Now, using the fact that j is a natural transformation, it is readily shown that

$$j_A(S) = \{B \xrightarrow{f} A : f^*S \in K(B)\},$$

so we get from (4):

(IV) for all $A \in \text{Ob}(\mathbf{C})$, all $S \in \Omega(A)$

$$\{B \xrightarrow{f} A : f^*S \in K(B)\} \in K(A) \Rightarrow S \in K(A).$$

In the presence of (**I**) it is easy to show that (**III**) and (**IV**) can be combined into the equivalent statement

(IV′) for all $A \in \text{Ob}(\mathbf{C})$, all $S \in K(A)$, if $S' \in \Omega(A)$ is such that, for each $B \xrightarrow{f} A$ in S, we have $f^*S' \in K(B)$, then $S' \in K(A)$.

All this leads to the following.

DEFINITION. Let $\mathbf{C}$ be a small category. A *covering system* (or *Grothendieck topology*) *on* $\mathbf{C}$ is a function K assigning to each $A \in \text{Ob}(\mathbf{C})$ a family of cosieves on A satisfying (**I**)–(**IV**), or equivalently (**I**), (**II**), (**IV′**). The members of $K(A)$ are called *K-covering cosieves* on A. The pair $(\mathbf{C}, K)$ is called a *site*.

We have seen that each modality in Th($\mathbf{E}(\mathbf{C})$) engenders a covering system on $\mathbf{C}$. Conversely, given a covering system K on $\mathbf{C}$, by (**I**) K is a subfunctor of Ω in $\mathbf{E}(\mathbf{C})$, and its characteristic arrow $j : \Omega \to \Omega$ has an associated function symbol $\mathbf{j}$ of signature $\mathbf{\Omega} \to \mathbf{\Omega}$ in $\mathscr{L}(\mathbf{E}(\mathbf{C}))$. It is now not hard to show that the formula $\mathbf{j}(\omega)$ is a modality in Th($\mathbf{E}(\mathbf{C})$). Thus *modalities in* Th($\mathbf{E}(\mathbf{C})$) *are in bijective correspondence with covering systems on* $\mathbf{C}$.

Given a covering system K on $\mathbf{C}$, we obtain the corresponding *modal operator* j in $\mathbf{E}(\mathbf{C})$ as the characteristic arrow of K: for $A \in \text{Ob}(\mathbf{C})$, $S \in \Omega(A)$,

$$j_A(S) = \{B \xrightarrow{f} A : f^*S \in K(B)\}.$$

Examples

(i) The covering systems $A \mapsto \Omega(A)$ and $A \mapsto \max^A$ correspond to the trivial modalities **true** and **fix**, respectively.

(ii) Let γ be a sentence in $\mathscr{L}(\mathbf{E}(\mathbf{C}))$. Its interpretation $[\![\gamma]\!]$ in $\mathbf{E}(\mathbf{C})$ is an arrow $1 \xrightarrow{[\![\gamma]\!]} \Omega$, i.e. an assignment, to each $A \in \text{Ob}(\mathbf{C})$, of an element $[\![\gamma]\!]_A \in \Omega(A)$ such that, for any $B \xrightarrow{f} A$, we have $[\![\gamma]\!]_B = f^*[\![\gamma]\!]_A$. Then the covering system K_γ associated with the modality μ_γ is given by

$$S \in K_\gamma(A) \quad \text{iff} \quad [\![\gamma]\!]_A \subseteq S,$$

and the covering system K^γ associated with the modality μ^γ by

$$S \in K_\gamma(A) \quad \text{iff} \quad S = \max{}^A \quad \text{or} \quad [\![\gamma]\!]_A = \max{}^A.$$

(iii) The covering system $K_{\neg\neg}$ associated with the double negation modality $\neg\neg$ in Th(**E**(**C**)) is given by

$$S \in K_{\neg\neg}(A) \text{ iff for each } B \xrightarrow{f} A \text{ there is } C \xrightarrow{g} B \text{ such that } f \circ g \in S.$$

Now let $(\mathbf{C},K)$ be a site, let μ be the corresponding modality in Th(**E**(**C**)) and let j be the corresponding modal operator in **E**(**C**). To emphasize the central role that is going to be played by K, we shall use the terms 'K-closed', 'K-sheaf', etc., for 'j-closed', 'j-sheaf', etc. We shall characterize the K-separated objects and the K-sheaves in **E**(**C**).

To begin with, we observe that, for $F \in \mathbf{E}(\mathbf{C})$, $x \in FA$ and α an **F**-formula, we have

$$A \Vdash^{\mathbf{C}} \mu(a)[x] \quad \text{iff} \quad [\![\alpha]\!]_A(x) \in K(A)$$

and it follows easily from this that the K-closure $\bar{G}$ of a subfunctor G of F is given by

$$\bar{G}(A) = \{x \in GA : \{B \xrightarrow{f} A : (Ff)x \in GB\} \in K(A)\}.$$

Now F is K-separated provided that Δ_F is K-closed in $F \times F$; but

$$\bar{\Delta}_F(A) = \{\langle x,y\rangle \in FA \times FA : \{B \xrightarrow{f} A : (Ff)x = (Ff)y\} \in K(A)\}.$$

Therefore F is K-separated iff, for all A and all $x,y \in FA$,

$$\{B \xrightarrow{f} A : (Ff)x = (Ff)y\} \in K(A) \Rightarrow x = y. \tag{5}$$

This condition may be informally expressed as: 'if x and y are identified by F sufficiently often, they are the same'.

Given $S \in \Omega(A)$, a set $\{a_f : f \in S\}$ with $a_f \in F(\text{dom}(f))$ for all $f \in S$ is called *(S,F)-compatible* provided that, for any $f \in S$ and any $C \xrightarrow{g} \text{dom}(f)$, we have

$$a_{f\circ g} = (Fg)a_f$$

This condition may also be put in the 'symmetric' form:

for any $B \xrightarrow{f} A$, $C \xrightarrow{g} A$ in S and any commutative diagram

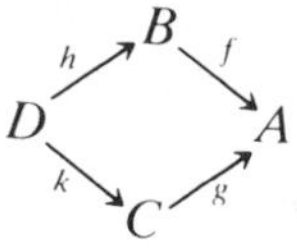

we have $(Fh)(a_f) = (Fk)(a_g)$.

It is now easily established, using (5), that a presheaf F is K-separated iff the following condition is satisfied:

(Sep) For any $A \in \mathrm{Ob}(\mathbf{C})$, any $S \in K(A)$ and any (S, F)-compatible set $\{a_f : f \in S\}$, there is *at most one* $a \in FA$ such that $(Ff)a = a_f$ for all $f \in S$.

We can provide another formulation of separatedness as follows. First, we observe that each cosieve $S \in \Omega(A)$ corresponds to the subfunctor S_* of H^A given by

$$S_*(B) = \{f \in S : \mathrm{dom}(f) = B\}$$

and conversely each subfunctor G of H^A corresponds to the cosieve

$$\bigcup_{B \in \mathrm{Ob}(\mathbf{C})} GB.$$

Next, any arrow $S_* \xrightarrow{\eta} F$ corresponds to the (S, F)-compatible set

$$\bigcup_{B \in \mathrm{Ob}(\mathbf{C})} \{\eta_B(f) : f \in S_*(B)\}$$

and conversely any (S, F)-compatible set $\{a_f : f \in S\}$ engenders the arrow $S_* \xrightarrow{\eta} F$ defined by

$$\eta_B(f) = a_f$$

for $f \in S_*(B)$. Thus (S, F)-compatible sets may be identified with arrows $S_* \to F$ in $\mathbf{E}(\mathbf{C})$.

Now the Yoneda lemma asserts that elements of FA correspond precisely to arrows $H^A \to F$ in $\mathbf{E}(\mathbf{C})$. Putting all these facts together, we (should) see that (Sep) may be restated as:

(Sep′) For each $A \in \mathrm{Ob}(\mathbf{C})$, and each $S \in K(A)$, any arrow $S_* \to F$ has at most one extension to an arrow $H^A \to F$.

The necessity of this condition for separatedness could also have been derived from the readily established observation that, for $S \in \Omega(A)$,

$$S \in K(A) \Leftrightarrow S_* \text{ is } K\text{-dense in } H^A.$$

Now we proceed to characterize the sheaves in $\mathbf{E}(\mathbf{C})$. Let F be a sheaf in $\mathbf{E}(\mathbf{C})$, and let $S \in K(A)$. Given an (S, F)-compatible set $\{a_f : f \in S\}$, let $\eta : S_* \to F$ be the corresponding arrow in $\mathbf{E}(\mathbf{C})$. Since $S \in K(A)$, S_* is, according to the observation immediately above, K-dense in H^A. Therefore, since F is a K-sheaf, there is a unique arrow $H^A \xrightarrow{\xi} F$ extending η. By the ever-reliable Yoneda lemma, ξ corresponds to an element $a \in FA$. For this element a it is now readily established that

$$(Ff)a = a_f$$

for all $f \in S$. It therefore follows that, if F is a K-sheaf, the condition stated below obtains.

(Shv) For any $A \in \mathrm{Ob}(\mathbf{C})$, any $S \in K(A)$ and any (S,F)-compatible set $\{a_f : f \in S\}$ there is a *unique* $a \in FA$ such that $(Ff)a = a_f$ for all $f \in S$.

Conversely, any presheaf F satisfying (Shv) is a K-sheaf. To see this, we show that F satisfies condition (5.13)(iv) in Th($\mathbf{E}(\mathbf{C})$). To do this it suffices to show that, for any $\mathbf{F}$-formula α and any $A \in \mathrm{Ob}(\mathbf{C})$,

$$A \Vdash^{\mathbf{C}} (\exists! u \in F.\alpha)^* \qquad \text{implies} \qquad A \Vdash^{\mathbf{C}} \exists u \in F.\alpha^*.$$

Suppose $A \Vdash^{\mathbf{C}} (\exists! u \in F.\alpha)^*$. Then

$$S = \{B \xrightarrow{f} A : B \Vdash^{\mathbf{C}} \exists! u \in F.\alpha\} \in K(A).$$

Thus for each $B \xrightarrow{f} A$ in S there is a unique element $a_f \in FB$ such that $B \Vdash^{\mathbf{C}} \alpha[a_f]$. It is now easy to see that $\{a_f : f \in S\}$ is an (S,F)-compatible set, so by (Shv) there is a (unique) $a \in FA$ such that $a_f = (Ff)a$. We then have

$$\{B \xrightarrow{f} A : B \Vdash^{\mathbf{C}} \alpha[(Ff)a]\} \supseteq S \in K(A),$$

so that $A \Vdash^{\mathbf{C}} \alpha^*[a]$, whence $A \Vdash^{\mathbf{C}} \exists u \in F.\alpha^*$ as required.

We conclude that (Shv) is a necessary and sufficient condition for a presheaf F to be a K-sheaf.

We shall write $\mathbf{Sh}_K(\mathbf{C})$ for the full subcategory of $\mathbf{E}(\mathbf{C})$ whose objects are the K-sheaves.

Sheaves over locales and topological spaces

The special case in which $\mathbf{C}$ is (the category associated with) a locale H is of particular interest. In this case we define the *canonical covering system* on $\mathbf{H}$ to be the covering system K given by

$$K(p) = \{S \in \Omega(p) : \bigvee S = p\}$$

for $p \in H$. We write $\mathbf{Sh}(H)$ for the category of K-sheaves in $\mathbf{E}(\mathbf{H})$. Objects of $\mathbf{Sh}(H)$ will be called *sheaves on H.*

More specifically still we can take H to be the locale $\mathcal{O}(X)$ of open sets in a topological space X. In this case the canonical covering system K in $\mathbf{E}(\mathcal{O}(X))$ is given by

$$K(U) = \{S \in \Omega(U) : \bigcup S = U\}$$

for $U \in \mathcal{O}(X)$. In this definition we can discern the origin of the term 'covering system', since here the covering cosieves on an open set U are those cosieves which cover U in the usual set-theoretical sense. The category $\mathbf{Sh}(\mathcal{O}(X))$ is written $\mathbf{Sh}(X)$ and called the *category of sheaves on* X.

A classical example of a sheaf on X is the sheaf C of real-valued continuous functions on X, where, for $U, V \in \mathcal{O}(X)$, $C(U)$ is the ring of continuous real-valued functions on U and, for $f \in C(U)$, $C_{UV}(f) = f \mid V$.

In the next chapter we shall obtain characterizations of toposes equivalent to $\mathbf{Sh}(H)$ for some locale H and to $\mathbf{Sh}(X)$ for some topological space X.

REMARK *The origins of the terms 'sheaf' and 'geometric morphism'.* A *displayed space* (from the French 'espace étalé') over a topological space X is a space Y equipped with a local homeomorphism to X, i.e. a continuous map $\pi : Y \to X$ such that, for any $y \in Y$, there is a neighbourhood U of y such that the restriction $\pi \mid U$ is a homeomorphism $U \to \pi[U]$. Given a displayed space (Y, π) over X, the *stalk* of (Y, π) at a point $x \in X$ is the set $Y_x = \pi^{-1}(x)$. Since $Y = \bigcup_{x \in X} Y_x$, we may think of the displayed space Y as the 'sheaf' or 'bundle' of stalks Y_x, as in the following picture:

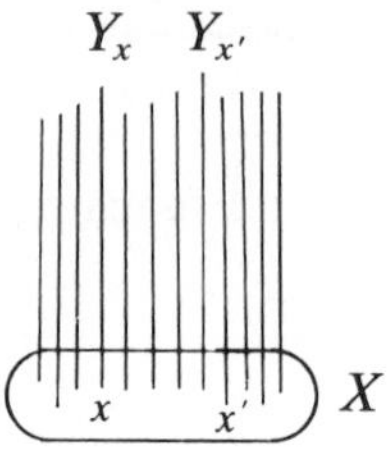

The displayed spaces over X form the objects of a category $\mathbf{Esp}(X)$ if we define an arrow between displayed spaces (Y, π) and (Z, ρ) over X to be a continuous map $h : Y \to Z$ such that the diagram

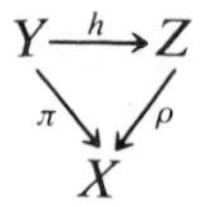

commutes.

It can be shown (see, e.g., Johnstone 1977) that the categories $\mathbf{Sh}(X)$ and $\mathbf{Esp}(X)$ are equivalent. It is the fact of this equivalence which justifies calling the objects of $\mathbf{Sh}(X)$ 'sheaves'.

So a sheaf on X may be regarded either as a presheaf satisfying a certain condition, or as a displayed space over X. Using these alternative descriptions, it can now be shown (see, e.g., Johnstone 1977, Section 0.26) that any continuous map $X \to Y$ to a topological space Y induces a geometric morphism $\mathbf{Sh}(X) \to \mathbf{Sh}(Y)$. These induced functors between categories of sheaves were the first examples of what were later to be called geometric morphisms: the term 'geometric' was assigned to emphasize their topological origins.

5.35 EXERCISES Let X be a topological space.

(1) If F is a presheaf on X, U and V are open sets in X such that $V \subseteq U$ and $a \in F(U)$, write $a \mid V$ for $F_{UV}(a)$. Show that F is a sheaf iff it satisfies the following condition: for any open covering $\{U_i : i \in I\}$ of an open set U and any family $\{a_i : i \in I\}$ such that $a_i \in F(U_i)$ for all $i \in I$ and, for any $i, j \in I$, we have $a_i \mid U_i \cap U_j = a_j \mid U_i \cap U_j$, there is a unique $a \in F(U)$ such that $a_i = a \mid U_i$ for all $i \in I$.

(2) A map f on X to a set I is said to be *locally constant* if each element of X has a neighbourhood on which f is constant. Let $I^\dagger$ be the presheaf on X whose object function is given by $I^\dagger(U) =$ set of locally constant functions $U \to I$, for $U \in \mathcal{O}(X)$, with the arrow function given by restriction. Show that $I^\dagger$ is a sheaf, and that, in $\mathbf{Sh}(X)$, it is the I-fold copower of the terminal object. Deduce that $I^\dagger \cong L\hat{I}$, where $\hat{I}$ is the constant presheaf with value I, and $L : \mathbf{Set}^{\mathbf{C}^{\mathrm{op}}} \to \mathbf{Sh}(X)$ is the sheafification functor.

6

Locale-valued sets

Given a locale H, we may think of the members of H as possible 'truth values' for propositions. The situation in which these propositions are assertions of equality among the elements of a set gives rise to the concept of *H-valued set.* We show in this chapter that the collection of H-valued sets can be organized into a category in a natural way, and that this category is a topos equivalent to the category of sheaves on H. Necessary and sufficient conditions are formulated for a topos to be equivalent to a topos of H-valued sets for some locale H, and hence to a category of sheaves on a locale. (In particular, we obtain a characterization of the category of sets.) We continue with a discussion of the related category of *H-fuzzy sets,* and conclude the chapter with a topos-theoretic analysis of Boolean extensions of the universe of sets.

Locale-valued sets

Let H be a locale. An *H-valued set*—or *H-set* for short—is an (intuitive) set equipped with an *H-valued equality relation,* that is, a pair (I, δ) consisting of a set I and a map $\delta : I \times I \to H$ satisfying

$$\delta(i,i') = \delta(i',i) \qquad \text{(symmetry)}$$

$$\delta(i,i') \wedge \delta(i',i'') \leqslant \delta(i,i'') \qquad \text{(transitivity)}$$

for all $i,i',i'' \in I$. Notice that we do *not* insist on the 'reflexivity' condition $\delta(i,i) = 1$.

We shall usually write $\delta_{ii'}$ for $\delta(i,i')$.

We define the *category* $\mathbf{Set}_H$ *of H-valued sets* as follows. Its *objects* are all H-valued sets. An *arrow* f between two H-valued sets (I, δ) and (J, ε) is an *H-valued functional relation between* (I, δ) and (J, ε), that is, a map $f : I \times J \to H$ satisfying

$$\left.\begin{array}{l} \delta(i,i') \wedge f(i,j) \leqslant f(i',j) \\ f(i,j) \wedge \varepsilon(j,j') \leqslant f(i,j') \end{array}\right\} \qquad \text{(substitutivity)}$$

$$f(i,j) \wedge f(i,j') \leqslant \varepsilon(j,j') \qquad \text{(functionality)}$$

$$\bigvee_{j \in J} f(i,j) = \delta(i,i) \qquad \text{(definition on } I\text{)}$$

for all $i,i' \in I$, $j,j' \in J$. We shall often write f_{ij} for $f(i,j)$. Given arrows $(I,\delta)\xrightarrow{f}(J,\varepsilon)\xrightarrow{g}(K,\eta)$ in $\mathbf{Set}_H$, the composition $(I,\delta)\xrightarrow{g\circ f}(K,\eta)$ is defined by

$$(g \circ f)(i,k) = \bigvee_{j\in J} f(i,j) \wedge g(j,k).$$

The identity arrow $1_{(I,\delta)}$ on (I,δ) is just δ.

Using the distributive law

$$a \wedge \bigvee_{i\in I} b_i = \bigvee_{i\in I} a \wedge b_i$$

in H, it is readily checked that these specifications do yield a category. We are later going to show that $\mathbf{Set}_H$ is actually a *topos*. Rather than tackling this problem head-on, however, we shall argue as follows. First, we determine sufficient (and also necessary) conditions on a (full) local set theory S for $\mathbf{C}(S)$ to be equivalent to $\mathbf{Set}_{\Omega(S)}$ where $\Omega(S)$ is the locale of sentences of S. Next, given a locale H, we construct a topos $\mathbf{E}$ for which $\Omega(\mathrm{Th}(\mathbf{E})) \cong H$, and such that $\mathrm{Th}(\mathbf{E})$ satisfies the aforesaid conditions. It will then follow that $\mathbf{Set}_H \simeq \mathbf{C}(\mathrm{Th}(E)) \simeq \mathbf{E}$, so that $\mathbf{Set}_H$ is a topos.

Let S be a full local set theory. Then by (4.35) $(\Omega(S), \leqslant)$ is a locale. We shall define a functor $F:\mathbf{Set}_{\Omega(S)} \to \mathbf{C}(S)$ which, under certain further conditions to be specified presently, we will show to be an equivalence of categories.

Let (I,δ) be an $\Omega(S)$-set. By the universal property of $(I\times I)\hat{}$ in S, there is a formula α such that $\vdash_S \alpha((i,j)\hat{}) = \delta_{ij}$ for all $i,j\in I$. So if ρ is the term given in (4.34) which establishes an isomorphism $\hat{I}\times\hat{I}\to(I\times I)\hat{}$, writing $\delta^*(x,y)$ for $\alpha(\rho\langle x,y\rangle))$, we have

$$\vdash_S \delta^*(\hat{\imath},\hat{\jmath}) = \delta_{ij}.$$

Let

$$I^* = \{x \in \hat{I} : \delta^*(x,x)\}$$

and

$$R = \{\langle x,y\rangle \in I\times I : \delta^*(x,y)\}.$$

It is now easily shown, using (4.34)(i), that

$\vdash_S$ *R is an equivalence relation on I^*.*

For instance, to show that $\vdash_S R \subseteq I^* \times I^*$ we note that $\delta_{ij} \leqslant \delta_{ii} \wedge \delta_{jj}$ for any $i,j\in I$, so

$$\vdash_S \delta^*(\hat{\imath},\hat{\jmath}) \Rightarrow \delta^*(\hat{\imath},\hat{\imath}) \wedge \delta^*(\hat{\jmath},\hat{\jmath})$$

and therefore, by (4.34)(i),

$$\vdash_S \forall x\, \forall y[\delta^*(x,y) \Rightarrow \delta^*(x,x) \wedge \delta^*(y,y)].$$

Hence

$$\vdash_S \forall x\, \forall y[\langle x,y\rangle \in R \Rightarrow x \in I^* \wedge y \in I^*].$$

The other conditions are established similarly.

We write $\tilde{x}$ for the term $\{y \in I^* : \langle x,y\rangle \in R\}$, $\bar{i}$ for $(\hat{\imath})^{\sim}$ and define (writing I for (I,δ))

$$FI = I^*/R = \{\tilde{x} : \delta^*(x,x)\}.$$

Given an arrow $(I,\delta) \xrightarrow{f} (J,\varepsilon)$ in $\mathbf{Set}_{\Omega(S)}$, we define $FI \xrightarrow{Ff} FJ$ in $\mathbf{C}(S)$ as follows. Using an argument similar to that employed in constructing δ^* we get a formula $\gamma(x,z)$ such that

$$\vdash_S \gamma(\hat{\imath},\hat{\jmath}) = f_{ij}$$

for $i \in I$, $j \in J$. Now define

$$Ff = \{\langle \tilde{x},\tilde{y}\rangle : \gamma(x,y)\}.$$

Then a straightforward argument using (4.34)(i) shows that Ff is an arrow $FI \to FJ$ in $\mathbf{C}(S)$. Using (4.36), one also shows that $F(g \circ f) = Fg \circ Ff$, and it is easy to see that $F1_{(I,\delta)} = 1_{FI}$. Therefore F is a functor.

Now we can prove

6.1 PROPOSITION. $F : \mathbf{Set}_{\Omega(S)} \to \mathbf{C}(S)$ is full and faithful.

Proof. *F is faithful.* Suppose that $(I,\delta) \underset{f'}{\overset{f}{\rightrightarrows}} (J,\varepsilon)$ and $Ff = Ff'$. Then

$$\vdash_S \langle \tilde{x},\tilde{y}\rangle \in Ff \Leftrightarrow \langle \tilde{x},\tilde{y}\rangle \in Ff'$$

whence for all $i \in I$, $j \in J$,

$$\vdash_S \langle \bar{i},\bar{j}\rangle \in Ff \Leftrightarrow \langle \bar{i},\bar{j}\rangle \in Ff'$$

so that

$$\vdash_S \delta_{ii} \wedge \varepsilon_{jj} \Rightarrow [f_{ij} \Leftrightarrow f'_{ij}].$$

Therefore, for all $i \in I$, $j \in J$,

$$\vdash_S [\delta_{ii} \wedge \varepsilon_{jj} \wedge f_{ij}] \Leftrightarrow [\delta_{ii} \wedge \varepsilon_{jj} \wedge f'_{ij}]$$

and so

$$\vdash_S f_{ij} = f'_{ij}.$$

It follows that $f = f'$, and so F is faithful.

F is full. Given $FI \xrightarrow{g} FJ$ in $\mathbf{C}(S)$, define $f : I \times J \to \Omega(S)$ by

$$f_{ij} = (\langle \bar{i},\bar{j}\rangle \in g).$$

It is easy to check that f is an arrow $(I,\delta)\to(J,\varepsilon)$ in $\mathbf{Set}_{\Omega(S)}$. Moreover, for $i\in I$, $j\in J$,

$$\begin{aligned}\vdash_S \langle\bar{i},\bar{j}\rangle \in Ff &\Leftrightarrow \delta_{ii}\wedge\varepsilon_{jj}\wedge\gamma(\hat{i},\hat{j})\\ &\Leftrightarrow \delta_{ii}\wedge\varepsilon_{jj}\wedge\langle\bar{i},\bar{j}\rangle\in g\\ &\Leftrightarrow \langle\bar{i},\bar{j}\rangle\in g.\end{aligned}$$

So by (4.34)(i)

$$\vdash_S \forall x\,\forall y\,[\langle\tilde{x},\tilde{y}\rangle\in Ff\Leftrightarrow\langle\tilde{x},\tilde{y}\rangle\in g],$$

whence $Ff=g$. Therefore F is full. □

We now determine a condition under which F is an equivalence. Given a type symbol $\mathbf{A}$ in the language $\mathscr{L}$ of S, let $\Delta(\mathbf{A})$ be the set of all $\mathbf{A}$-singletons (4.29(iv)). We turn $\Delta(\mathbf{A})$ into an $\Omega(S)$-set by defining, for U, $V\in\Delta(\mathbf{A})$,

$$\delta(U,V)=\exists x.x\in U\cap V.$$

Now each $U\in\Delta(\mathbf{A})$ is a closed term of type $\mathbf{PA}$, so by the universal property of $\boldsymbol{\Delta}(\mathbf{A})\hat{}$ there is a term $\tau(u)$ of type $\mathbf{PA}$ such that, for all $U\in\Delta(\mathbf{A})$,

$$\vdash_S \tau(\hat{U})=U.$$

Define

$$k_{\mathbf{A}}=\{\langle\tilde{u},x\rangle : x\in\tau(u)\}.$$

It is readily checked, using (4.34)(i), that $k_{\mathbf{A}}$ is a monic arrow $F(\Delta(\mathbf{A}))\to A$ in $\mathbf{C}(S)$. We can now prove the

6.2 PROPOSITION. $k_{\mathbf{A}}:F(\Delta(\mathbf{A}))\to A$ is an isomorphism iff $\mathbf{A}$ is S-singleton-complete (4.29(iv)).

Proof. Since $k_{\mathbf{A}}$ is monic, it is an isomorphism iff it is epic, i.e. if

$$\vdash_S \forall x\,\exists u.x\in\tau(u). \tag{1}$$

Assume (1) and suppose α is a formula such that $\vdash_S\forall x\in U.\alpha$ for any $U\in\Delta(\mathbf{A})$. Then $\vdash_S\forall x\in\tau(\hat{U}).\alpha$ for all $U\in\Delta(\mathbf{A})$, so by (4.34(i))

$$\vdash_S \forall u\,\forall x\in\tau(u).\alpha. \tag{2}$$

From (1) and (2) we deduce $\vdash_S\forall x\alpha$. So $\mathbf{A}$ is S-singleton-complete.

Conversely, suppose $\mathbf{A}$ is S-singleton-complete. Let α be the formula $\exists v.x\in\tau(v)$. Then clearly

$$\vdash_S \forall u\,\forall x\in\tau(u).\alpha,$$

whence

$$\vdash_S \forall x\in\tau(\hat{U}).\alpha$$

for every $U \in \Delta(\mathbf{A})$. Hence

$$\vdash_S \forall x \in U.\alpha$$

for all $U \in \Delta(\mathbf{A})$, so since $\mathbf{A}$ is S-singleton-complete,

$$\vdash_S \forall x \alpha,$$

i.e. (1). Therefore $k_{\mathbf{A}}$ is an isomorphism. □

As an immediate consequence of this proposition we get the following

6.3 THEOREM. If S is full and singleton-complete, then $\mathbf{C}(S) \simeq \mathbf{Set}_{\Omega(S)}$. □

Now if $\mathbf{E}$ is a topos defined over $\mathbf{Set}$, then $\mathrm{Th}(\mathbf{E})$ is full and so $\Omega(\mathrm{Th}(\mathbf{E}))$ is a locale. Writing $\Omega(\mathbf{E})$ for $\mathbf{E}(1,\Omega)$, we have $\Omega(\mathbf{E}) \cong \Omega(\mathrm{Th}(\mathbf{E}))$. Moreover, $\mathbf{E}$ is subextensional iff $\mathrm{Th}(\mathbf{E})$ is singleton-complete (4.32). Since $\mathbf{E} \simeq \mathbf{C}(\mathrm{Th}(\mathbf{E}))$ it now follows immediately from (6.3) that

6.4 THEOREM. If $\mathbf{E}$ is a subextensional topos defined over $\mathbf{Set}$, then $\mathbf{E} \simeq \mathbf{Set}_{\Omega(\mathbf{E})}$. □

Hence

6.5 COROLLARY. Let $\mathbf{E}$ and $\mathbf{E}'$ be subextensional toposes defined over $\mathbf{Set}$. Then

$$\mathbf{E} \simeq \mathbf{E}' \quad \text{iff} \quad \Omega(\mathbf{E}) \cong \Omega(\mathbf{E}'). \quad \square$$

We now show that, for any locale H, we have $H \cong \Omega(\mathbf{Sh}(H))$. For if K is the canonical covering system on H, j the associated modal operator in $\mathbf{E}(H)$, and Ω^* the subobject classifier in $\mathbf{Sh}(H)$, then for $p \in H$, S a cosieve on p, we have

$$j_p(S) = \{q \leq p : q \leq \bigvee S\}$$

so that

$$S \in \Omega^*(p) \Leftrightarrow j_p(S) = S \Leftrightarrow \bigvee S \in S.$$

It follows that

$$\Omega^*(p) = \{\max^q : q \leq p\}. \qquad (*)$$

Now define $\Omega^{**} \in \mathbf{E}(H)$ by

$$\Omega^{**}(p) = \max^p;\ \Omega^{**}_{pq}(r) = q \wedge r \qquad \text{for} \quad q,r \leq p.$$

It is then easy to check, using (*), that the map $p \mapsto \max^p$ for $p \in H$ establishes a natural isomorphism between Ω^{**} and Ω^* in $\mathbf{E}(H)$. *It follows that Ω^{**} is a subobject classifier in* $\mathbf{Sh}(H)$.

It is now readily seen that there is a bijective correspondence between $\Omega(\mathbf{Sh}(H))$, that is, the set of arrows $1 \rightarrow \Omega^{**}$ in $\mathbf{Sh}(H)$, and H, which yields the required isomorphism between $\Omega(\mathbf{Sh}(H))$ and H.

Accordingly, in order to prove that $\mathbf{Set}_H \simeq \mathbf{Sh}(H)$ we need only show that $\mathbf{Sh}(H)$ satisfies the conditions imposed on $\mathbf{E}$ in (6.4), viz., that it is subextensional and defined over **Set**.

To see that $\mathbf{Sh}(H)$ is subextensional, we first recall that $\mathbf{E}(H)$ is subextensional (Example (ii) following (4.32)) and so it suffices to prove the following.

6.6 LEMMA. If $\mathbf{E}$ is a subextensional topos, and j a modal operator in $\mathbf{E}$, then $\mathbf{Sh}_j(\mathbf{E})$ is subextensional.

Proof. Since the sheafification functor $L: \mathbf{E} \rightarrow \mathbf{Sh}_j(\mathbf{E})$ is left exact (5.24), it preserves monics and terminal objects. Now suppose that $X \underset{g}{\overset{f}{\rightrightarrows}} Y$ is a diagram in $\mathbf{Sh}_j(\mathbf{E})$ with $f \neq g$. Then since $\mathbf{E}$ is subextensional, there is $U \rightarrowtail 1$ and $U \xrightarrow{h} X$ in $\mathbf{E}$ such that $f \circ h \neq g \circ h$. Then $LU \rightarrowtail L1 \cong 1$ and since X is a sheaf there is an arrow $h^0: LU \rightarrow X$ such that the diagram

$$\begin{array}{ccc} U & \xrightarrow{h} & X \\ {\scriptstyle l_U}\downarrow & \nearrow{\scriptstyle h^0} & \\ LU & & \end{array}$$

commutes. Clearly $f \circ h^0 \neq g \circ h^0$, and the lemma follows. □

To show that $\mathbf{Sh}(H)$ is defined over **Set**, we recall that $\mathbf{E}(H)$ is defined over **Set**, so we need only establish the following

6.7 LEMMA. If $\mathbf{E}$ is a topos defined over **Set** and j is a modal operator in $\mathbf{E}$, then $\mathbf{Sh}_j(\mathbf{E})$ is defined over **Set**.

Proof. Since the sheafification functor $L: \mathbf{E} \rightarrow \mathbf{Sh}_j(\mathbf{E})$ has a right adjoint (cf. (5.22)), it preserves colimits. Therefore, since $\coprod_I 1$ exists in $\mathbf{E}$ for any set I,

$$L\Big(\coprod_I 1\Big) \cong \coprod_I L(1) \cong \coprod_I 1 \quad \text{in } \mathbf{Sh}_j(\mathbf{E}).$$

That is, $\coprod_I 1$ also exists in $\mathbf{Sh}_j(\mathbf{E})$, so the latter is defined over **Set**. □

Putting all these things together, we have proved

6.8 THEOREM. For any locale H, $\mathbf{Set}_H \simeq \mathbf{Sh}(H)$ and so $\mathbf{Set}_H$ is a topos. □

6.9 COROLLARY. Let $\mathbf{E}$ be a topos. Then the following conditions are equivalent.

(i) $\mathbf{E}$ is subextensional and defined over **Set**.
(ii) $\mathbf{E} \simeq \mathbf{Set}_H \simeq \mathbf{Sh}(H)$ for some locale H.

Proof. (i)$\Rightarrow$(ii) follows from (6.4) and (6.8), and (ii)$\Rightarrow$(i) from (6.6) and (6.7). □

In view of (6.9), a topos which is subextensional and defined over **Set** will be called *localic.*

6.10 COROLLARY. The following conditions on a topos **E** are equivalent:

(i) **E** satisfies **AC** and is defined over **Set**.
(ii) **E** is Boolean and localic;
(iii) $\mathbf{E} \simeq \mathbf{Set}_B \simeq \mathbf{Sh}(B)$ for some complete Boolean algebra B.

Proof. (i)$\Rightarrow$(ii) follows from (4.38). Assuming (ii), it is clear that $\Omega(\mathbf{E})$ is a complete Boolean algebra, so (iii) follows from (6.4) and (6.8). Finally, assuming (iii), **E** is then Boolean, subextensional, and defined over **Set**, and so satisfies **AC** by (4.38). Hence (i). □

As a consequence we obtain the following characterization of the category of sets.

6.11 COROLLARY. The following conditions on a topos **E** are equivalent:

(i) $\mathbf{E} \simeq \mathbf{Set}$;
(ii) **E** is extensional and defined over **Set**;
(iii) **E** is bivalent, satisfies **AC** and is defined over **Set**.

Proof. (i)$\Rightarrow$(ii) is obvious. Assuming (ii), it follows from 4.31(iii) that Th(**E**) is classical, singleton-complete and complete, so that **E** is Boolean, subextensional and bivalent. So by (6.10) **E** also satisfies **AC**, and (iii) follows.

Finally, assume (iii). Then by (6.10) $\mathbf{E} \simeq \mathbf{Set}_B$ for some complete Boolean algebra B, and since **E** is bivalent, B has just two elements 0,1, that is, B is isomorphic to the two-element Boolean algebra 2. Therefore $\mathbf{E} \simeq \mathbf{Set}_2$. But (6.5) implies that $\mathbf{Set}_2 \simeq \mathbf{Set}$, so (i) follows. □

EXERCISE Let S be a full, well-termed local set theory. (1) Show that $\Delta(1)$ may be identified with the $\Omega(S)$-set $(\Omega(S), \Leftrightarrow)$ and deduce that this is the terminal object in $\mathbf{Set}_{\Omega(S)}$. Hence obtain a description of the terminal object in $\mathbf{Set}_H$.
(2) Show that $\Delta(\Omega)$ may be identified with the $\Omega(S)$-set

$$\Omega' = (\{(\alpha,\beta) \in \Omega(S) \times \Omega(S) : \alpha \leqslant \beta\},\ \delta)$$

where δ is given by

$$\delta((\alpha,\beta),\ (\alpha',\beta')) = (\alpha \Leftrightarrow \alpha') \wedge \beta \wedge \beta'.$$

Deduce that Ω' is the subobject classifier in $\mathbf{Set}_{\Omega(S)}$ and hence obtain a description of the subobject classifier in $\mathbf{Set}_H$.

The topos of sheaves on a topological space

We now determine necessary and sufficient conditions on a topos for it to be equivalent to a topos of sheaves on a topological space, or to a topos of presheaves on a partially ordered set.

Let L be a lattice. An element $a \in L$ is called *large* if whenever $x,y \in L$ and $x \wedge y \leqslant a$ then $x \leqslant a$ or $y \leqslant a$. An element $a \in L$ is said to *distinguish* the elements $x,y \in L$ if ($x \leqslant a$ iff $y \nleqslant a$). L is said to be *spatial* if it is complete and, for any pair of distinct elements of L, there is a large element distinguishing them.

6.12 LEMMA. A lattice L is isomorphic to a topology $\mathcal{O}(X)$ on some topological space X if and only if it is spatial.

Proof. For any topological space X, it is easy to show that $\mathcal{O}(X)$ is spatial by observing that all open sets of the form $X - \overline{\{x\}}$ are large in $\mathcal{O}(X)$.

Conversely, suppose that L is spatial and let A be the set of large elements of L. Define the map $\phi: L \to P(A)$ by

$$\phi(x) = \{a \in A : x \nleqslant a\}.$$

It is now easy to see that the image of L under ϕ is closed under arbitrary unions and finite intersections; i.e. it is a topology on A. And since L is spatial, ϕ establishes an order isomorphism of L with M. □

As a result, we have

6.13 THEOREM. Let $\mathbf{E}$ be a topos. Then the following conditions are equivalent:

(i) $\mathbf{E}$ is localic and $\Omega(\mathbf{E})$ is spatial;
(ii) $\mathbf{E} \simeq \mathbf{Sh}(X)$ for some topological space X.

Proof. (ii)$\Rightarrow$(i) is evident. Conversely, if (i) holds, then by (6.9) $\mathbf{E} \simeq \mathbf{Sh}(H)$ for some locale $H \cong \Omega(\mathbf{E})$. But then H is spatial, and so by (6.12) $H \cong \mathcal{O}(X)$ for some topological space X. Thus $E \simeq \mathbf{Sh}(\mathcal{O}(X)) = \mathbf{Sh}(X)$. □

6.14 EXERCISE. Let $(P, \leqslant)$ be a partially ordered set. The *order topology* Ord(P) on P is the family of all *cosieves* in P, i.e. subsets U of P such that $x \in U$, $y \leqslant x \Rightarrow y \in U$.

(1) Show that, with the order topology, P is a T_0-space (i.e. for any pair

of distinct points of P there is an open set containing one of the points but not the other) in which the intersection of any family of open sets is open. A space satisfying these conditions is called an *Alexandrov space*.
(2) Conversely, show that, for any Alexandrov space (X, T) there is a (unique) partial ordering $\leqslant$ on X such that $T = \mathrm{Ord}(X)$. (Define $x \leqslant y$ to be $y \in \overline{\{x\}}$.)
(3) Let L be a complete lattice. An element $a \in L$ is called *huge* if whenever $\bigwedge_{i \in I} x_i \leqslant a$ in L, we have $x_i \leqslant a$ for at least one $i \in I$. L is called an *Alexandrov lattice* if for any pair of distinct elements of L there is a huge element distinguishing them. Show that L is an Alexandrov lattice iff $L \cong \mathrm{Ord}(P)$ for some partially ordered set $(P, \leqslant)$.
(4) Let $\mathbf{E(P)}$ be the topos of presheaves on $\mathbf{P}$. Show that $\Omega(\mathbf{E(P)}) \cong \mathrm{Ord(P)}$, and deduce that $\mathbf{E(P)} \simeq \mathbf{Sh}(X)$, where $X = (P, \mathrm{Ord}(P))$.
(5) Show that the following conditions on a topos $\mathbf{E}$ are equivalent:
(a) $\mathbf{E}$ is localic and $\Omega(\mathbf{E})$ is an Alexandrov lattice;
(b) $\mathbf{E} \simeq \mathbf{E(P)}$ for some partially ordered set $(P, \leqslant)$.

Decidable, subconstant, and fuzzy sets

Let S be a local set theory. An S-set X is said to be *decidable* if the equality predicate on X is decidable, that is, if

$$x \in X, x' \in X \vdash_S x = x' \vee \neg(x = x').$$

If Y is an S-set and $\vdash_S X \subseteq Y$, then X is said to be *complemented* in Y if there is an S-set Z, called a *complement* of X in Y, such that

$$\vdash_S X \cup Z = Y \wedge \vdash_S X \cap Z = \varnothing.$$

It is evident that S is classical iff every S-set is decidable or, in particular, iff Ω is decidable. Moreover, an S-set X is decidable iff the diagonal Δ_X is complemented in $X \times X$. Clearly, also, any subset of a decidable set is decidable.

Now suppose that S is full. An S-set of the form $\hat{I}$ will be called a *constant* (S-)set, and a subset of a constant set a *subconstant* set. Let $\mathbf{C}^-(S)$ be the full subcategory of $\mathbf{C}(S)$ whose objects are all subconstant sets.

6.15 PROPOSITION. Any subconstant set is decidable.

Proof. It suffices to show that $\hat{I}$ is decidable for any I. Let $i, j \in I$. If $i = j$, then evidently $\vdash_S \hat{\imath} = \hat{\jmath}$. Now suppose that $i \neq j$. For $k \in I$ define the terms τ_k by $\tau_i = 0$, $\tau_j = 1 = \tau_k$ for $k \in I - \{i, j\}$. Then by the universal property of $\hat{\mathbf{I}}$ there is a term τ such that $\vdash_S \tau(\hat{k}) = \tau_k$ for $k \in I$. It follows that†

† We write $x \neq y$ for $\neg(x = y)$.

$\vdash_S \tau(\hat{i}) = 0 \neq 1 = \tau(\hat{j})$, whence $\vdash_S \hat{i} \neq \hat{j}$. Hence $\vdash_S \hat{i} = \hat{j}$ or $\vdash_S \hat{i} \neq \hat{j}$ so that $\vdash_S \hat{i} = j \vee i \neq j$. Therefore, by (4.34)(i),

$$\vdash_S \forall x \in \hat{I}\, \forall y \in \hat{I}\ (x = y \vee x \neq y),$$

so that $\hat{i}$ is decidable. □

We now proceed to give an alternative description of $\mathbf{C}^-(S)$.

Given a locale H, we define the category $\tilde{\mathbf{H}}$ of *H-fuzzy sets* as follows. The *objects* of $\tilde{\mathbf{H}}$ are all functions $I \xrightarrow{a} H$, $a = \langle a_i \rangle_{i \in I}$ for all sets I. If $I \xrightarrow{a} H$, $J \xrightarrow{b} H$ are $\tilde{\mathbf{H}}$-objects, an $\tilde{\mathbf{H}}$-*arrow* is a function $p: I \times J \to H$, $p = \langle p_{ij} \rangle_{i \in I, j \in J}$ such that

(6.16) $$p_{ij} \leqslant b_j \qquad (i \in I, j \in J)$$

(6.17) $$p_{ij} \wedge p_{ij'} = 0 \qquad (i \in I,\ j \neq j' \in J)$$

(6.18) $$\bigvee_{j \in J} p_{ij} = a_i \qquad (i \in I)$$

If $c = \langle c_k \rangle_{k \in K}$ is an object of $\tilde{\mathbf{H}}$ and $q: J \times K \to H$ is an arrow $b \to c$ in $\tilde{\mathbf{H}}$, the composition $q \circ p = r$ of p and q in $\tilde{\mathbf{H}}$ is defined by

$$r_{ik} = \bigvee_{j \in J} p_{ij} \wedge q_{jk}.$$

It is easy to check that composition as just defined is associative and that the identity arrow $1: a \to a$ is given by the map $1: I \times I \to H$ such that

$$1_{ii'} = 0 \quad (i \neq i'), \qquad 1_{ii} = a_i.$$

If S is full we know that $\Omega(S)$ is a locale. For this locale we have:

6.19 THEOREM. $\mathbf{\Omega(S)}^{\sim} \simeq \mathbf{C}^-(S)$. If in addition S is well-typed and AC holds in S, then $\Omega(S)$ is a complete Boolean algebra and $\mathbf{\Omega(S)}^{\sim} \simeq \mathbf{C}(S)$.

Proof. We define a functor $F: \mathbf{\Omega(S)}^{\sim} \to \mathbf{C}^-(S)$ as follows. Given an object $I \xrightarrow{a} \Omega(S)$ in $\mathbf{\Omega(S)}^{\sim}$ let τ be a term (whose existence is guaranteed by the universal property of $\hat{\mathbf{I}}$ in S) such that $\vdash_S \tau(\hat{i}) = a_i$ for all $i \in I$: τ will be called the term *associated* with a. Now define

$$F(a) = \{x \in \hat{I} : \tau(x)\}.$$

Clearly $F(a)$ is a subconstant S-set.

Given an $\tilde{\mathbf{H}}$-object $J \xrightarrow{b} \Omega(S)$ and an $\tilde{\mathbf{H}}$-arrow $p: a \to b$, we define $F(p): F(a) \to F(b)$ in $\mathbf{C}^-(S)$ as follows. Let σ be the term associated with b and let ξ be a term (whose existence is again guaranteed by the universal property of $\hat{\mathbf{I}}$) such that $\vdash_S \xi(\hat{i}, \hat{j}) = p_{ij}$ for all $i \in I$, $j \in J$: ξ will be called the term *associated* with p. Then, using (4.34)(i), condition (6.16) yields

$$\vdash_S \forall x \in \hat{I}\, \forall y \in \hat{J}[\xi(x, y) \Rightarrow \sigma(y)] \qquad (*)$$

Using (4.34)(i) and the fact established in (6.15) that $\vdash_S \hat{j} \neq \hat{j}'$ whenever $j \neq j'$, (6.17) yields

$$\vdash_S \forall x \in \hat{I}\, \forall y, y' \in \hat{J}[\xi(x,y) \wedge \xi(x,y') \Rightarrow y = y']. \qquad (**)$$

Finally, using (4.36) and (4.34)(i), (6.18) yields

$$\vdash_S \forall x \in \hat{I}[\exists y \in \hat{J}\xi(x,y) \Leftrightarrow \tau(x)]. \qquad (***)$$

Now (*), (**), (***) jointly assert that the S-set

$$\{\langle x,y \rangle \in \hat{I} \times \hat{J} : \xi(x,y)\}$$

is an S-map $F(a) \to F(b)$. We take this map for $F(p)$.

It is now easy to show, using (4.36) and (4.34)(i), that F defined in this way is a functor from $\mathbf{\Omega(S)}^{\sim}$ to $\mathbf{C}^{-}(S)$. Moreover, clearly every subconstant S-set is in the range of F. So for F to be an equivalence it suffices that it be full and faithful, properties we now verify.

To verify the fidelity of F, suppose $a \overset{p}{\underset{q}{\rightrightarrows}} b$ are arrows in $\mathbf{\Omega(S)}^{\sim}$. Let ξ, η be the terms associated with p,q. If $F(p) = F(q)$, then $\vdash_S \xi(x,y) \Leftrightarrow \eta(x,y)$, whence $\vdash_S \xi(\hat{i},\hat{j}) = \eta(\hat{i},\hat{j})$, so that $p_{ij} = q_{ij}$ for all $i \in I$, $j \in J$, and so $p = q$.

To verify the fullness of F, given an arrow $F(a) \xrightarrow{f} F(b)$, let $\xi(x,y)$ be the formula $\langle x,y \rangle \in f$ and define

$$p_{ij} = \xi(\hat{i},\hat{j}).$$

It is now readily shown that $p = \langle p_{ij} \rangle_{i \in I, j \in J}$ is an arrow $a \to b$ in $\mathbf{\Omega(S)}^{\sim}$ and that $F(p) = f$.

Finally, suppose that the axiom of choice holds in S. Then by (4.31)(iv), S is singleton-complete. Given any type symbol $\mathbf{A}$, let τ be a term of type $\mathbf{PA}$ such that, for all $U \in \Delta(\mathbf{A})$, $\vdash_S \tau(\hat{U}) = U$, and define $f : \Delta(\mathbf{A})\hat{} \to A$ by

$$f = \{\langle u,x \rangle \in \Delta(\mathbf{A})\hat{} \times A : x \in \tau(u)\}.$$

Then the proof of (6.2) shows that f is epic. Since AC holds in S, there is a map $g : A \to \Delta(\mathbf{A})\hat{}$ such that $f \circ g = 1_A$. The map f is then necessarily monic and establishes an isomorphism between A and its image $f[A]$ which is a subconstant set. Therefore every type, and hence—if S is well-typed—every S-set, is isomorphic to a subconstant set, so that $\mathbf{C}(S) \simeq \mathbf{C}^{-}(S) \simeq \mathbf{\Omega(S)}^{\sim}$. □

Now let $\mathbf{E}$ be a topos defined over **Set** by a geometric morphism $\mu : \mathbf{E} \to \mathbf{Set}$. Then under the canonical equivalence between $\mathbf{C}(\mathrm{Th}(\mathbf{E}))$ and $\mathbf{E}$, constant sets in the former correspond to objects of the form $\mu^* I$ in the latter, and subconstant sets to subobjects of these objects. So let us call any subobject of an object of the form $\mu^* I$ a *subconstant* object, and write $\mathbf{E}^-$ for the full subcategory of subconstant objects of $\mathbf{E}$. Then $\mathbf{E}^- \simeq \mathbf{C}^-(\mathrm{Th}(\mathbf{E}))$ and (6.19) has the immediate consequence

6.20 COROLLARY. Let $\mathbf{E}$ be a topos defined over **Set**. Then $\mathbf{E}^- \simeq \boldsymbol{\Omega}(\mathbf{E})^{\tilde{}}$. If in addition the axiom of choice holds in $\mathbf{E}$ then $\Omega(\mathbf{E})$ is a complete Boolean algebra and $\mathbf{E} \simeq \boldsymbol{\Omega}(\mathbf{E})^{\tilde{}}$. □

We next establish the equivalence of a number of conditions on $\mathbf{C}^-(S)$ and $\boldsymbol{\Omega}(\mathbf{S})^{\tilde{}}$.

6.21 THEOREM. Let S be a full, well-termed, well-typed local set theory. Consider the conditions:

(i) AC holds in S;
(ii) $\mathbf{C}^-(S) \hookrightarrow \mathbf{C}(S)$ is an equivalence;
(iii) Ω is isomorphic to an object in $\mathbf{C}^-(S)$;
(iv) S is classical;
(v) $\mathbf{C}^-(S) \simeq \boldsymbol{\Omega}(\mathbf{S})^{\tilde{}}$ is a topos;
(vi) $\mathbf{C}^-(S) \simeq \boldsymbol{\Omega}(\mathbf{S})^{\tilde{}}$ has a subobject classifier;
(vii) $\Omega(S)$ is a Boolean algebra.

Then (i)$\Leftrightarrow$(ii)$\Rightarrow$(iii)$\Leftrightarrow$(iv)$\Rightarrow$(v)$\Leftrightarrow$(vi)$\Leftrightarrow$(vii). If, moreover, S is singleton-complete, then all the conditions are equivalent.

Proof. (i)$\Rightarrow$(ii) follows from (6.19).

(ii)$\Rightarrow$(iii) is trivial.

(iii)$\Rightarrow$(iv). By (6.15), any object of $\mathbf{C}^-(S)$ is decidable, so that if (iii) holds, Ω is decidable and S is classical.

(iv)$\Rightarrow$(iii). If S is classical, then $\Omega \cong 1+1 \cong \hat{2} \in \mathbf{C}$.

(ii)$\Rightarrow$(i). Suppose that $\mathbf{C}^-(S) \hookrightarrow \mathbf{C}(S)$ is an equivalence. We show that S is singleton-complete. Since by assumption every object of $\mathbf{C}(S)$ is isomorphic to a subconstant set, for any type $\mathbf{A}$ there is a set I and a term τ of type $\hat{\mathbf{I}}$ such that $x \mapsto \tau$ is a monic map $A \to \hat{I}$. For each $i \in I$ let

$$U_i = \{x \in A : \tau(x) = \hat{\imath}\}.$$

Then since $x \mapsto \tau$ is monic, U_i is an $\mathbf{A}$-singleton. Therefore, if $\vdash_S \forall x \in U. \alpha(x)$ for any $\mathbf{A}$-singleton U, we have $\vdash_S \forall x \in U_i \alpha(x)$, i.e.

$$\vdash_S \forall x \in A[\tau(x) = \hat{\imath} \Rightarrow \alpha(x)].$$

So

$$\vdash_S \forall x \in A[\tau(x) = \tau(x) \Rightarrow \alpha(x)],$$

i.e. $\vdash_S \forall x \in A \alpha(x)$. Thus S is singleton-complete.

Now since (ii)$\Rightarrow$(iii)$\Rightarrow$(iv), S is also classical. So by (4.37) S satisfies AC.

(iii)$\Rightarrow$(vi). It is readily shown that the product of any pair of objects of

$\mathbf{C}^{-}(S)$, as well as any subobject of an object of $\mathbf{C}^{-}(S)$, is isomorphic to an object of $\mathbf{C}^{-}(S)$. Therefore the pullback of any pair of maps in $\mathbf{C}^{-}(S)$ is isomorphic to a diagram in $\mathbf{C}^{-}(S)$. The implication in question follows easily from this.

(v)$\Rightarrow$(vi) is trivial.

(vi)$\Rightarrow$(vii). Let $(\Omega^{-}, \top^{-})$ be the subobject classifier in $\mathbf{C}^{-}(S)$. For any $\alpha \in \Omega(S)$, $\{*:\alpha\}$ is a subset of $1 \in \mathrm{Ob}(\mathbf{C}^{-}(S))$, so there must be a map $1 \xrightarrow{f} \Omega^{-}$ such that the diagram

$$\begin{array}{ccc} \{*:\alpha\} & \rightarrow & 1 \\ \downarrow & & \downarrow \top^{-} \\ 1 & \xrightarrow{f} & \Omega^{-} \end{array}$$

is a pullback. Writing $true^{-}$ for $\top^{-}(*)$ and β for $f(*)$, it follows that

$$\vdash_S \alpha \Leftrightarrow (\beta = true^{-}).$$

But Ω^{-}, as a subconstant set, is decidable, so

$$\vdash_S \beta = true^{-} \vee \beta \neq true^{-}.$$

Hence $\vdash_S \alpha \vee \neg\alpha$, so that $\Omega(S)$ is a Boolean algebra.

(vii)$\Rightarrow$(v). Suppose that $\Omega(S)$ is a Boolean algebra and let $\mathbf{E} = \mathbf{Sh}(\Omega(S))$. Then by (6.10) **AC** holds in $\mathbf{E}$, so AC holds in Th($\mathbf{E}$). Hence by (6.19)

$$\mathbf{C}(\mathrm{Th}(\mathbf{E})) \simeq \boldsymbol{\Omega}(\mathbf{Th}(\mathbf{E}))^{\sim} \simeq \boldsymbol{\Omega}(\mathbf{E})^{\sim} \simeq \boldsymbol{\Omega}(\mathbf{S})^{\sim}.$$

So $\boldsymbol{\Omega}(\mathbf{S})^{\sim}$ is a topos.

Finally, if S is singleton-complete, then (vii)$\Rightarrow$(iv), so that all the conditions are equivalent. For assume (vii) and let U be an $\mathbf{\Omega}$-singleton. Then

$$true \in U,\ \omega \in U \vdash_S \omega = true \vdash_S \omega \tag{1}$$

and

$$\begin{aligned} \neg(true \in U),\ \omega \in U,\ \omega &\vdash_S \neg(true \in U) \wedge \omega \in U \wedge \omega = true \\ &\vdash_S \neg(true \in U) \wedge true \in U \\ &\vdash_S false. \end{aligned}$$

Hence

$$\neg(true \in U),\ \omega \in U \vdash_S \neg\omega. \tag{2}$$

Now since $\Omega(S)$ is Boolean, we have

$$\vdash_S true \in U \vee \neg(true \in U),$$

and this, together with (1) and (2), yields

$$\omega \in U \vdash_S \omega \vee \neg\omega,$$

whence

$$\vdash_S \forall \omega \in U. \omega \vee \neg\omega,$$

Since S is singleton-complete, it follows that

$$\vdash_S \forall \omega. \omega \vee \neg\omega,$$

so that S is classical, i.e. (iv).

So, by (4.37), S satisfies AC, and we have (i). □

By taking $S = \mathrm{Th}(\mathbf{E})$ in (6.21), we obtain

6.22 COROLLARY. Let **E** be a topos defined over **Set**. Consider the following conditions:

(i) **E** satisfies **AC**;
(ii) $\mathbf{E}^- \hookrightarrow \mathbf{E}$ is an equivalence;
(iii) $\Omega_{\mathbf{E}}$ is isomorphic to an object in $\mathbf{E}^-$;
(iv) **E** is Boolean;
(v) $\mathbf{E}^- \simeq \tilde{\boldsymbol{\Omega}(\mathbf{E})}$ is a topos;
(vi) $\mathbf{E}^- \simeq \tilde{\boldsymbol{\Omega}(\mathbf{E})}$ has a subobject classifier;
(vii) $\Omega(\mathbf{E})$ is a Boolean algebra.

Then (i)⇔(ii)⇒(iii)⇔(iv)⇒(v)⇔(vi)⇔(vii). If, moreover, **E** is *localic,* then all these conditions are equivalent. □

6.23 COROLLARY. The following conditions are equivalent for a locale H:

(i) $\tilde{\mathbf{H}}$ is a topos;
(ii) $\tilde{\mathbf{H}}$ has a subobject classifier;
(iii) H is a Boolean algebra.

Proof. This follows from (6.22), taking $\mathbf{E} = \mathbf{Sh}(H)$. □

We may paraphrase (6.23) by saying that *a category of fuzzy sets is a topos iff it is Boolean.*

Boolean extensions as toposes

In this final section we discuss Boolean extensions of the universe of sets from the topos-theoretic standpoint. Our discussion will be based on the presentation of Boolean-valued set theory in Bell (1985).

Let B be a complete Boolean algebra. We write $V^{(B)}$ for the B-extension of the universe V of sets (Bell (1985), Chapter 1). For each sentence σ of the first-order language of set theory augmented by names for elements of $V^{(B)}$ we write $[\![\sigma]\!]^B$ or $[\![\sigma]\!]$ for the *B-value* (sometimes called the *probability*) of σ. Writing 0_B, 1_B for the least and largest elements of B, we have a canonical map $V \xrightarrow{\wedge} V^{(B)}$ satisfying $x \in y$ iff $[\![\hat{x} \in \hat{y}]\!] = 1_B$, $x \notin y$ iff $[\![\hat{x} \in \hat{y}]\!] = 0_B$.

We obtain a category $\mathbf{Set}^{(B)}$ from $V^{(B)}$ in the following way. First, we *identify* elements of $V^{(B)}$ which are equal with probability 1_B. The objects of $\mathbf{Set}^{(B)}$ are the (thus identified) elements of $V^{(B)}$ and its arrows are those (identified) elements of $V^{(B)}$ which are functions with probability 1_B. Composition and identity arrows are defined in the obvious way.

6.24 PROPOSITION. $\mathbf{Set}^{(B)}$ is a topos with terminal object $\hat{1}$ and subobject classifier $\hat{2}$.

Proof. This follows immediately from the fact (Bell 1985, 1.33) that with probability 1_B all the axioms of Zermelo–Fraenkel set theory hold in $V^{(B)}$. □

6.25 LEMMA.

(i) $X \xrightarrow{f} Y$ is monic in $\mathbf{Set}^{(B)}$ iff $[\![f \text{ is one–one}]\!] = 1_B$;
(ii) $X \xrightarrow{f} Y$ is epic in $\mathbf{Set}^{(B)}$ iff $[\![f \text{ is onto}]\!] = 1_B$.

Proof. This is left as a straightforward exercise to the reader. □

6.26 PROPOSITION. $\mathbf{Set}^{(B)}$ satisfies **AC**.

Proof. We know (Bell 1985, 1.43) that $[\![\mathbf{AC}]\!] = 1_B$. Therefore if $X \xrightarrow{f} Y$ is epic in $\mathbf{Set}^{(B)}$, by (6.25) we have $[\![f \text{ is onto}]\!] = 1_B$, so there is $g \in V^{(B)}$ such that

$$[\![g : Y \to X \wedge f \circ g = 1_Y]\!] = 1_B,$$

whence $f \circ g = 1_Y$ in $\mathbf{Set}^{(B)}$ and so **AC** holds in $\mathbf{Set}^{(B)}$. □

6.27 PROPOSITION. $\mathbf{Set}^{(B)}$ is Boolean and $\Omega(\mathbf{Set}^{(B)}) \cong B$.

Proof. Since any topos satisfying **AC** is Boolean, $\mathbf{Set}^{(B)}$ is so by (6.26). Any $\alpha \in \Omega(\mathbf{Set}^{(B)})$ is a map $1 \xrightarrow{\alpha} \Omega$ in $V^{(B)}$, so we can define $\tilde{\alpha} \in B$ by $\tilde{\alpha} = [\![\alpha(0) = 1]\!]$. It is easy to verify that $\alpha \mapsto \tilde{\alpha}$ establishes an isomorphism $\Omega(\mathbf{Set}^{(B)}) \cong B$. □

6.28 PROPOSITION. $\mathbf{Set}^{(B)}$ is defined over $\mathbf{Set}$. The geometric morphism $\mathbf{Set}^{(B)} \xrightarrow{\gamma} \mathbf{Set}$ may be obtained as follows. For I in $\mathbf{Set}$, $\gamma^* I = \hat{I}$ and for $I \xrightarrow{f} J$ in $\mathbf{Set}$, $\gamma^* f = \hat{f}$. For X in $\mathbf{Set}^{(B)}$, $\gamma_* X = \{x \in V^{(B)} : [\![x \in X]\!] = 1_B\}$ and, for $X \xrightarrow{g} Y$ in $\mathbf{Set}^{(B)}$, $\gamma_* X \xrightarrow{\gamma_* g} \gamma_* Y$ is given by $[\![(\gamma_* g)(x) = g(x)]\!] = 1_B$ for $x \in \gamma_* X$. Finally, the counit ε of the adjunction $\gamma^* \dashv \gamma_*$ is defined as follows. For X in $\mathbf{Set}^{(B)}$, $\varepsilon_X : \gamma^* \gamma_* X \to X$ is the unique function in $V^{(B)}$ such that $[\![\varepsilon_X(\hat{x}) = x]\!] = 1_B$ for all $x \in \gamma_* X$.

Proof. Most of this is (or should be) clear. We verify only that $\gamma^* \dashv \gamma_*$ and that ε is the counit of the adjunction. First, it is easy to see that ε is a natural transformation $\gamma^*\gamma_* \to 1_{\mathbf{Set}}(B)$. It thus suffices to show that, for any X in $\mathbf{Set}^{(B)}$, I in $\mathbf{Set}$ and $\gamma^* I \xrightarrow{f} X$ in $\mathbf{Set}^{(B)}$, there is a unique $I \xrightarrow{g} \gamma_* X$ such that the diagram

$$\begin{array}{ccc} \gamma^* I & \xrightarrow{f} & X \\ {\scriptstyle \gamma^* g}\downarrow & \nearrow_{\varepsilon_X} & \\ \gamma^*\gamma_* X & & \end{array}$$

commutes. But clearly the map $I \xrightarrow{g} \gamma_* X$ defined by: $g(i) =$ (unique $x \in \gamma_* X$ such that $[\![f(\hat{i}) = x]\!] = 1_B$) is the required unique function. □

6.29 COROLLARY. (i) $\mathbf{Set}^{(B)} \simeq \mathbf{Set}_B \simeq \mathbf{Sh}(B) \simeq \tilde{\mathbf{B}}$ for any complete Boolean algebra B.

(ii) Let $\mathbf{E}$ be a topos defined over $\mathbf{Set}$. Then $\mathbf{E}$ satisfies $\mathbf{AC}$ iff $\mathbf{E} \simeq \mathbf{Set}^{(B)}$ for some complete Boolean algebra B. In fact, B may be taken to be $\Omega(\mathbf{E})$. □

REMARK. Let P be a partially ordered set and let B be its regular open algebra, that is, the complete Boolean algebra of all subsets of P which are equal to the interiors of their closures with respect to the order topology on P ((6.14); see also Bell 1985, Chapter 2). It is not hard to show, using (6.14)(iv) and (5.28), that $\Omega(\mathbf{Sh}_{\neg\neg}(\mathbf{E}(\mathbf{P}))) \cong B$ and hence by (6.29) $\mathbf{Set}^{(B)} \simeq \mathbf{Sh}_{\neg\neg}(\mathbf{E}(\mathbf{P}))$. That is, the 'Boolean-valued model' $\mathbf{Set}^{(B)}$ may be obtained by taking $\neg\neg$-sheaves in the Kripke model $\mathbf{E}(\mathbf{P})$. We have seen at the end of Chapter 4 that truth in $\mathbf{E}(\mathbf{P})$ is governed by the forcing relation $\Vdash^{\mathbf{P}}$. Thus the process of passing from $\mathbf{E}(\mathbf{P})$ to $\mathbf{Set}^{(B)}$ by taking $\neg\neg$-sheaves corresponds to the process—familiar to set-theorists—of replacing the forcing relation by Boolean values.

We now return to the counit ε of the geometric morphism $\mathbf{Set}^{(B)} \xrightarrow{\gamma} \mathbf{Set}$. Let us determine the action of ε_Ω in $\mathbf{Set}^{(B)}$. We know that $\Omega = \hat{2}$ in $\mathbf{Set}^{(B)}$, and so $\gamma_*\Omega$ is the set of objects of $\mathbf{Set}^{(B)}$ which are members of $\hat{2}$ with probability 1_B. Now these objects are in natural bijective correspondence with elements of B. Specifically, this correspondence sends $x \in \gamma_*\Omega$ to $[\![x = \hat{1}]\!]$ in B. Thus, identifying x and $[\![x = \hat{1}]\!]$ for $x \in \gamma_*\Omega$ leads to an identification of $\gamma_*\Omega$ with B. But (6.28) tells us that $[\![\varepsilon_\Omega(\hat{x}) = x]\!] = 1_B$ for $x \in \gamma_*\Omega$. Hence

$$[\![\varepsilon_\Omega(\hat{x}) = \hat{1}]\!] = [\![x = \hat{1}]\!]$$

for $x \in \gamma_*\Omega$ and so, under the above identification of $\gamma_*\Omega$ with B, ε_Ω is that arrow from $\hat{B}$ to $\hat{2}$ such that

$$[\![\varepsilon_\Omega(\hat{x}) = \hat{1}]\!] = x$$

for all $x \in B$. Let U_* be the subobject of B classified by ε_Ω. Then

$$[\![\hat{x} \in U_*]\!] = [\![\varepsilon_\Omega(\hat{x}) = \hat{1}]\!] = x.$$

But this is precisely the definition of the *canonical generic ultrafilter* in $V^{(B)}$ (Bell 1985, (4.17)). Accordingly, we have proved the

6.30 PROPOSITION. The canonical generic ultrafilter in $V^{(B)}$ is—up to isomorphism—the subobject of B classified by the counit arrow $\varepsilon_\Omega: \hat{B} \to \hat{2}$ determined by the geometric morphism $\mathbf{Set}^{(B)} \xrightarrow{\gamma} \mathbf{Set}$. □

We conclude this chapter with a topos-theoretic account of iterated Boolean extensions (Bell 1985, Chapter 6).

We shall write $V^{(B))} \models \sigma$ for $[\![\sigma]\!]^B = 1_B$.

Let $C, \leqslant_C$ be elements of $V^{(B)}$ such that

$$V^{(B)} \models (C, \leqslant_C) \text{ is a complete Boolean algebra.}$$

Now form the Boolean extension $V^{(C)}$ inside $V^{(B)}$, thus obtaining the iterated extension $V^{(B)(C)}$. We extract a category $\mathbf{Set}^{(B)(C)}$ from $V^{(B)(C)}$ as follows. First, we identify elements x,y of $V^{(B)(C)}$ whenever

$$V^{(B)} \models [\![x = y]\!]^C = 1_C.$$

The objects of the category $\mathbf{Set}^{(B)(C)}$ are the (thus identified) elements of $V^{(B)(C)}$ and its arrows are all objects f of $V^{(B)(C)}$ such that

$$V^{(B)} \models [\![f \text{ is a function}]\!]^C = 1_C.$$

Composition of arrows in $\mathbf{Set}^{(B)(C)}$ is defined in the obvious way. Using the fact that

$$V^{(B)} \models [\![V^{(C)} \text{ is a model of } ZFC]\!]^C = 1_C,$$

it follows that $\mathbf{Set}^{(B)(C)}$ is a topos satisfying **AC**.

Now let $\phi(x,y,X)$ be a set-theoretical formula expressing 'X is a complete Boolean algebra, $y \in V^{(X)}$ and $y = \hat{x}$'. Then we can define a canonical map $\mu^*: V^{(B)(C)}$ by

$$\mu^*(x) = \text{the unique } y \in V^{(B)} \text{ such that } [\![\phi(x,y,C)]\!]^B = 1_B.$$

Thus μ^* is 'the functor $\mathbf{Set} \hat{\to} \mathbf{Set}^{(C)}$ constructed inside $V^{(B)}$' and it follows that, for each set I, $\mu^* I$ is $\coprod_I 1$ in $\mathbf{Set}^{(B)(C)}$. Therefore $\mathbf{Set}^{(B)(C)}$ is defined over $\mathbf{Set}$ and so, by 6.29(ii), $\mathbf{Set}^{(B)(C)} \simeq \mathbf{Set}^{(A)}$ where A is the complete Boolean algebra $\Omega(\mathbf{Set}^{(B)(C)})$.

We can now formulate an 'external' description of the algebra A. Let $\leqslant_A$ be the partial ordering on A and let $(A', \leqslant_{A'})$ be 'the algebra $\Omega(\mathbf{Set}^{(C)})$ constructed in $V^{(B)}$'. Then clearly we have, for $u \in V^{(B)}$,

(6.31) $$u \in A \Leftrightarrow V^{(B)} \models u \in A',$$

and for $u, v \in A$,

$$u \leqslant_A v \Leftrightarrow V^{(B)} \models u \leqslant_{A'} v.$$

We also know that, in $V^{(B)}$, A' is naturally isomorphic to C. Thus we can find $f \in V^{(B)}$ such that

$$V^{(B)} \models f \text{ is an isomorphism of } (A, \leqslant_{A'}) \text{ with } (C, \leqslant_C).$$

This gives rise to a bijection

$$\gamma_* A' \xrightarrow{\gamma_* f} \gamma_* C.$$

Now it follows immediately from (6.31) that $\gamma_* A' = A$. Thus if we put $\gamma_* f = g$, we have a bijection

$$A \xrightarrow{g} \gamma_* C.$$

We use g to transfer the ordering $\leqslant_A$ to $\gamma_* C$. That is, we define $\leqslant$ on $\gamma_* C$ by

$$x \leqslant y \Leftrightarrow g^{-1}(x) \leqslant g^{-1}(y),$$

for $x, y \in \gamma_* C$. Then we have, for $x, y \in \gamma_* C$,

$$\begin{aligned} x \leqslant y &\Leftrightarrow g^{-1}(x) \leqslant g^{-1}(y) \\ &\Leftrightarrow V^{(B)} \models g^{-1}(x) \leqslant_{A'} g^{-1}(y) \\ &\Leftrightarrow V^{(B)} \models f^{-1}(x) \leqslant_A f^{-1}(y) \\ &\Leftrightarrow V^{(B)} \models x \leqslant_C y. \end{aligned}$$

Therefore $(\gamma_* C, \leqslant)$ is a complete Boolean algebra isomorphic to $(A, \leqslant_A)$ and $\mathbf{Set}^{(\gamma_* C)} \simeq \mathbf{Set}^{(B)(C)}$. Accordingly we have proved the following theorem, which amounts to a topos-theoretic description of iterated Boolean extensions (cf. the Remarks following (6.20) of Bell 1985).

6.32 THEOREM. Let B be a complete Boolean algebra and let $(C, \leqslant_C)$ be a complete Boolean algebra (with probability 1_B) in $V^{(B)}$. Let $\mathbf{Set}^{(B)} \xrightarrow{\gamma} \mathbf{Set}$ be a geometric morphism and define $\leqslant$ on $\gamma_* C$ by

$$x \leqslant y \Leftrightarrow [\![x \leqslant_C y]\!]^B = 1_B.$$

Then $(\gamma_* C, \leqslant)$ is a complete Boolean algebra and

$$\mathbf{Set}^{(B)(C)} \simeq \mathbf{Set}^{(\gamma_* C)}. \quad \square$$

7

Natural numbers and real numbers

In this brief chapter we introduce the natural numbers and real numbers into local set theories, and formulate their interpretation in toposes. Some of the arguments here will only be sketched.

Natural numbers in local set theories

Let S be a local set theory in a local language $\mathscr{L}$. A *natural number system* in S is a triple $(\mathbf{N},\mathbf{s},\bar{0})$ consisting of a type symbol $\mathbf{N}$, a function symbol $\mathbf{s}$ of signature $\mathbf{N}\rightarrow\mathbf{N}$, and a closed term $\bar{0}$ of type $\mathbf{N}$, satisfying the following *Peano axioms*:

(P1) $$\vdash_S \neg(\mathbf{s}n=\bar{0})$$

(P2) $$\mathbf{s}m=\mathbf{s}n \vdash_S m=n$$

(P3) $$\bar{0}\in u,\ \forall n(n\in u \Rightarrow \mathbf{s}n\in u)\vdash_S \forall n.n\in u$$

Here m and n are variables of type $\mathbf{N}$, the variable u is of type $\mathbf{PN}$, and we have written $\mathbf{s}n$ for $\mathbf{s}(n)$. P3 is the *axiom of induction.*

A local set theory with a natural number system will be called *naturalized.*

In any naturalized local set theory S, $\bar{0}$ is called the *zeroth numeral.* For each natural number $n\geqslant 1$ the nth *numeral* $\bar{n}$ in S is defined recursively by putting $\bar{n}=\mathbf{s}\bar{k}$, where $k=n-1$. Numerals are closed terms of type $\mathbf{N}$ which may be regarded as the *formal representatives* in S of the natural numbers.

Our first proposition furnishes another condition equivalent to P3.

7.1 PROPOSITION. The following conditions are equivalent for any triple $(\mathbf{N},\mathbf{s},\bar{0})$ as specified above:

(i) P3

(ii) the *induction scheme:* for any formula α with exactly one free variable n of type $\mathbf{N}$, if $\vdash_S \alpha(\bar{0})$ and $\alpha(n)\vdash_S \alpha(\mathbf{s}n)$, then $\vdash_S \forall n\alpha(n)$.

Proof. (i)$\Rightarrow$(ii). To get (ii) from (i), take $\{n:\alpha(n)\}$ for u in P3.

(ii)$\Rightarrow$(i). Assuming (ii), define $\alpha(n)$ to be the formula

$$\forall u[[\bar{0}\in u \wedge \forall m(m\in u\Rightarrow \mathbf{s}m\in u)]\Rightarrow n\in u].$$

Then clearly $\vdash_S \alpha(\bar{0})$ and $\alpha(n) \vdash_S \alpha(\mathbf{s}n)$. So by (ii) $\vdash_S \forall n \alpha(n)$, which evidently yields P3. □

We now show that functions may be defined on N by the usual process of *simple recursion*.

7.2 THEOREM. Let S be a naturalized local set theory. Then we have the following *simple recursion principle* (*SRP*): for any S-set X of type $\mathbf{PA}$, any closed term a of type $\mathbf{A}$ and any S-map $g: X \to X$, there is a unique S-map $f: N \to X$ such that

$$\vdash_S f(\bar{0}) = a \wedge \forall n[f(\mathbf{s}n) = g(f(n))].$$

Proof. Define

$$U = \{u \in P(N \times X) : \langle \bar{0}, a \rangle \in u \wedge \forall n \forall x(\langle n,x \rangle \in u \Rightarrow \langle \mathbf{s}n, g(x) \rangle \in u)\}.$$

Put $f = \bigcap U$. We claim that f satisfies the conditions of the theorem; its uniqueness is left as an exercise for the reader. It suffices to show that f is a map from N to X.

Clearly $\vdash_S f \subseteq N \times X$. Let $V = \{n : \exists x. \langle n,x \rangle \in f\}$. Then $\vdash_S V \subseteq N$ and $\vdash_S \bar{0} \in V$ since $\vdash_S \langle \bar{0}, a \rangle \in f$. Moreover

$$n \in V \vdash_S \exists x. \langle n,x \rangle \in f \vdash_S \exists x \langle \mathbf{s}n, g(x) \rangle \in f \vdash_S \mathbf{s}n \in V.$$

Hence by P3 we conclude that

$$\vdash \forall n. n \in V.$$

It remains to show that f is single-valued. To this end define

$$W = \{n : \forall n \forall y(\langle n,x \rangle \in f \wedge \langle n,y \rangle \in f \Rightarrow x = y)\}.$$

We need to show that $\vdash_S W = N$.

First, we show that $\vdash_S \bar{0} \in W$. Define

$$f' = \{\langle n,x \rangle \in f : n = \bar{0} \Rightarrow x = a\}.$$

It is easily verified that $\vdash_S f' \in U$, whence $\vdash_S f = f'$. Therefore

$$\langle \bar{0},x \rangle \in f \vdash_S \langle \bar{0},x \rangle \in f' \vdash_S x = a,$$

so that

$$\vdash_S \forall x(\langle \bar{0},x \rangle \in f \Rightarrow x = a)$$

and from this it immediately follows that $\vdash_S \bar{0} \in W$.

Finally, we need to show that $n \in W \vdash_S \mathbf{s}n \in W$. To do this we first prove the auxiliary result

$$n \in W, \langle n,x \rangle \in f, \langle \mathbf{s}n,y \rangle \in f \vdash_S y = g(x). \qquad (*)$$

To establish this it will be easier to argue informally. Given $n \in W$, $\langle n,x \rangle \in f$ and $\langle \mathbf{s}n,y \rangle \in f$, write f_{nx} for

$$\{\langle m,y \rangle \in f : m = \mathbf{s}n \Rightarrow y = g(x)\}.$$

We claim that $f_{nx} \in U$. Clearly $\langle \bar{0},a \rangle \in f_{nx}$ (using P1). Moreover, if $\langle m,y \rangle \in f_{nx}$, then $\langle m,y \rangle \in f$ and $[m = \mathbf{s}n \Rightarrow y = g(x)]$. We need to show that $\langle \mathbf{s}m, g(y) \rangle \in f_{nx}$, and for this it suffices to show that $\langle \mathbf{s}m, g(y) \rangle \in f$ and $\mathbf{s}m = \mathbf{s}n \Rightarrow g(y) = g(x)$. The first of these assertions follows from the assumption that $\langle m,y \rangle \in f$. As for the second, if $\mathbf{s}m = \mathbf{s}n$, then $m = n$ by P2, so from $\langle m,y \rangle \in f$ it follows that $\langle n,y \rangle \in f$. But we are assuming that $n \in W$ and $\langle n,x \rangle \in f$, so we conclude from the defining property of W that $x = y$. Hence, certainly, $g(x) = g(y)$ proving the second assertion.

We conclude that $f_{nx} \in U$. Therefore (reverting to formal notation)

$$n \in W,\ \langle n,x \rangle \in f,\ \langle \mathbf{s}n,y \rangle \in f \vdash_S f_{nx} \in U$$

whence

$$\vdash_S f_{nx} = f,$$

$$n \in W,\ \langle n,x \rangle \in f,\ \langle \mathbf{s}n,y \rangle \in f \vdash_S \langle \mathbf{s}n,y \rangle \in f_{nx}$$

$$\vdash_S y = g(x),$$

that is (*).

Now, from the fact that $\vdash_S V = N$, it follows that

$$n \in W \vdash_S \exists! x.\langle n,x \rangle \in f,$$

so, using (*), we have

$$n \in W,\ \langle \mathbf{s}n,y \rangle \in f,\ \langle \mathbf{s}n,z \rangle \in f$$

$$\vdash_S \exists! x[y = g(x) \wedge z = g(x)]$$

$$\vdash_S y = z.$$

Therefore $n \in W \vdash_S \mathbf{s}n \in W$, as required.

So by P3 we have $\vdash_S W = N$ as claimed. □

EXERCISE. Show that a natural number system in a local set theory is uniquely determined up to isomorphism in the evident sense.

We now show that *SRP* yields the Peano axioms.

7.3 THEOREM. Let S be a local set theory and $(\mathbf{N},\mathbf{s},\bar{0})$ a triple satisfying *SRP* in S. Then $(\mathbf{N},\mathbf{s},\bar{0})$ is a natural number system in S.

Proof. In order to prove (7.3) we need first to show that *SRP* implies the *primitive recursion principle*:

(*PRP*) for any S-set X of type $\mathbf{PA}$, any closed term a of type $\mathbf{A}$ such that $\vdash_S a \in X$ and any S-map $g : X \times N \to X$, there is a unique S-map $f : N \to X$ such that $\vdash_S f(\bar{0}) = a \wedge \forall n(f(\mathbf{s}n) = g(f(n), n))$.

To get *PRP* from *SRP*, we take $Y = X \times N$ and $h: Y \to Y$ defined by

$$h = (\langle x,n \rangle \mapsto \langle g(\langle x,n \rangle), n \rangle).$$

Applying *SRP* to h, Y and $\langle a, \bar{0} \rangle$ yields a unique $k: N \to Y$ such that

$$\vdash_S k(\bar{0}) = \langle a, \bar{0} \rangle \wedge \forall n(k(\mathbf{s}n) = h(k(n))).$$

It is now easily checked that $f = \pi_1 \circ k: N \to X$ is the unique map such that

$$\vdash_S f(\bar{0}) = a \wedge \forall n[f(\mathbf{s}n) = g(\langle f(n), n \rangle)].$$

PRP follows.

Using this we can prove (7.3). We need to verify P1, P2 and the induction scheme (7.1).

(P1) $\vdash_S \neg(\mathbf{s}n = \bar{0})$. Define $2 = \{0,1\}$ and $g: 2 \to 2$ by

$$g = \{\langle 0,1 \rangle, \langle 1,1 \rangle\}.$$

Using *SRP*, there is $f: N \to 2$ such that

$$\vdash_S f(\bar{0}) = 0 \wedge \forall n[f(\mathbf{s}n) = g(f(n))].$$

Then

$$\mathbf{s}n = \bar{0} \vdash_S 0 = f(\bar{0}) = f(\mathbf{s}n) = g(f(n)) = 1$$

and P1 follows.

(P2) $\mathbf{s}m = \mathbf{s}n \vdash_S m = n$. Consider $\pi_2: N \times N \to N$. By *PRP* there is a map $f: N \to N$ such that

$$\vdash_S f(\bar{0}) = \bar{0} \wedge \forall n[f(\mathbf{s}n) = \pi_2(\langle f(n), n \rangle) = n].$$

Then

$$\mathbf{s}m = \mathbf{s}n \vdash_S m = f(\mathbf{s}m) = f(\mathbf{s}n) = n$$

and P2 follows.

Induction scheme 7.1. Suppose $\vdash_S \alpha(\bar{0})$ and $\alpha(n) \vdash_S \alpha(\mathbf{s}n)$. Let $X = \{n : \alpha(n)\}$. Then $n \mapsto \mathbf{s}n : X \to X$ and so *SRP* yields a map $f: N \to X$ such that

$$\vdash_S f(\bar{0}) = \bar{0} \wedge \forall n[f(\mathbf{s}n) = \mathbf{s}(f(n))].$$

Writing j for the insertion map $X \xrightarrow{n \mapsto n} N$, we get

$$\vdash_S (j \circ f)(\bar{0}) = \bar{0} \wedge \forall n[(j \circ f)(\mathbf{s}n) = \mathbf{s}((j \circ f)(n))] \qquad (*)$$

But the identity map $1_N: N \to N$ also satisfies (*), so by the uniqueness condition in *SRP*, $j \circ f = 1_N$. It follows easily from this that $\vdash_S X = N$, i.e. $\vdash_S \forall n \alpha(n)$. This proves the induction scheme, and hence also (7.3). □

Given a naturalized local set theory S, if we denote the map $n \mapsto \mathbf{s}(n): N \to N$ by s and the map $* \mapsto \bar{0}: 1 \to N$ by o, it is easy to see, using P1 and P2, that the map

$$N+1 \xrightarrow{\binom{s}{o}} N$$

is an isomorphism in $\mathbf{C}(S)$. Conversely, the presence of an S-set X and an isomorphism

$$X+1 \xrightarrow{f} X$$

yields a natural number system:

7.4 PROPOSITION. Suppose $X+1 \xrightarrow{f} X$ is an isomorphism in $\mathbf{C}(S)$. Define

$$U = \bigcap \{u \subseteq X : f(*) \in u \wedge \forall x \in u . f(x) \in u\},$$

where we have identified X and 1 with their canonical images in $X+1$. Then, writing $\bar{0}$ for $f(*)$, the triple $(U, f, \bar{0})$ satisfies the Peano axioms

$$x \in U \vdash_S \neg(f(x) = \bar{0})$$
$$x \in U, y \in U, f(x) = f(y) \vdash_S x = y$$
$$u \subseteq U, \bar{0} \in u, \forall x[x \in u \Rightarrow f(x) \in u] \vdash_S \forall x \in U . x \in u.$$

Proof. This is a straightforward consequence of the definition of U. □

Assume throughout the remainder of this section that S is a local set theory with natural number system $(\mathbf{N}, \mathbf{s}, \bar{0})$.

We want now to define the binary relation 'less than' on N. To do this we apply primitive recursion with $X = PN$, $a = \varnothing_{\mathbf{N}}$, and

$$g = PN \times N \xrightarrow{\langle u,x\rangle \mapsto u \cup \{x\}} PN.$$

We get a (unique) $f: N \to PN$ such that

$$\vdash_S f(\bar{0}) = \varnothing_{\mathbf{N}} \wedge f(\mathbf{s}n) = f(n) \cup \{n\}.$$

Now write

$$m < n \qquad \text{for} \qquad m \in f(n).$$

Then we have

7.5 PROPOSITION. *$\vdash_S <$ is a strict total ordering of N with least element* $\bar{0}$. In particular, we have

(i) $\vdash_S \neg(m < \bar{0})$;
(ii) $\vdash_S m < \mathbf{s}n \Leftrightarrow m = n \vee m < n$;
(iii) $\vdash_S \neg(m < m)$;
(iv) $m < n, n < p \vdash_S m < p$;
(v) $\vdash_S m < n \vee m = n \vee n < m$.

Proof. (i) and (ii) are obvious and (iii) and (iv) are easily proved by induction. We establish (v). First, it is readily shown by induction that $\vdash_S \bar{0} < \mathbf{s}n$ and hence that $\vdash_S \bar{0} < n \vee \bar{0} = n$. Therefore, writing $\alpha(m)$ for $\forall n[m < n \vee m = n \vee n < m]$, we have $\vdash_S \alpha(\bar{0})$. A straightforward inductive argument shows that $m < n \vdash_S \mathbf{s}m = n \vee \mathbf{s}m < n$. Using this and (ii), we see that

$$\alpha(m) \vdash_S \mathbf{s}m < n \vee \mathbf{s}m = n \vee n < \mathbf{s}m$$

and so $\alpha(m) \vdash_S \alpha(\mathbf{s}m)$. Hence $\vdash_S \forall m \alpha(m)$ by induction, giving (v). □

As an immediate consequence of this we have

7.6 PROPOSITION. N is decidable, i.e.

$$\vdash_S m = n \vee \neg(m = n). \quad \square$$

The operations of *addition* and *multiplication* on N are defined essentially as usual. We indicate how this is done in the case of addition.

Given a map $f: X \to Y$, and a set A, write f^A for the map $g \mapsto f \circ g : X^A \to Y^A$. Now we apply simple recursion with $X = N^N$, $a = 1_N$, $g = s^N$, where $s = (n \mapsto \mathbf{s}n)$. We get $f: N \to N^N$ such that

$$\vdash_S f(\bar{0}) = 1_N \wedge f(\mathbf{s}m) = s^N(f(m)) = (s \circ f)(m). \qquad (*)$$

We define $+ : N \times N \to N$ to be the exponential transpose of f and write $m + n$ for $+(m,n)$, so that $\vdash_S m + n = f(m)(n)$. We see from (*) that

$$\vdash_S \bar{0} + n = n, \qquad \vdash_S \mathbf{s}m + n = \mathbf{s}(m + n),$$

which are the usual defining equations for $+$.

Similarly, now using $+$ in place of s, we get a map $\bullet : N \times N \to N$ satisfying the usual defining equations for multiplication, viz

$$\vdash_S \bar{0}.n = \bar{0}, \qquad \vdash_S (\mathbf{s}m).n = m.n + n.$$

The standard properties of $+$ and $\bullet$, e.g., commutativity and associativity, can now be established inductively as usual.

We next turn to the interpretation of these ideas in toposes. Let **E** be a topos and let $(\mathbf{N}, \mathbf{s}, o)$ be a triple consisting of an **E**-object N and **E**-arrows $N \xrightarrow{s} N$, $1 \xrightarrow{o} N$. Let **s**,**o** be the function symbols in $\mathscr{L}(\mathbf{E})$ associated with s,o respectively and let $\bar{0}$ be the closed term $\mathbf{o}(*)$. Then clearly $(\mathbf{N}, \mathbf{s}, \bar{0})$ satisfies the simple recursion principle in Th(**E**) iff the following condition, known as the *Peano–Lawvere axiom*, holds. For any diagram $1 \xrightarrow{a} X \xrightarrow{g} X$ in **E** there exists a unique $N \xrightarrow{f} X$ such that the diagram

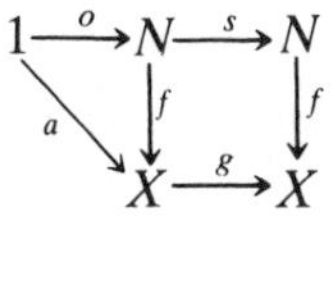

commutes. A triple $(\mathbf{N},\mathbf{s},o)$ satisfying this condition will be called a *natural number system,* and N a *natural number object,* respectively, in $\mathbf{E}$.

Upon interpreting (7.2), (7.3), and (7.4) in Th($\mathbf{E}$), we obtain:

7.7 THEOREM. The following are equivalent for any topos $\mathbf{E}$:

(i) $\mathbf{E}$ has a natural number system;

(ii) there exists a pair (A,f) consisting of an $\mathbf{E}$-object A and an isomorphism $A+1 \xrightarrow{f} A$. □

Calling an object A of $\mathbf{E}$ *infinite* if $A+1 \cong A$, (7.7) asserts that *a topos has a natural number system iff it contains an infinite object.* It is remarkable that this result, familiar from classical set theory, can be translated to arbitrary toposes.

7.8 EXERCISES

(1) Let $\mathbf{E}$ be a topos with a natural number object N and let $\mathbf{E}' \xrightarrow{\gamma} \mathbf{E}$ be a geometric morphism. Show that γ^*N is a natural number object in $\mathbf{E}'$. (*Hint*: given $1 \xrightarrow{a} X \xrightarrow{g} X$ in $\mathbf{E}'$, apply γ_* to get $\gamma_*1 \cong 1 \xrightarrow{\gamma_* a} \gamma_* X \xrightarrow{\gamma_* g} \gamma_* X$ in $\mathbf{E}$, thence obtaining $N \xrightarrow{f} \gamma_* X$ and hence $\gamma^*N \to X$ by transposition across $\gamma^* \dashv \gamma_*$.)

(2) Deduce from (1) that, for any site $(\mathbf{C},K)$, $\mathbf{Sh}_K(\mathbf{C})$ has a natural number system.

(3) Let X be a topological space. Show that $\omega^\dagger$ (5.35(2)) is a natural number object in $\mathbf{Sh}(X)$.

Real numbers in local set theories

We now turn to the construction of the real numbers in a local set theory with a natural number system. This will be achieved in essentially the classical manner: first the integers are constructed, next, the rationals, and thence the reals obtained as Dedekind cuts or Cauchy sequences. In the classical context it is of course well known that these two methods of constructing the reals lead to identical (i.e. isomorphic) results. However, this is *not* the case for local set theories. Although we shall not be establishing this fact here (the interested reader may consult Johnstone 1977, Section 6.6) we will show, for example, that in general the reals constructed as Dedekind cuts within a local set theory need not be (conditionally) order-complete.

Throughout this section let S be a local set theory with a natural number system $(\mathbf{N},\mathbf{s},\bar{0})$.

We construct within S the system Z of integers (positive and negative) in the usual way (see, e.g., MacLane and Birkhoff (1967), Chapter II,

Section 4). In this context Z is defined as the coproduct

$$\{n:0<n\}+N.$$

One shows in the standard way that Z may be turned into an ordered ring $\langle Z, +, \bullet, <\rangle$.

The customary procedure for obtaining the rational field as the ordered field of quotients of Z (MacLane and Birkhoff 1967, Chapter V, Section 2) now yields, in S, the *ordered field of rationals* $\langle Q, +, \bullet, <\rangle$. We shall use letters p,q to denote variables ranging over Q.

For any (intuitive) rational number q, its representative $\bar{q}$ in S is readily constructed from the numerals in S: we leave the details to the reader. We then have $\vdash_S \bar{q} \in Q$ for any rational q.

Now we can define the (*Dedekind*) *real numbers* in S as 'cuts' in Q. That is, we define the S-set

$$\begin{aligned} R = \{\langle u,v\rangle \in PQ \times PQ : \exists p\, \exists q(p \in u \wedge q \in v) \\ \wedge\, u \cap v = \varnothing \wedge \forall p[p \in u \Leftrightarrow \exists q \in u.p<q] \\ \wedge\, \forall p[p \in v \Leftrightarrow \exists q \in v.q<p] \\ \wedge\, \forall p\, \forall q[p<q \Rightarrow p \in u \vee q \in v]. \end{aligned}$$

We use letters r,s as variables ranging over R.

The ordering $\leqslant$ on R is defined as usual, viz.

$$r \leqslant s \Leftrightarrow \pi_1(r) \subseteq \pi_1(s),$$

where π_1 is the projection $PQ \times PQ \to PQ$.

In the classical case one now proceeds to show that R is conditionally order-complete, i.e. every bounded subset has a supremum and an infimum. This, as we have already remarked, is *not* in general the case within a local set theory. In fact, we shall see that R is conditionally complete iff S satisfies the logical rule

$$\vdash_S \forall \omega[\neg\omega \vee \neg\neg\omega].$$

This surprising result (due to Johnstone) illustrates the close connection between mathematics and logic in local set theories.

Define $i: Q \to R$ by

$$i = (p \mapsto \langle \{q:q<p\}, \{q:p<q\}\rangle).$$

The map i is clearly monic and embeds Q in R. We shall often identify p with $i(p)$.

We define the *strict ordering* $<$ on R by

$$r<s \Leftrightarrow_{\mathrm{df}} \exists p.p \in \pi_2(r) \cap \pi_1(s).$$

It is easy to see that

$$\vdash_S p \in \pi_1(r) \Leftrightarrow p<r.$$

We remark that it is *not* the case that

$$\vdash_S r \leqslant s \Leftrightarrow r < s \vee r = s.$$

However, we have

(7.9) $$\vdash_S r \leqslant s \Leftrightarrow \neg(s < r).$$

To prove this, note that

$$\begin{aligned} r \leqslant s,\ s < r \vdash_S & \exists p.p \in \pi_2(s) \cap \pi_1(s) \\ \vdash_S & \mathit{false} \end{aligned}$$

so

$$r \leqslant s \vdash_S \neg(s < r).$$

Conversely,

$$\begin{aligned} \neg(s < r), p \in \pi_1(r) \vdash_S & \exists q[p < q \wedge q \in \pi_1(r)] \\ \vdash_S & \exists q[p < q \wedge \neg(q \in \pi_2(s))] \\ \vdash_S & \exists q[p < q \wedge \neg(q \in \pi_2(s)) \\ & \wedge (p \in \pi_1(s) \vee q \in \pi_2(s))] \\ \vdash_S & p \in \pi_1(s). \end{aligned}$$

Hence

$$\begin{aligned} \neg(s < r) \vdash_S & p \in \pi_1(r) \Rightarrow p \in \pi_1(s) \\ \vdash_S & \pi_1(r) \subseteq \pi_1(s) \\ \vdash_S & r \leqslant s. \end{aligned}$$

This proves (7.9).

Now define (using evident informal notations)

$$\begin{aligned} u^+ &= \{p : \exists q < p\ \forall q' \in u.q' < q\} \\ u^- &= \{p : \exists q > p\ \forall q' \in u.q < q'\}. \end{aligned}$$

The *extended reals* $^*R$ are then defined by

$$^*R = \{\langle u,v \rangle \in PQ \times PQ : \exists p\ \exists q(p \in u \wedge q \in v) \wedge u = v^- \wedge v = u^+\}.$$

We define $<$, $\leqslant$ on $^*R$ as we did for R. Now, using the fact that Q satisfies trichotomy, i.e.

$$\vdash_S \forall p\ \forall q(p < q \vee p = q \vee q < p),$$

it is not hard to show that

$$\vdash_S R \subseteq {}^*R.$$

A partially ordered set is called *conditionally complete* if every non-empty subset with an upper bound has a least upper bound. It is quite easy to show that

$$\vdash_S \langle {}^*R, \leqslant \rangle \quad \textit{is conditionally complete.}$$

(*Sketch of proof*: given a family $\{\langle U_i, V_i \rangle : i \in I\}$ (bounded above) of members of $^*R$, one shows that, putting $\bigcup_{i \in I} U_i = U$, the pair $\langle U^{+-}, U^+ \rangle$ is the least upper bound of the family.)

However, the situation in respect of R is quite different. For we have

7.10 THEOREM. The following are equivalent for any naturalized local set theory S:

(i) $\vdash_S \forall \omega[\neg \omega \vee \neg\neg \omega]$;
(ii) $\vdash_S R = {}^*R$;
(iii) $\vdash_S \langle R, \leqslant \rangle$ *is conditionally complete*;
(iv) $\vdash_S \langle 2, \leqslant \rangle$ *is complete.* (Here $2 = \{0,1\}$ and $\leqslant = \{\langle 0,0 \rangle, \langle 0,1 \rangle, \langle 1,1 \rangle\}$.)

Proof. (i)$\Rightarrow$(ii). We have already noted that $\vdash_S R \subseteq {}^*R$, so it remains to prove the reverse inclusion. To do this it suffices to show that

$$p < q,\ \langle u,v \rangle \in {}^*R \vdash_S p \in u \vee q \in v. \tag{1}$$

Now we have

$$\langle u,v \rangle \in {}^*R, q \in u, p \leqslant q \vdash_S p \in u.$$

Hence

$$\langle u,v \rangle \in {}^*R, \neg(p \in u),\ q \in u \vdash_S \neg(p \leqslant q) \vdash_S q < p,$$

since Q satisfies trichotomy. Therefore

$$\langle u,v \rangle \in {}^*R, \neg(p \in u) \vdash_S \forall q \in u . q < p,$$

whence

$$\langle u,v \rangle \in {}^*R, \neg(p \in u),\ p < q \vdash_S q \in u^+ \vdash_S q \in v. \tag{2}$$

Similarly, we get

$$\langle u,v \rangle \in {}^*R, \neg(q \in v),\ p < q \vdash_S p \in u. \tag{3}$$

Write q' for $\frac{1}{2}(p+q)$. Then, by (i),

$$\vdash_S \neg(q' \in u) \vee \neg\neg(q' \in u). \tag{4}$$

Now, using (2),

$$\langle u,v \rangle \in {}^*R, p < q, \neg(q' \in u) \vdash_S q \in v, \tag{5}$$

and since

$$\langle u,v \rangle \in {}^*R, q' \in v \vdash_S q' \in u,$$

we get, using (3),

$$\langle u,v\rangle \in {}^*R, p<q, \neg\neg(q' \in u) \vdash_S \neg(q' \in v) \vdash_S p \in u. \tag{6}$$

(4), (5) and (6) now yield (1).

(ii)$\Rightarrow$(iii). We have already remarked that $\vdash_S \langle {}^*R, \leqslant\rangle$ *is conditionally complete.*

(iii)$\Rightarrow$(iv). Identify 2 with the subset $\{\bar{0},\bar{1}\}$ of R. Assuming (iii), write $\bigvee u$ for the supremum of u in $\langle R, \leqslant\rangle$. It suffices to show that

$$u \subseteq 2, \bar{0} \in u \vdash_S \bigvee u \in u. \tag{7}$$

For simplicity write a for $\bigvee u$. Then clearly

$$u \subseteq 2, \bar{1} \in u \vdash_S a = \bar{1}$$
$$u \subseteq 2, \bar{0} \in u, \neg(\bar{1} \in u) \vdash_S a = \bar{0}.$$

But since† $\vdash_S \neg\neg((\bar{1} \in u) \vee \neg(\bar{1} \in u))$ it follows that

$$\begin{aligned} u \subseteq 2, \bar{0} \in u &\vdash_S \neg\neg(a = \bar{0} \vee a = \bar{1}) \\ &\vdash_S \neg\neg(a \in 2). \end{aligned} \tag{8}$$

Now by (7.9) we have

$$\begin{aligned} &\vdash_S \neg(a < \bar{0}) \Leftrightarrow a \geqslant \bar{0}, \\ &\vdash_S \neg(a > \bar{1}) \Leftrightarrow a \leqslant \bar{1}. \end{aligned} \tag{9}$$

Also, again using (7.9)

$$\begin{aligned} \neg(a > \bar{0} \wedge a < \bar{1}), a > \bar{0} \vdash_S \neg(a < \bar{1}) \vdash_S a \geqslant \bar{1}. \\ \neg(a > \bar{0} \wedge a < \bar{1}), a < \bar{1} \vdash_S \neg(a > \bar{0}) \vdash_S a \leqslant \bar{0}. \end{aligned} \tag{10}$$

However, it is easily seen that

$$r \in R \vdash_S r > \bar{0} \vee r < \bar{1},$$

so that, combining this with (10), we get

$$\neg(a > \bar{0} \wedge a < \bar{1}) \vdash_S a \leqslant \bar{0} \vee a \geqslant \bar{1}. \tag{11}$$

Conversely, it is easy to show that

$$a \leqslant \bar{0} \vee a \geqslant \bar{1} \vdash_S \neg(a > \bar{0} \wedge a < \bar{1}).$$

Combining this with (11) gives

$$\vdash_S \neg(a > \bar{0} \wedge a < \bar{1}) \Leftrightarrow (a \leqslant \bar{0} \vee a \geqslant \bar{1}).$$

† Recall that $\vdash \neg\neg(\omega \vee \neg\omega)$.

Therefore, writing α for

$$a<\bar{0} \vee a>\bar{1} \vee (a>\bar{0} \wedge a<\bar{1})$$

we have, using (9),

$$\begin{aligned}\vdash_S \neg\alpha &\Leftrightarrow \neg(a<\bar{0}) \wedge (a>\bar{1}) \wedge \neg(a>\bar{0} \vee a<\bar{1})\\ &\Leftrightarrow a\geqslant\bar{0} \wedge a\leqslant\bar{1} \wedge (a\leqslant\bar{0} \vee a\geqslant\bar{1})\\ &\Leftrightarrow a=\bar{0} \vee a=\bar{1} \Leftrightarrow a\in 2.\end{aligned}$$

Hence, by (8),

$$\begin{aligned}u\subseteq 2, \bar{0}\in u \vdash_S \neg\neg(a\in 2) \vdash_S \neg\neg\neg\alpha \vdash_S \neg\alpha\\ \vdash_S a\in 2.\end{aligned}$$

This is (7).

(iv)$\Rightarrow$(i). Suppose that 2 is complete. Then since $2\cong\{\mathit{false}, \mathit{true}\}$, we may assume the latter complete. Given a formula α, let $X=\{\omega:\omega=\mathit{true}\wedge\alpha\}$ and let β be the supremum of X in $\{\mathit{false}, \mathit{true}\}$. Then $\vdash_S \beta\in\{\mathit{false}, \mathit{true}\}$ so that $\vdash_S \beta=\mathit{false} \vee \beta=\mathit{true}$, whence $\vdash_S \neg\beta \vee \beta$, so that

$$\vdash_S \neg\beta \vee \neg\neg\beta. \tag{12}$$

Also $\alpha \vdash_S \beta=\mathit{true} \vdash_S \beta$, so that

$$\neg\beta \vdash_S \neg\alpha. \tag{13}$$

Moreover

$$\neg\alpha \vdash_S (\omega=\mathit{true}\wedge\alpha)\Rightarrow \mathit{false} \vdash_S X=\varnothing \vdash_S \beta=\mathit{false} \vdash_S \neg\beta$$

so that

$$\neg\neg\beta \vdash_S \neg\neg\alpha. \tag{14}$$

(12)–(14) now yield $\vdash_S \neg\alpha \vee \neg\neg\alpha$, which gives (i). □

7.11 EXAMPLE. *The object of real numbers in* **Sh**(X). Let X be a topological space, and construct R in Th(**Sh**(X)) as above. Now suppose that U, V are Th(**Sh**(X))-sets such that

$$\vdash_{\mathrm{Th}(\mathbf{Sh}(X))} \langle U,V\rangle \in R.$$

We know that $\Omega(\mathrm{Th}(\mathbf{Sh}(X)))\cong\mathcal{O}(X)$: for each sentence α of $\mathcal{L}(\mathbf{Sh}(X))$ let $|\alpha|$ be the member of $\mathcal{O}(X)$ associated with α under this isomorphism. For $t\in X$ define

$$\begin{aligned}U_t&=\{q\in\mathbb{Q}:t\in|\bar{q}\in U|\}\\ V_t&=\{q\in\mathbb{Q}:t\in|\bar{q}\in V|\},\end{aligned}$$

where $\mathbb{Q}$ is the set of (intuitive) rational numbers. It can now be readily checked that the pair $(U_t, V_t) = r_t$ is a Dedekind cut in the rationals, i.e. a real number. Moreover, the map $t \mapsto r_t$ is continuous from X to $\mathbb{R}$, the set of (intuitive) real numbers. To see this, note that for $q \in \mathbb{Q}$ we have $q \in U_t$ iff $q < r_t$ and $q \in V_t$ iff $q > r_t$. Given $t \in X$ and $\varepsilon > 0$ choose $q, q' \in \mathbb{Q}$ such that $q < r_t < q'$ and $r_t - q < \varepsilon$, $q' - r_t < \varepsilon$. For any $s \in |\bar{q} \in U \wedge \bar{q}' \in V|$, a neighbourhood of t, we have $q < r_s < q'$, so that $|r_s - r_t| < \varepsilon$. Thus the continuity of $t \mapsto r_t$ is established.

Conversely, let $f: X \to \mathbb{R}$ be continuous. We can identify the canonical interpretation $[Q]_{\mathbf{Sh}(X)}$ with the sheaf $Q^\dagger$ of locally constant Q-valued functions on X (cf. 5.35(2)). The assignments

$$W \mapsto \{g \in Q^\dagger(W) : \forall x \in W . g(x) < f(x)\}$$
$$W \mapsto \{g \in Q^\dagger(W) : \forall x \in W . f(x) < g(x)\}$$

for $W \in \mathcal{O}(X)$ then define subsheaves U, V of $Q^\dagger$. It is now not hard to show that

$$\vdash_{\mathrm{Th}(\mathbf{Sh}(X))} \langle U, V \rangle \in R.$$

In this way we establish a bijective correspondence between pairs U, V of Th(**Sh**(X))-sets such that $\vdash_{\mathrm{Th}(\mathbf{Sh}(X))} \langle U, V \rangle \in R$ and continuous real-valued functions on X. Calling the canonical interpretation $[R]_{\mathbf{Sh}(X)}$ the *object of real numbers* in **Sh**(X), and its **Sh**(X)-elements *real numbers* in **Sh**(X), it follows that the latter may be identified with continuous real-valued functions on X. And from this fact it can be shown that the object of real numbers in **Sh**(X) may be identified with the sheaf of real-valued continuous functions on X.

EXERCISES

(1) Show that condition (i) of (7.10) is equivalent to either of the following conditions:

(i) $\vdash_S \forall\omega[\neg(\omega \wedge \omega') \Leftrightarrow (\neg\omega \vee \neg\omega']$ (de Morgan's law)
(ii) $\{$*false*$\}$ has a complement in Ω.

(2) Let **C** be a small category, and as usual write **E(C)** for the topos of presheaves on **C**. Show that Th(**E**(**C**)) satisfies condition (ii) of Exercise (1) iff every diagram in **C** of the form

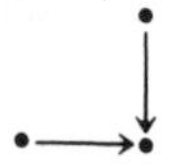

can be completed to a commutative square.

(3) Let X be a topological space. Show that condition (i) of (7.10) holds in Th(**Sh**(X)) iff X is *extremally disconnected*, i.e., for any open set U in X, its closure $\bar{U}$ is also open.

The free topos

Let $\mathscr{L}_{\mathbf{N}}$ be the local language with just one ground type symbol **N**, one function symbol **s** of signature $\mathbf{N} \to \mathbf{N}$ and one function symbol **0** of signature $\mathbf{1} \to \mathbf{N}$. Write $\bar{0}$ for $\mathbf{0}(*)$. Let P be the local set theory in $\mathscr{L}_{\mathbf{N}}$ whose axioms are

$$:\neg(\mathbf{s}n = \bar{0})$$

$$\mathbf{s}m = \mathbf{s}n : m = n$$

$$\bar{0} \in u, \forall n(n \in u \Rightarrow \mathbf{s}n \in u) : \forall n. n \in u$$

where m,n are variables of type **N** and u is a variable of type **PN**. The triple $(\mathbf{N},\mathbf{s},\bar{0})$ is then a natural number system in P, so that P is a naturalized local set theory: it is called the *free* naturalized local set theory.

P is particularly important because it is an *initial object* in the category of naturalized local set theories. Given two naturalized local set theories S, S', a *natural translation* of S into S' is a translation $K: S \to S'$ which preserves **N**, **s** and $\bar{0}$. Let **Natloc** be the category of naturalized local set theories with natural translations as arrows. It should now be obvious that P is an initial object in **Natloc**. The associated topos **C**(P) is called the *free topos*.

P has some features which make it attractive from a constructive standpoint: for instance, it is witnessed and has the *disjunction property*, viz., for sentences α, β,

$$\vdash_P \alpha \vee \beta \quad \text{implies} \quad \vdash_P \alpha \quad \text{or} \quad \vdash_P \beta.$$

(For proofs establishing these and other properties of P see Lambek and Scott (1986)).

These facts lead one to suggest that P is the *ideal theory*, and its model the free topos the *ideal universe*, for the constructively minded mathematician.

If to the axioms of P we add the law of excluded middle

$$: \forall \omega(\omega \vee \neg \omega)$$

we get the theory P^c—the *free classical naturalized local set theory*—which is the classical counterpart of P. The associated topos $\mathbf{C}(P^c)$ is called the *free Boolean topos*. It would seem natural to regard this topos as the ideal universe for the classically minded mathematician; however, the incompleteness of first-order set theory implies that P^c is not complete, so that $\mathbf{C}(P^c)$ is not bivalent, an evident drawback from the classical standpoint.

8

Epilogue: the wider significance of topos theory

In this concluding chapter we attempt to place topos theory in a wider context. We start with an account of the relationship between set theory and topos theory, contrasting the 'monism' of the former with the 'pluralism' of the latter. Next, an analogy is drawn between topos theory and relativity theory, based on the fact that both employ the notion of a reference frame in an essential way. Finally, we show how the concept of variable set underlies two important developments in the foundations of mathematics: Robinson's nonstandard analysis and Cohen's independence proofs in set theory.

From set theory to topos theory

With the creation and refinement of set theory by the early decades of this century, the problem of providing a suitable foundation for the 'abstract postulational' formulation of mathematics (as championed, for example, by Hilbert) appeared to be solved. The universe of sets ('Cantor's paradise') came to be regarded as the source of raw material for building the structures called for by the abstract postulational program. So, although the concepts of (abstract) *structure* and *operation* on structures had come to play a fundamental role in most mathematical disciplines, these concepts were not taken as primitive; instead they were explicated by reducing them to the ostensibly more fundamental concept of *set*. On the other hand, the transparency and apparent reliability of the set-theoretical framework enabled mathematicians not primarily interested in set theory—in other words, the vast majority—to take the set-theoretic reduction for granted and get on with mathematical business as usual.

With the rise of abstract algebra in the 1930s it came to be discerned that the concepts and constructions (for example, isomorphism, quotient, product, etc.) most frequently arising in connection with mathematical structures possess a universality that is in some sense independent of their set-theoretical origin. Furthermore, it was observed that these concepts and constructions can usually be formulated in terms of the single concept of structure-preserving map, or *morphism* (arrow). Thus gradually arose the view that the essence of a mathematical structure is to be sought not in its internal constitution as a set-theoretical entity, but rather in the form of its relationship with other structures through the network of morphisms. (In particular, it came to be seen that the notion of equality appropriate for structures is not numerical identity, but *isomorphism,* an idea going back in principle to Dedekind and Klein.) This view, strikingly reminiscent of the operational structuralism associated with, for example, linguistics and psychology, was espoused most resolutely by the Bourbaki school in France, which had proposed a 'structuraliste' account of mathematics in the 1930s. However, although the account of mathematics they give in their justly famed *Éléments de Mathématique* is manifestly structuralist in intention, their uncritical employment of (axiomatic) set theory in their formulation of the concept of mathematical structure prevented them from achieving the structuralist objective of treating structures as autonomous forms with no specified substance. In fact, it was not until the 1940s that an axiomatic framework for mathematics was developed that was more in keeping with the spirit of operational structuralism, viz., the theory of *categories and functors* devised by Eilenberg and MacLane. Here for the first time we have a theory which takes the notions of structure and morphism as primitive (as *objects* and *arrows,* respectively) and which is indifferent to any particular set-theoretic construction that structures may possess.

Category theory may be said to bear the same relation to abstract algebra as does the latter to elementary algebra. Elementary algebra results from the replacement of *constant quantities* (i.e. numbers) by *variables,* keeping the operations on these quantities fixed. Abstract algebra, in its turn, carries this a stage further by allowing the *operations* to vary while ensuring that the resulting mathematical structures (groups, rings, etc) remain of a certain prescribed kind. Finally, category theory allows even the *kind* of structure to vary: it is concerned with *structure in general.*

In category theory the morphisms (arrows) between structures (objects) play an autonomous role which is in no way subordinate to that played by the structures themselves. So category theory is like a language in which the 'verbs' are on an equal footing with the 'nouns'. In this respect category theory differs crucially from set theory in which the corresponding notion of *function* is reduced to the concept of *set* (of points). This difference arises more generally in the respective accounts that set theory and category theory provide of mathematical structure. Both set theory and category theory transcend the *particularity* of mathematical structures. Set theory *strips away* structure from the ontology of mathematics leaving pluralities of structureless individuals open to the imposition of new structure. Category theory, by contrast, transcends particular structure not by doing away with it, but by taking it as given and *generalizing* it. It may be said that the success of category theory as a unifying language for mathematics is due to the fact that it, and it alone, gives direct expression to the centrality of form and structure in mathematics.

The emergence of category theory has tended to undermine the received idea that the meaning of any mathematical concept is fixed by referring it to the context of a unique absolute universe of sets. By contrast, category theory suggests that the interpretation of a mathematical concept may vary with the choice of 'category of discourse'. To indicate what is meant by this, consider, for example, the concept 'group'. From the set-theoretical standpoint, the term 'group' signifies a set (equipped with a couple of operations) satisfying certain elementary axioms expressed in terms of the elements of the set. Thus the set-theoretical interpretation directly refers to the absolute universe of sets. Now consider the category-theoretic formulation of the group concept. Here the reference to the 'elements' of the group has been replaced by an 'arrows only' formulation, thereby enabling the concept to become interpretable not merely in the universe (category) of sets but in virtually *any* category. The possibility of varying the domain of interpretation afforded by category theory confers on the group concept a truly protean generality. Indeed, its interpretation within the category of topological spaces is *topological group,* within the category of differentiable manifolds it is *Lie group,* and within the category of sheaves over a topological space it is *sheaf of groups.*

We see, then, that the category-theoretic meaning of a mathematical concept is determined only in relation to a 'category of discourse' which can itself *vary.* Thus the effect of casting a mathematical concept in category-theoretic terms is to confer a degree of *ambiguity of reference* on the concept.

In fact, a similar kind of referential ambiguity is already present within classical set theory, since its axioms are formulated in *first-order* terms and therefore admit many essentially *different* interpretations. Indeed, as far back as 1922, Skolem remarked that, for this reason, set-theoretical notions—in particular, infinite cardinalities—have no absolute meaning. He concluded that axiomatized set theory 'is not a satisfactory ultimate foundation for mathematics'. Skolem's strictures were largely ignored by mathematicians, but a new challenge to the absoluteness of the set-theoretical framework arose in 1963 when Paul Cohen constructed models of set theory (i.e. Zermelo–Fraenkel set theory ZF) in which important mathematical *propositions* such as the axiom of choice and the continuum hypothesis are falsified (Gödel having already in the 1930s produced models in which these propositions are validated). The resulting ambiguity in the *truth values* of mathematical propositions was regarded by many set-theorists (and even by more 'orthodox' mathematicians) as a much more serious matter than the ambiguity of reference of mathematical concepts already pointed out by Skolem. The techniques of Cohen and his successors have led to a vast proliferation of models of set theory with essentially different mathematical properties, which has in turn engendered a disquieting uncertainty in the minds of set-theorists as to the identity of the 'real' universe of sets, or at least as to precisely what mathematical properties it should possess. Notwithstanding the entrenched Platonist attitudes of many mathematicians, the set concept has turned out to be *radically underdetermined.* As Mostowski said in 1965, in response to Cohen's discoveries:

> Probably we shall have in the future essentially different intuitive notions of sets just as we have different notions of space, and will base our discussions of sets on axioms which correspond to the kind of sets which we want to study.

In this event, it becomes natural, indeed mandatory, to seek for the set concept a formulation that takes account of its underdetermined character, that is, one that does not bind it so tightly to the absolute universe of sets with its rigid hierarchical structure. Category theory furnishes such a formulation through the concept of *topos,* and its formal counterpart *local set theory*.

We have seen in Chapter 3 that each local set theory S gives rise to a topos $\mathbf{C}(S)$ in such a way that validity or truth in $\mathbf{C}(S)$ corresponds precisely to provability in S. And conversely, given a topos $\mathbf{E}$ we can define a local set theory Th($\mathbf{E}$) in such a way that truth in $\mathbf{E}$ corresponds precisely to provability in Th($\mathbf{E}$). Moreover, the equivalence theorem asserts that Th($\mathbf{E}$) determines $\mathbf{E}$ up to equivalence.

Any topos may be regarded as a mathematical domain of discourse or 'world' in which mathematical concepts can be interpreted and mathematical constructions performed. In this event, the associated local set theory may be treated as a 'chart' mapping that world. Just as all the charts in an atlas share a common geometry, so all local set theories share a common internal *logic,* which we have established in Chapter 3 to be the intuitionist logic of types.

In a topos $\mathbf{E}$, arrows $1 \to \Omega$ are *truth values* of propositions interpreted in $\mathbf{E}$: there are always at least the two truth values $1 \xrightarrow{\top} \Omega$ and $1 \xrightarrow{\bot} \Omega$ corresponding to truth and falsehood, but, as we have seen, there may be more than just these. We have called a topos in which these are the only truth values *bivalent,* and we have observed that bivalence of a topos corresponds to *completeness* of the associated local set theory. Now since truth in a topos corresponds to provability in a local set theory, and many local set theories are *incomplete,* we must be prepared to accept the phenomenon of *polyvalence,* that is, the fact that mathematical propositions (formulated in local set theories and interpreted in toposes) will in general possess truth values *different* from truth or falsehood.

We may think of a topos (with a natural number system) and its associated theory as dual aspects of the same entity—let us call it a (*mathematical*) *framework*—the topos constituting, so to speak, its *substance* and the theory its *form.* Like a work of fiction, the substance of a framework is entirely determined by the form: any (meaningful) question which one can ask about the substance is either answerable within the form or, if not (as in the case of an incomplete theory), must be regarded as having no determinate answer within the given framework. Just as, for instance, in Kafka's *The Castle,* the remaining letters of its protagonist *K*'s name must forever remain undetermined since no scrutiny of the text will ever reveal them, so, analogously, the 'truth' or 'falsity' of the continuum hypothesis (suitably formulated within a naturalized local language) will always be indeterminate within the pure classical naturalized local set theory. (This follows from the fact that there exist models of set theory in which the continuum hypothesis is true, and others in which it is false.)

Topos theory yields the concordance shown in Table 8.1 between *formal* and *substantive* notions. Confining attention to the left-hand side of this scheme, we get *formalism*; to the right, *realism* (or platonism). The concordance between the left- and right-hand sides of this scheme suggests the possibility of a rapprochement between these two opposed doctrines, a rapprochement extending over those parts of mathematics formulable within toposes or local set theories.

Some analogies with the theory of relativity

The notion of *geometric morphism* is of central importance in topos theory. We may think of a geometric morphism $\mathbf{E} \to \mathbf{E}'$ between toposes $\mathbf{E}, \mathbf{E}'$ as a nexus between the mathematical worlds represented by $\mathbf{E}$

Table 8.1

Formal	Substantive
Local set theories	Toposes
Pure naturalized local set theory	Free topos
Provability	Truth
Completeness	Bivalence

and **E**′, or as a method of shifting from **E** to **E**′ and vice versa. There is an analogy here with the physical geometric notion of *change of coordinate system.* Indeed, just as in astronomy one effects a change of coordinate system to simplify the description of the motion of, e.g., a planet, so it also proves possible to simplify the formulation of a mathematical concept by effecting a shift of mathematical framework. Consider, for example, the concept 'real-valued continuous function on a topological space X'. Any such function may be regarded as a real number (or quantity) *varying continuously* over X. Now consider the topos **Sh**(X) of sheaves on X. Here *everything* is varying (continuously) over X, so shifting from **Set** to **Sh**(X) essentially amounts to placing oneself in a framework which is, so to speak, itself 'co-moving' with the variation over X of any given variable real number. This causes its variation not to be 'noticed' in **Sh**(X); it is accordingly regarded as being a *constant* real number. In this way the concept 'real-valued continuous function on X' is transformed into the concept 'real number' when interpreted in **Sh**(X). (To be precise, the objects in **Sh**(X) satisfying the condition of being real numbers correspond, via the geometric morphism **Sh**(X)→**Set**, to the real-valued continuous functions on X: cf. (7.11)). Putting it the other way around, the concept 'real number', interpreted in **Sh**(X) corresponds to the concept 'real-valued continuous function on X' interpreted in **Set**.

Thus, we see again that a mathematical concept may possess a fixed *sense,* but a variable *reference.* Indeed, we may think of the *sense* of the concept 'real number' as being fixed by its definition within a (naturalized) local set theory, while, as we have seen, its *reference* will vary with the mathematical framework of interpretation. That is, the reference of this concept is determined only *relative* to a mathematical framework.

Another instance of this relativity of mathematical concepts is the phenomenon of *cardinal collapse,* familiar to all set-theorists. Suppose we are given an uncountable set I (in **Set**). Then we can shift (via a geometric morphism) to a new mathematical framework **E** (a Boolean extension of **Set**) in which the cardinality of I is *countable.* Thus the cardinality of an infinite set is not absolute or intrinsic but determined only relative to the mathematical framework with respect to which the cardinality is 'measured'.

Now, the recognition that properties originally held to be absolute or intrinsic must instead be treated as relative has arisen frequently in the theory of thought. For example, Leibniz recognized that a state of rest or motion of a material body is not an intrinsic state of the body but is determined only relative to other bodies. One of the profoundest instances of this phenomenon arose in the transition from *classical* (Newtonian) to *relativistic physics,* when physical concepts such as simultaneity of events and mass of a body formerly ascribed an absolute meaning were seen to possess meaning only in relation to *local coordinate systems.*

There is an evident *analogy* between mathematical frameworks and the local coordinate systems of relativity theory: each serve as the appropriate *reference frames* for fixing the meaning of mathematical or physical concepts respectively. Pursuing this analogy suggests certain further parallels.

For example, consider the concept of *invariance.* In relativistic physics, *invariant physical laws* are statements of mathematical physics (e.g. Maxwell's equations) that, suitably formulated, hold universally, i.e. in every local coordinate system. Analogously, *invariant mathematical laws* are mathematical assertions that again hold universally, i.e. in every mathematical framework. These are the theorems of the free naturalized local set theory, or, if you prefer, of typed intuitionistic logic with a natural number system. Thus the invariant mathematical laws are those provable *constructively*: so, for example, one would count as invariant mathematical laws the theorems of intuitionistic arithmetic but not the law of excluded middle. Notice in this connection that a theorem of classical logic that is not constructively provable will not in general hold universally until it has been transformed into its intuitionistic correlate (which, e.g. in the case of the excluded middle $\omega \vee \neg\omega$, is $\neg\neg(\omega \vee \neg\omega)$). The procedure of translating classical into intuitionistic logic is thus the logical counterpart of casting a physical law in invariant form.

The physical concept of *inertial coordinate system* also has its counterpart in the context of mathematical frameworks. An inertial coordinate system is one in which undisturbed bodies undergo no accelerations, i.e. in which *Newton's first law of motion* holds. Thus inertial coordinate systems act as surrogates for Newtonian absolute space. Analogously a *constant* mathematical framework is one in which objects undergo no variations, in other words, one which resembles the absolute universe of (constant) sets as much as possible. This resemblance can be ensured by the satisfaction of the *strong axiom of choice.* This is because the truth of the strong axiom of choice in a topos **E** implies not only that **E** is Boolean and bivalent, but also that it is extensional, i.e. its arrows resemble set-theoretic mappings in that they are determined by their action on points of their domains. These features may be taken as distinguishing toposes of constant sets among arbitrary toposes. Since the strong axiom of choice ensures the presence of these features, we may *define* a constant topos, or mathematical framework, to be one in which the strong axiom of choice is satisfied (equivalently: the associated local set theory is Hilbertian). Then: *constant mathematical frameworks correspond to inertial coordinate systems, and the* (*strong*) *axiom of choice to Newton's first law of motion.*

These observations suggest that the topos-theoretic interpretation of mathematical concepts bears the same relation to classical set theory as relativity theory does to classical physics.

The negation of constancy

Suppose we are given a topos **E** defined over the topos **Set** (and not equivalent to it). In general, we may regard **E** as a topos which results when the constant objects of **Set** are allowed to *vary* in some manner. (For example, when **E** is **Sh**(X), the objects of **E** are those varying continuously over X.) In passing from **Set** to **E** (via the geometric morphism **E** $\rightarrow$ **Set**), thereby obtaining the 'variation' in the objects of **E**. In passing from **Set** to a topos defined over it we are, in short, *negating constancy*.

Now in certain important cases we can proceed in turn to negate the 'variation' in **E** to obtain a new (constant) topos **Set*** in which constancy again prevails. **Set*** may be regarded as arising from **Set** through the process of *negating negation*. (In general, **Set*** is not equivalent to **Set** and is therefore (by (6.11)) not defined over **Set**.)

As we shall see, this process of 'negating negation' underlies two key developments in the foundations of mathematics: *Robinson's nonstandard analysis* and *Cohen's independence proofs in set theory*.

Given a set I, each element $i \in I$ may be identified with the principal ultrafilter $U_i = \{Z \subseteq T : i \in Z\}$ over I. This identification suggests that we think of *arbitrary* ultrafilters over I as *generalized points* of I. The collection of generalized points of I forms a new set βI (which topologists call the Stone–Čech compactification of I). Identifying I as a subset of βI, we call elements of I *standard* points of I, and elements of $\beta I - I$ *ideal* points of I. If I is infinite, it always has ideal points.

Now consider the topos **Set**^{I} of sets varying over I. Objects of **Set**^{I}—which we shall call *variable sets*—are I-indexed families of sets $X = \langle X_i : i \in I \rangle$. An 'element' of a variable set X is an I-indexed family $x = \langle x_i : i \in I \rangle$ such that $x_i \in X_i$ for $i \in I$, i.e. a 'choice function' on X. Thus the Cartesian product $\prod_{i \in I} X_i$ is the set of 'elements' of the variable set X.

Each (constant) set A is associated with the variable set $\hat{A}$ given by the function on I with constant value A. The set of 'elements' of the variable set $\hat{A}$ is A^I.

We note that the topos **Set**^{I} is defined over **Set** via the geometric morphism μ given by $\mu^*(A) = \hat{A}$, with $\mu_*(x) = \prod_{i \in I} X_i$.

Given an element $i_0 \in I$, we can arrest the variation of any variable set X by *evaluating* at i_0, i.e. by considering X_{i_0}. If we apply this in particular to the set A^I of 'elements' of the variable set $\hat{A}$, i.e. if we evaluate each such 'element' at i_0, we just retrieve A. So, in this case, if we negate the constancy of (the elements of) A by passing to the set A^I of (variable) 'elements' of A, and then negate the variation of these 'elements' by evaluating at a *standard* point of I, we come full circle. This instance of 'negation of negation' is, accordingly, *trivial*. The situation is decidedly different, however, when the evaluation is made at an *ideal* point of I.

Given an ideal point U of I, i.e. a nonprincipal ultrafilter over I, how shall we 'evaluate' functions in A^I at U? To this end, observe that the result of evaluating at a standard point i_0 of I is essentially the same as *identifying* functions in A^I when their values at i_0 coincide, i.e. stipulating that, for $f,g \in A^I$,

$$f \sim_{i_0} g \quad \text{iff} \quad f(i_0)=g(i_0) \quad \text{iff} \quad \{i \in I : f(i)=g(i)\} \in U_{i_0}.$$

We use this last equivalence as the basis for evaluating functions in A^I at an ideal point U of I. That is, we define

$$f \sim_U g \quad \text{iff} \quad \{i \in I : f(i)=g(i))\} \in U.$$

Then, the result of 'evaluating' all the functions in A^I at U is the set A^* of equivalence classes of A^I modulo $\sim_U$. It is well known that if I is infinite (and U an ideal point of I), then A^* cannot be the same as A. In particular if, for example, A is the real line R, then R^* will have the same elementary properties as R but will in addition contain new 'infinite' and 'infinitesimal' elements. That is, R^* will be a *nonstandard model of the real line*. This, in essence, is the basis of Robinson's nonstandard analysis.

To sum up, we get Robinson's infinitesimals by first negating the constancy of the classical real line, and then negating the resulting variation by arresting it at an ideal point.

Now, if we arrest the variation of all the objects of $\mathbf{Set}^I$ *simultaneously* at an ideal point of I we obtain a new constant topos $\mathbf{Set}^*$ (actually an ultrapower or enlargement of **Set**) which has the same elementary properties as **Set**. So this instance of 'negating negation' leads to a constant topos which, although not identical with **Set**, is nonetheless *internally indistinguishable* from it. By contrast, the whole purpose of *Cohen's* techniques in set theory is to obtain new constant toposes which are internally *distinguishable* from **Set**.

Let P be a partially ordered set: think of the elements of P as *stages of knowledge* and $p \leqslant q$ as meaning that q is a deeper (or later!) stage of knowledge than p. Consider now the topos $\mathbf{Set}^{\mathbf{P}}$ of sets varying over $\mathbf{P}$ (cf. Chapter 2). The objects of $\mathbf{Set}^{\mathbf{P}}$ may be thought of as sets varying over the stages of knowledge assembled in P. Within $\mathbf{Set}^{\mathbf{P}}$ we consider objects X such that, for $p \leqslant q$, $X(p) \subseteq X(q)$ and X_{pq} is the insertion map $X(p) \rightarrow X(q)$. Such an object we call a set *varying steadily* over $\mathbf{P}$. If we think of $X(p)$ as the collection of elements of the variable set X *secured* at stage p, then the 'steadiness' condition means that no secured elements are ever lost. For $p \in P$ and sets X, Y varying steadily over $\mathbf{P}$, we write

$$p \Vdash X \subseteq Y$$

for

$$\forall q \geqslant p \forall x \in X(q) \exists r \geqslant q \, [x \in Y(r)]$$

that is, at stage p, X is *eventually contained* in Y. We also write

$$p \Vdash X = Y$$

for

$$p \Vdash X \subseteq Y \,\&\, p \Vdash Y \subseteq X,$$

that is, at stage p, X *eventually coincides* with Y.

Two elements $p, q \in P$ are *compatible* if $\exists r \in P[p \leqslant r \,\&\, q \leqslant r]$. A set of compatible elements of P is called a (coherent) *body of knowledge*. A body of knowledge K is said to be *complete* if whenever $p \in P$ is compatible with every member of K, then p belongs to K.

Given a complete body of knowledge K, define the equivalence relation $\sim_K$ on the collection of sets varying steadily over P by

$$X \sim_K Y \quad \text{iff} \quad \exists p \in K \, [p \Vdash X = Y].$$

Thus $X \sim_K Y$ means that our body of knowledge K yields the assertion that X and Y eventually coincide. The collection **Set*** of equivalence classes modulo $\sim_K$ of steadily varying sets forms a *new* constant topos, which is, in general, internally *distinguishable* from **Set** in the sense that it does not share all the elementary properties of **Set**: **Set*** is in fact a (possibly nonstandard) *Cohen extension* of **Set**.

To recapitulate: the topos $\mathbf{Set}^{\mathbf{P}}$ was obtained by negating constancy in allowing variation ('growth') over stages of knowledge, and the Cohen extension **Set*** obtained from (the steadily varying objects in) $\mathbf{Set}^{\mathbf{P}}$ by using a complete body of knowledge to determine the 'eventual' identities among the variable sets, in other words, to negate their variation.

In the 'Cohen extension' case, passage from **Set** to **Set*** preserves some of the principles associated with constancy of sets (e.g. axiom of choice, classical logic) but, as Cohen showed, the partially ordered set P may be chosen in such a way—now familiar to every set theorist—as to ensure that other such principles (e.g., axiom of constructibility, continuum hypothesis) are *violated* in this passage. In passing from **Set** to $\mathbf{Set}^{\mathbf{P}}$ (negation of constancy), the classical bivalent logic of **Set** is replaced by the intuitionistic polyvalent logic of $\mathbf{Set}^{\mathbf{P}}$. And passage from $\mathbf{Set}^{\mathbf{P}}$ to **Set*** ('negation of negation') restores classical logic and constancy but, as we have remarked, not all principles associated with constancy.

Let us sum up. As we have said, the topos-theoretical viewpoint suggests that the absolute universe of sets be replaced by a plurality of 'toposes of discourse', each of which may be regarded as a possible 'world' in which mathematical activity may (figuratively) take place. The mathematical activity that takes place within such 'worlds' is codified within local set theories; it seems appropriate, therefore, to call this codification *local mathematics,* to contrast it with the *absolute* (i.e. classical) mathematics associated with the absolute universe of sets. *Constructive provability* of a mathematical assertion now means that it is *invariant,* i.e. valid in *every* local mathematics. Thus, from the standpoint of local mathematics, the use of constructive proof procedures, far from hobbling mathematical activity, has instead the opposite effect of extending the validity of mathematical reasoning to the widest possible spectrum of contexts.

APPENDIX: GEOMETRIC THEORIES AND CLASSIFYING TOPOSES

In this Appendix we discuss, in abbreviated fashion, the rudiments of the model theory of *first-order* languages in local set theories and toposes. We single out a particular class of first-order theories—the so-called *geometric theories*—whose models in toposes have an especially simple category-theoretic description. This description enables us to construct, for any geometric theory Σ, a topos—the so-called *classifying topos* of Σ—which contains a canonical or *generic* model of Σ validating exactly those (geometric) formulae which hold in every model of Σ.

A (many-sorted) *first-order language* L has the following ingredients:

(i) A set of *sorts* $A, B, C, \ldots$

(ii) A set of *function symbols* $f, g, h, \ldots$. Each function symbol is assigned a *signature* which is a pair $(\vec{A}, B)$ where $\vec{A} = (A_1, \ldots, A_n)$ is a finite (possibly empty) sequence of sorts and B is a sort.

(iii) A set of *predicate symbols* $\alpha, \beta, \ldots$ each of which is assigned a *signature* $\vec{A}$ which is a finite sequence of sorts.

(iv) Logical symbols $\wedge, \vee, \Rightarrow, \forall, \exists$.

(v) Propositional constants $\top$ (true) $\bot$ (false).

(vi) Equality symbol $=$

(vii) For each sort A, variables $x_A, y_A, \ldots$.

The *terms* of L and their corresponding *sorts* are defined inductively as follows:

(i) Each variable x_A is a term of sort A.

(ii) If f is a function symbol of signature $(\vec{A}, B)$ and $\vec{\tau} = (\tau_1, \ldots, \tau_n)$ is a sequence of terms of sorts $A_1, \ldots, A_n$, then $f(\vec{\tau})$ is a term of sort B.

Finally, the *formulae* of L are given inductively by the following clauses

(i) $\top$, $\bot$ are formulae.

(ii) If σ, τ are terms of the same sort, then $\sigma = \tau$ is a formula.

(iii) If α is a predicate symbol of signature $\vec{A}$ and $\vec{\tau}$ is a sequence of terms of sorts $A_1, \ldots, A_n$, then $\alpha(\vec{\tau})$ is a formula.

(iv) If ϕ, ψ are formulae, so are $\phi \wedge \psi$, $\phi \vee \psi$, $\phi \Rightarrow \psi$.

(v) If ϕ is a formula and x is a variable, then $\forall x \phi$ and $\exists x \phi$ are formulae.

We write $\neg\phi$, $\phi \Leftrightarrow \psi$ for $\phi \Rightarrow \bot$ and $(\phi \Rightarrow \psi) \wedge (\psi \Rightarrow \phi)$ respectively.

We assume that L is equipped with the usual deductive machinery of intuitionistic predicate logic (cf. Bell and Machover 1977, Chapter 9,

Section 9), *except* that the inference rule *modus ponens* is replaced by the rule *restricted modus ponens*, viz., from ϕ, $\phi \Rightarrow \psi$, deduce ψ *provided that every variable free in* ψ *is also free in* ϕ. (This is to allow for the possibility that an interpretation of L may contain an empty domain.)

A set of formulae of L is called a *theory* in L. If Σ is a theory in L, and ϕ a formula of L, we write $\Sigma \vdash \phi$ for 'ϕ is deducible from Σ using the axioms of intuitionistic predicate logic and restricted modus ponens', and $\Sigma \vdash_C \phi$ for 'ϕ is deducible from Σ in classical logic (i.e. with the addition of the axiom of excluded middle $\phi \vee \neg\phi$)'. We write $\vdash \phi$, $\vdash_C \phi$ for $\varnothing \vdash \phi$, $\varnothing \vdash_C \phi$.

Now let $\mathscr{L}$ be a local language and S a local set theory in $\mathscr{L}$. An *interpretation* I of a first-order language L in S assigns:

(i) to each sort A of L, a type symbol $\mathbf{A}_I^*$ of $\mathscr{L}$ and an S-set A_I of type $\mathbf{PA}_I^*$;
(ii) to each function symbol f of L of signature $(\vec{A}, B)$, an S-map $f_I : (A_1)_I \times \cdots \times (A_n)_I \rightarrow B_I$;
(iii) to each predicate symbol α of L of signature $\vec{A}$, a formula $[\![\alpha]\!]_I$ with free variables $x_1, \ldots, x_n$ of types $(\mathbf{A}_1^*)_I, \ldots, (\mathbf{A}_n^*)_I$ such that $\vdash_S [\![\alpha]\!]_I \Rightarrow x_1 \in (A_1)_I \wedge \cdots \wedge x_n \in (A_n)_I$.

The interpretation I can now be extended to all the terms of L recursively as follows:

(i) for each variable x of sort A, x_I is a variable of type $\mathbf{A}_I^*$;
(ii) if f is a function of signature $(\vec{A}, B)$ and $\vec{\tau}$ is a sequence of terms of sorts $A_1, \ldots, A_n$, then

$$(f(\vec{\tau}))_I = f_I(\langle (\tau_1)_I, \ldots, (\tau_n)_I \rangle).$$

It is readily shown that if τ is a term of sort B, then $\vdash_S \tau_I \in B_I$.

Each formula ϕ of L is assigned an *interpretation* $[\![\phi]\!]_I$ in S as follows:

(i) $[\![\top]\!]_I = \mathit{true}$, $[\![\bot]\!]_I = \mathit{false}$;
(ii) $[\![\sigma = \tau]\!]_I = (\sigma_I = \tau_I)$;
(iii) $[\![\alpha(\vec{\tau})]\!]_I = [\![\alpha]\!]_I(\langle (\tau_1)_I, \ldots, (\tau_r)_I \rangle$;
(iv) $[\![\phi \wedge \psi]\!]_I = [\![\phi]\!]_I \wedge [\![\psi]\!]_I$, $[\![\phi \vee \psi]\!]_I = [\![\phi]\!]_I \vee [\![\psi]\!]_I$,
$[\![\forall x \phi]\!]_I = \forall x_I \in A_I . [\![\phi]\!]_I$, $[\![\exists x \phi]\!]_I = \exists x_I \in A_I . [\![\phi]\!]_I$,

where x is if sort A. Clearly $[\![\phi]\!]_I$ is a formula of $\mathscr{L}$.

Now let Σ be a theory in L. An interpretation I of L in S is said to be a *model* of Σ *in* S is $\vdash_S [\![\phi]\!]_I$ for all $\phi \in \Sigma$. We write $\vDash \phi$ for

$\vdash_S [\![\phi]\!]_I$ *for any interpretation* I *of* L *in any local set theory* S.

By induction on the formation of formulae of L one now easily proves the

A.1 FIRST-ORDER SOUNDNESS THEOREM. For any formula ϕ of L,

$$\vdash \phi \quad \text{implies} \quad \models \phi \quad \square$$

A.2 COROLLARY. If Σ is a theory in L, and I a model of Σ in a local set theory S, then

$$\Sigma \vdash \phi \quad \text{implies} \quad \vdash_S [\![\phi]\!]_I. \quad \square$$

We now define what is meant by an interpretation of a first-order language in a *topos*.

Let $\mathbf{E}$ be a topos, and L a first-order language. An *interpretation* J of L in $\mathbf{E}$ assigns:

(i) to each sort A, an $\mathbf{E}$-object A_J;
(ii) to each function symbol f of L of signature $(\vec{A}, B)$, an $\mathbf{E}$-arrow

$$(A_1)_J \times \cdots \times (A_n)_J \xrightarrow{f_J} B_J;$$

(iii) to each predicate symbol α of L of signature $\vec{A}$, an $\mathbf{E}$-arrow

$$(A_1)_J \times \cdots \times (A_n)_J \xrightarrow{[\![\alpha]\!]_J} \Omega_E.$$

Recall that Th($\mathbf{E}$), the theory of $\mathbf{E}$, is a local set theory formulated within the canonical language $\mathscr{L}(\mathbf{E})$ of $\mathbf{E}$. Each interpretation J of L in $\mathbf{E}$ induces an interpretation J^* of L in Th($\mathbf{E}$) in the obvious way.

Now, for each term τ of $\mathscr{L}(\mathbf{E})$, we recall that we have a natural interpretation $[\![\tau]\!]_{E,x}$ of τ in $\mathbf{E}$ (cf. Chapter 3). We define the interpretation of terms and formulae of L in $\mathbf{E}$ under J by composing J^* with the natural interpretation:

$$\tau_J = [\![\tau_{J^*}]\!]_{E,\mathbf{x}}$$
$$[\![\phi]\!]_J = [\![[\![\phi]\!]_{J^*}]\!]_{E,\mathbf{x}}$$

where $\mathbf{x}$ lists the free variables of τ_{J^*} or $[\![\phi]\!]_{J^*}$.

If ϕ is a formula of L, and J an interpretation of L in $\mathbf{E}$, we say that ϕ is *true* under J and write $\models_J \phi$ if $[\![\phi]\!]_J = T$. J is said to be a *model* of a theory Σ in $\mathbf{E}$ if, for any formula ϕ of L,

$$\phi \in \Sigma \quad \text{implies} \quad \models_J \phi.$$

If $\models_J \phi$ for any interpretation J, we write $\models_{\mathbf{Top}} \phi$ and say that ϕ is *valid*..

An an immediate consequence of the soundness theorem A.1, we have

A.3 TOPOS FIRST-ORDER SOUNDNESS THEOREM. For any formula ϕ of L,

$$\vdash \phi \quad \text{implies} \quad \models_{\mathbf{Top}} \phi. \quad \square$$

A.4 COROLLARY. If Σ is a theory in L and J a model of Σ in a topos, then

$$\Sigma \vdash \phi \quad \text{implies} \quad \models_J \phi. \quad \square$$

Given a theory Σ, we construct a category $\mathbf{Syn}(\Sigma)$—the *syntactic category* derived from Σ.

We shall write $\vec{x},\vec{y}$, etc. for finite lists of variables of L, and $\phi(\vec{x})$, etc. to indicate that the free variables of ϕ are among $\vec{x}$.

The *objects* of the category $\mathbf{Syn}(\Sigma)$ are to be *formal class terms*, i.e. symbols either of the form $\{\vec{x} \mid \phi(\vec{x})\}$, where $\phi(\vec{x})$ is a formula of L whose free variables occur in the list $\vec{x}$ or $\{\phi\}$ where ϕ has no free variables. We adopt the convention that a formula or class term is not changed if its bound variables (including $\vec{x}$ in $\{\vec{x} \mid \phi(\vec{x})\}$) are renamed subject to the usual precautions for avoiding clashes.

To define the arrows in $\mathbf{Syn}(\Sigma)$ between two objects $\{\vec{x} \mid \phi(\vec{x})\}$ and $\{\vec{y} \mid \psi(\vec{y})\}$ we may assume without loss of generality that the lists $\vec{x}$ and $\vec{y}$ are disjoint. An *arrow* between $\{\vec{x} \mid \phi(\vec{x})\}$ and $\{\vec{y} \mid \phi(\vec{y})\}$ is then defined to be an equivalence class, with respect to Σ-provable equivalence, of formulae $\theta(\vec{x},\vec{y})$ which are *functional* with domain $\{\vec{x} \mid \phi(\vec{x})\}$ and codomain $\{\vec{y} \mid \phi(\vec{y})\}$, i.e. such that

$$\Sigma \vdash \forall\vec{x}\,\forall\vec{y}[\theta(\vec{x},\vec{y}) \rightarrow \phi(\vec{x}) \wedge \psi(\vec{y})],$$
$$\Sigma \vdash \forall\vec{x}[\phi(\vec{x}) \rightarrow \exists\vec{y}\theta(\vec{x},\vec{y})],$$
$$\Sigma \vdash \forall\vec{x}\,\forall\vec{y}\,\forall\vec{z}[\theta(\vec{x},\vec{y}) \wedge \theta(\vec{x},\vec{z}) \rightarrow \vec{y} = \vec{z}].$$

(If $\vec{y}$ is $(y_1, \ldots, y_n)$ and $\vec{z}$ is $(z_1, \ldots, z_n)$, then $\forall\vec{y}$ is $\forall y_1 \cdots \forall y_n$, $\exists\vec{y}$ is $\exists y_1 \cdots \exists y_n$ and $\vec{y} = \vec{z}$ is the conjunction of the n formulae $y_i = z_i$.) The arrow defined by $\theta(\vec{x},\vec{y})$ will be denoted by $[\vec{x} \mapsto \vec{y} \mid \theta(\vec{x},\vec{y})]$. The *composition* of $[\vec{x} \mapsto \vec{y} \mid \theta(\vec{x},\vec{y})]$ and $[\vec{y} \mapsto \vec{z} \mid \eta(\vec{y},\vec{z})]$ is defined to be $[\vec{x} \mapsto \vec{z} \mid \exists\vec{y}(\theta(\vec{x},\vec{y}) \wedge \eta(\vec{y},\vec{z}))]$.

It is now easily checked that this recipe defines a category $\mathbf{Syn}(\Sigma)$. Moreover, $\mathbf{Syn}(\Sigma)$ has finite limits: the object $\{\top\}$ is terminal (since for any $\phi(x)$, the arrow defined by ϕ is easily seen to be the unique arrow from $\{x : \phi(x)\}$ to $\{T\}$), and the pullback of the diagram

$$\begin{array}{ccc} & & \{\vec{y} : \psi(\vec{y})\} \\ & & \big\downarrow {\scriptstyle [\vec{y} \mapsto \vec{z} \mid \eta(\vec{y},\vec{z})]} \\ \{\vec{x} : \phi(\vec{x})\} & \xrightarrow{[\vec{x} \mapsto \vec{z} \mid \theta(\vec{x},\vec{z})]} & \{\vec{z} : \chi(\vec{z})\} \end{array}$$

is the object $\{\overrightarrow{xy} : \exists\vec{z}[\theta(\vec{x},\vec{y}) \wedge \eta(\vec{y},\vec{z})]\}$.

We define a covering system K_Σ on $\mathbf{Syn}(\Sigma)$ in the following way. First, a finite family of arrows

$$[\vec{x}_i \mapsto \vec{y} \mid \theta_i(\vec{x}_i,\vec{y})] : \{\vec{x}_i \mid \phi_i(\vec{x}_i)\} \rightarrow \{\vec{y} \mid \psi(\vec{y})\}$$

$(i = 1, \ldots, n)$ is said to be *Σ-probably epic* on $\{\vec{y} \mid \psi(\vec{y})\}$ if

$$\Sigma \vdash \forall\vec{y}\left[\psi(\vec{y}) \Rightarrow \bigvee_{i=1}^{n} \exists\vec{x}_i\theta_i(\vec{x}_i,\vec{y})\right].$$

And now, for each object $\{\vec{y} \mid \psi(\vec{y})\} = A$ we define $K_\Sigma(A)$ to consist of all cosieves U on A such that U contains a provably epic family. It is readily checked that K_Σ as just defined is a covering system on $\mathbf{Syn}(\Sigma)$. The site $(\mathbf{Syn}(\Sigma), K_\Sigma)$ is called the *syntactic site associated with* Σ.

A formula of L is called *geometric* if it does not contain $\Rightarrow$ or $\forall$ (and hence not $\neg$ *either*). *A geometric implication* is a sentence of the form $\forall\vec{x}(\phi \Rightarrow \psi)$, where ϕ and ψ are geometric formulae. A *geometric theory* is a theory consisting of geometric implications.

If Σ is a geometric theory, we construct the *geometric category* associated with Σ in the same way as we constructed the syntactic category, except that throughout the construction we confine attention to *geometric* formulae. Thus, for example, an arrow $\{\vec{x} \mid \phi(\vec{x})\} \to \{\vec{y} \mid \psi(\vec{y})\}$ with ϕ and ψ geometric is an equivalence class of *geometric* formulae $\theta(\vec{x},\vec{y})$ which are functional with domain $\{\vec{x} \mid \phi(\vec{x})\}$ and codomain $\{\vec{y} \mid \psi(\vec{y})\}$.

We write $\mathbf{Geom}(\Sigma)$ for the geometric category associated with Σ, and K_Σ for the corresponding covering system, defined as before. The site $(\mathbf{Geom}(\Sigma), K_\Sigma)$ is called the *geometric site* associated with Σ.

We shall need the following definitions. If A is an object of a category $\mathbf{D}$, a family $\{B_i \xrightarrow{f_i} A : i \in I\}$ of $\mathbf{D}$-arrows with common codomain A is said to be *jointly epic* if, for any arrows $A \underset{h}{\overset{g}{\rightrightarrows}} C$, $g \circ f_i = h \circ f_i$ for all $i \in I$ implies $g = h$. If the coproduct $\coprod_{i \in I} B_i$ exists in $\mathbf{D}$, it is easy to see that $\{f_i : i \in I\}$ is jointly epic if and only if the arrow

$$\coprod_{i \in I} B_i \xrightarrow{(f_i)_{i \in I}} A$$

is epic.

If $(\mathbf{C},K)$ is a site, a functor $F: \mathbf{C} \to \mathbf{D}$ to a category $\mathbf{D}$ is said to *preserve* (K-) *coverings* if, for any $\mathbf{C}$-object A and any cosieve $S \in K(A)$, the family $\{Ff : f \in S\}$ is jointly epic in $\mathbf{D}$. A functor $F: \mathbf{C} \to \mathbf{E}$ to a topos $\mathbf{E}$ is called *geometric* if it preserves K-coverings and is in addition *left exact*.

Now we can establish an important connection between models of a geometric theory and geometric functors on the associated geometric site.

A.5 THEOREM. Let Σ be a geometric theory and S a local set theory. Then models of Σ in S are in bijective correspondence with geometric functors $\mathbf{Geom}(\Sigma) \to \mathbf{C}(S)$.

Sketch of proof. Let $F: \mathbf{Geom}(\Sigma) \to \mathbf{C}(S)$ be a geometric functor. We define an interpretation I of L in S by setting:

$$A_I = F(\{x_A : \mathsf{T}\}) \text{ for each sort } A \text{ of } \mathsf{L}$$

$$f_I = F([\vec{x} \mapsto y \mid f(\vec{x}) = y]) \text{ for each function symbol } f \text{ of } \mathsf{L}$$

$$[\![\alpha]\!]_I = \chi(F([\vec{x} \mapsto \vec{y} \mid \alpha(\vec{x}) \wedge \vec{x} = \vec{y}])$$

for each predicate symbol α of L, where χ denotes the characteristic arrow in $\mathbf{C}(S)$.

By induction on the complexity of formulae (using the geometricity of F) one shows that for any geometric formula ϕ

$$[\![\phi]\!]_I = \chi(F(\vec{x} \mapsto \vec{y} \mid \phi(\vec{x}) \wedge \vec{x} = \vec{y}]).$$

If $\beta \in \Sigma$ then β is of the form $\forall \vec{x}(\phi \Rightarrow \psi)$ with ϕ and ψ geometric. Now define

$$m = [\vec{x} \mapsto \vec{z} \mid \phi(\vec{x}) \wedge \vec{x} = \vec{z}]$$
$$n = [\vec{x} \mapsto \vec{y} \mid \phi(\vec{x}) \wedge \vec{x} = \vec{y}]$$
$$p = [\vec{x} \mapsto \vec{y} \mid \psi(\vec{x}) \wedge \vec{x} = \vec{y}].$$

Then since $\Sigma \vdash \phi \Rightarrow \psi$, m is an arrow $\{\vec{x} \mid \phi(\vec{x})\} \to \{\vec{x} \mid \psi(\vec{x})\}$ in $\mathbf{Geom}(\Sigma)$. Clearly, also, $n = p \circ m$. Hence $F(n) = F(p) \circ F(m)$, so that $F(n) \subseteq F(p)$. Therefore

$$[\![\phi]\!]_I \leqslant \chi(F(n)) \leqslant \chi(F(p)) = [\![\psi]\!]_I$$

and so $[\![\phi]\!]_I \vdash_S [\![\psi]\!]_I$. Thus $\vdash_S \beta$ and accordingly I is a model of Σ.

Conversely, given a model I of Σ in S, define a functor $F\colon \mathbf{Geom}(\Sigma) \to \mathbf{C}(S)$ by

$$F(\{\vec{x} \mid \phi(\vec{x})\}) = \{\langle x_{1i}, \ldots, x_{nI}\rangle : [\![\phi]\!]_I\}$$
$$F([\vec{x} \mapsto \vec{y} \mid \theta(\vec{x}, \vec{y})]) = \{\langle\langle x_{1I}, \ldots, x_{nI}\rangle, \langle y_{1I}, \ldots, y_{nI}\rangle\rangle : [\![\theta]\!]_I\}.$$

It is readily checked that F is geometric. □

By taking S to be Th($\mathbf{E}$) for a topos $\mathbf{E}$, we obtain from this the

A.6 COROLLARY. For any geometric theory Σ and any topos $\mathbf{E}$ there is a bijective correspondence between models of Σ in $\mathbf{E}$ and geometric functors $\mathbf{Geom}(\Sigma) \to \mathbf{E}$. □

In view of this correspondence, a geometric functor $\mathbf{Geom}(\Sigma) \to \mathbf{E}$ will be called a *functorial model*—or just a *model*—of Σ in $\mathbf{E}$. Given a model F of Σ in $\mathbf{E}$, a geometric implication ϕ is said to be *true* in F if it is true under the associated interpretation I in $\mathbf{E}$. Clearly, if F is a model of Σ, and $\Sigma \vdash \phi$, then ϕ is true in F.

It is readily shown that, if $\mathbf{Geom}(\Sigma) \xrightarrow{F} \mathbf{E}$ is a geometric functor and $\mathbf{E} \xrightarrow{G} \mathbf{E}'$ an exact functor, then $G \circ F$ is geometric. In particular, if F is a model of Σ in a topos $\mathbf{E}$ and $\gamma\colon \mathbf{E}' \to \mathbf{E}$ is a geometric morphism, then $\gamma^* \circ F$ is a model of Σ in $\mathbf{E}'$. Thus, if ϕ is a geometric implication true in F, then F is a model of $\Sigma \cup \{\phi\}$, so that $\gamma^* \circ F$ is a model of $\Sigma \cup \{\phi\}$, whence ϕ is true in $\gamma^* \circ F$. (And conversely, if γ^* is *faithful*, the truth of ϕ in $\gamma^* \circ F$ implies that of ϕ in F.) In this sense γ^* *preserves the truth of geometric implications*, a property not shared by arbitrary formulae.

Now, given a geometric theory Σ, consider the Yoneda embedding

$$Y : \mathbf{Geom}(\Sigma) \to \mathbf{Set}^{\mathbf{Geom}(\Sigma)^{\mathrm{op}}}.$$

Y is left exact (by (1.28)) but *not* in general a model of Σ. In order to 'force' Y to be a model of Σ, we need the following result.

A.7 LEMMA. Let $(\mathbf{C},K)$ be a site; write $\mathbf{Sh}_K(\mathbf{C})$ for the topos of K-sheaves on $\mathbf{C}$, and $L : \mathbf{Set}^{\mathbf{C}^{\mathrm{op}}} \to \mathbf{Sh}_K(\mathbf{C})$ for the sheafification functor. Then the composition

$$C \xrightarrow{Y} \mathbf{Set}^{\mathbf{C}^{\mathrm{op}}} \xrightarrow{L} \mathbf{Sh}_K(\mathbf{C})$$

is geometric.

Proof. We know that both Y and L are left exact, hence so is $L \circ Y$. It remains to show that it preserves K-coverings. To do this, it suffices to show that, for any $\mathbf{C}$-object A and any K-covering cosieve $S = \{B_i \xrightarrow{f_i} A : i \in I\}$ on A, the induced arrow

$$\coprod_{i \in I} LYB_i \to LYA$$

is epic.

Let S_* be the subfunctor of the representable functor YA corresponding to the cosieve S on A. (That is, $S_*(B) = \{f \in S : \mathrm{dom}(f) = B\}$ for any $\mathbf{C}$-object B.) By the Yoneda lemma, each $f_i \in S$ corresponds to an arrow $YB_i \xrightarrow{\eta_i} S_*$. Let $\coprod_{i \in I} YB_i \xrightarrow{\eta} S_*$ be the induced arrow; we claim that η is epic. For this it suffices to show that, for any $\mathbf{C}$-object B, $\coprod_{i \in I} YB_i(B) \xrightarrow{\eta_B} S_*B$ is epic; but this is evident from the definitions of η and S_*.

Therefore $\coprod_{i \in I} B_i \xrightarrow{\eta} S_*$ is epic. Now since $S \in KA$, S_* is K-dense in YA and L converts the inclusion $S_* \hookrightarrow YA$ into an isomorphism. Since L is a left adjoint, it preserves epics and coproducts, so that the composition

$$\coprod_{i \in I} LYB_i \cong L\Big(\coprod_{i \in I} YB_i\Big) \xrightarrow{L(\eta)} L(S_*) \cong LYA$$

is epic, as required. □

For any geometric theory Σ, write $\mathbf{Set}[\Sigma]$ for $\mathbf{Sh}_{K_\Sigma}(\mathbf{Geom}(\Sigma))$. As an immediate consequence of (A.7) we have the

A.8 COROLLARY. For any geometric theory Σ, the composition

$$\mathbf{Geom}(\Sigma) \xrightarrow{Y} \mathbf{Set}^{\mathbf{Geom}(\Sigma)^{\mathrm{op}}} \xrightarrow{L} \mathbf{Set}[\Sigma]$$

is a model of Σ. □

Although we shall not prove it here, it can be shown (see, e.g. Johnstone (1977), Section 7.4) that the model $M(\Sigma) = L \circ Y$ of Σ in $\mathbf{Set}[\Sigma]$ has the following universal property. Call a topos of the form $\mathbf{Sh}_K(C)$, where $(\mathbf{C},K)$ is a site, a *Grothendieck topos*. (In particular, $\mathbf{Set}[\Sigma]$ is a Grothendieck topos.) Then for any Grothendieck topos $\mathbf{E}$ and any model F of Σ in $\mathbf{E}$, there is a unique (up to natural isomorphism) geometric morphism $\gamma : \mathbf{E} \to \mathbf{Set}[\Sigma]$ such that the diagram

$$\begin{array}{ccc} \mathbf{Geom}(\Sigma) & \xrightarrow{M(\Sigma)} & \mathbf{Set}[\Sigma] \\ & {\scriptstyle F}\searrow & \downarrow{\scriptstyle \gamma^*} \\ & & \mathbf{E} \end{array}$$

commutes. $M(\Sigma)$ is called the *generic model* of Σ and $\mathbf{Set}[\Sigma]$ the *classifying topos* of Σ.

This explains the notation $\mathbf{Set}[\Sigma]$: for we may think of $\mathbf{Set}[\Sigma]$ as the topos obtained by adjoining a generic (or universal) model of Σ to $\mathbf{Set}$.

Since functors of the form γ^* preserve the truth of geometric formulas, we see immediately from the universal property of $M(\Sigma)$ that *the geometric implications true in $M(\Sigma)$ are precisely those which are true in every model of Σ.* Moreover, it is not hard to show that a geometric implication ϕ is true in $M(\Sigma)$ if and only if $\Sigma \vdash \phi$. Accordingly, we have the

A.9 COMPLETENESS THEOREM FOR GEOMETRIC THEORIES. Given a geometric theory Σ and a geometric implication ϕ, $\Sigma \vdash \phi$ if and only if ϕ is true in every model of Σ. □

Somewhat surprisingly, the completeness theorem can be strengthened to assert that, for a geometric theory Σ and a geometric implication ϕ, $\Sigma \vdash_C \phi$ if and only if ϕ is true in every model of Σ. It follows in particular from this that $\Sigma \vdash_C \phi$ implies $\Sigma \vdash \phi$, in other words, *a geometric implication classically deducible from a geometric theory is also intuitionistically deducible from it.*

In order to prove the strengthened version of the completeness theorem, we use a result known as

BARR'S THEOREM. For any Grothendieck topos $\mathbf{E}$, there is a Boolean topos $\mathbf{B}$ and a geometric morphism $\gamma : \mathbf{B} \to \mathbf{E}$ such that γ^* is faithful.

For a proof of this result, see Johnstone (1977), Chapter 7. Now suppose $\Sigma \vdash_C \phi$, and let F be any model of Σ in a Grothendieck topos $\mathbf{E}$. Using Barr's theorem, let $\gamma : \mathbf{B} \to \mathbf{E}$ be a geometric morphism from a Boolean topos $\mathbf{B}$ with γ^* faithful. Since Σ is geometric, $\gamma^* \circ F$ is a model of Σ in $\mathbf{B}$. Now the laws of classical logic hold in the Boolean topos $\mathbf{B}$, so since $\Sigma \vdash_C \phi$, ϕ must be true in $\gamma^* \circ F$. And since γ^* is faithful, it follows that ϕ is true in F.

We conclude with some examples of geometric theories and their classifying toposes.

(i) *The theory* Ob *of objects.* The appropriate language has a single sort and no predicate or function symbols, and the theory itself has no axioms. A model of Ob in a topos **E** is then just an object of **E**. It can be shown (cf. Johnstone 1977, Section 6.33) that the classifying topos **Set**[Ob] is the functor category $\mathbf{Set}^{\mathbf{finset}}$, where **finset** is a skeleton of the category **Finset** of finite sets and functions. The generic model here is the insertion functor $\mathbf{finset} \hookrightarrow \mathbf{Set}$.

(ii) *The theory* D *of decidable objects.* Here the appropriate language has a single sort and a binary predicate $\neq$ and D has axioms

$$\forall x \, \forall y (\top \Rightarrow x = y \vee x \neq y)$$
$$\forall x (x \neq x \Rightarrow \bot).$$

A model of D in a topos **E** is a decidable object of **E** (i.e., a decidable set in Th(**E**)). The classifying topos **Set**[D], it turns out, may be taken to be the functor category $\mathbf{Set}^{\mathbf{finset}'}$, where **finset**′ is a skeleton of the category of finite sets and injections between them.

(iii) *The theory* Rng *of rings.* Here the appropriate language has one sort A, two function symbols $+$, $\cdot$ of signatures $((A,A),A)$ representing addition and multiplication on A and one function symbol 0 of type $((\ ),A)$. The theory Rng has as axioms geometric implications asserting that $(A, +, \cdot, 0)$ is a ring. A model of Rng in a topos **E** is a ring object in **E** (i.e. a ring in Th(**E**)). It can be shown that the classifying topos **Set**[Rng] of Rng is the functor category $\mathbf{Set}^{\mathbf{FPRng}}$, where **FPRng** is the category of *finitely presented rings,* i.e. quotient rings of the form $Z[t_1, \ldots, t_n]/I$, where $Z[t_1, \ldots, t_n]$ is the ring of polynomials over Z in indeterminates $t_1, \ldots, t_n$.

(iv) *The theory* Re *of a Dedekind real number.* Here the language has sorts N and Q corresponding to the natural numbers and rationals, and two relations U, V of signature (Q). The axioms of Re are geometric implications asserting that the pair $\langle U, V \rangle$ is a Dedekind cut in Q (cf. Chapter 7). A model of Re in a (Grothendieck) topos **E** is thus a Dedekind real in **E**. It can be shown that the classifying topos **Set**[Re] of Re is **Sh**($\mathbb{R}$), the category of sheaves over the real numbers with the usual topology. Thus **Sh**($\mathbb{R}$) is the topos obtained by adjoining a generic real number to **Set**.

HISTORICAL AND BIBLIOGRAPHICAL NOTES

Chapter 1. The basic concepts of category theory were invented and developed by Eilenberg and MacLane (1945). The Yoneda lemma (1.7) first appears in Yoneda (1954). The concept of an adjunction makes its first explicit appearance in Kan (1958); its central importance for category theory is emphasized in Freyd (1964) and Lawvere (1963, 1964). The account I have given of Galois connections is based on Lambek (1981).

The standard treatise on category theory is MacLane (1971). Useful accounts are also to be found in Herrlich and Strecker (1973), and, at a more elementary level, in Arbib and Manes (1975).

Chapter 2. The term 'topos' introduced here should rightly be '*elementary* topos' since the axioms governing the concept are formulable entirely within the elementary (= first-order) language of categories. The *original* concept of topos is that which came subsequently to be known as *Grothendieck* topos, viz., a category of sheaves over a site (= category with a Grothendieck topology). This concept arose with the French school of algebraic geometers centred around Grothendieck in the early 1960s: for an account of their work, see Grothendieck and Verdier (1972).

It was in 1969–70 that Lawvere and Tierney developed the concept of elementary topos, building on the former's previous insight that any Grothendieck topos contains a truth-value object. Early accounts of the theory of elementary toposes include Freyd (1972) and Kock and Wraith (1971). The first book on topos theory, and what must be regarded as the standard reference on the subject, is Johnstone (1977). (The introduction to this book is a wide-ranging and most informative account of the origins of topos theory.) Goldblatt (1979) contains an exposition of the first-order aspects of the theory. The recent book by Lambek and Scott (1986) presents a systematic account of topos theory via higher-order languages in a manner similar to that adopted in the present book. Barr and Wells (1985) approaches topos theory without the use of logical tools.

Chapter 3. The fact that logic can be formulated within a topos was known to Lawvere. However, it appears to be Mitchell (1972) who gave the first explicit description of a formal language designed to be interpreted within a topos; this idea seems also to have occurred independently to Bénabou and Joyal (among others). The concept of local set theory I employ here, and most of the material in the chapter, is due to Zangwill (1977). Zangwill's system is similar to that of Boileau and Joyal: cf. Boileau (1975), Boileau and Joyal (1981) (also Coste 1974). Lambek and Scott (1986) contains a systematic exposition of a type theory similar to all these. Fourman (1974, 1977) develops a typed language with a description operator.

The equivalence theorem (3.37) seems to have been proved independently by several people: at any rate, the result appears in Volger (1975), Fourman (1977), Zangwill (1977) and Boileau and Joyal (1981).

For intuitionistic (first-order) logic, consult Dummett (1977), Bell and Machover (1977), or Kleene (1952).

Chapter 4. The systematic use of the equivalence theorem in establishing topos-theoretic facts was explicitly advocated by Fourman (1974, 1977) and Zangwill (1977), although the idea undoubtedly occurred to others. Finite cocompleteness of toposes (4.2) was first proved, using category-theoretic methods, by Mikkelsen (1976). That the axiom of choice implies Booleanness for toposes is a result of Diaconescu (1975). The corresponding argument for local set theories given here is adapted from that of Zangwill (1980).

The definition of the forcing relation $\Vdash_S$ (in toposes) and the use of generalized elements seems to be due to Joyal. The first appearance of the forcing relation in print occurs in Osius (1975) where it is called 'Kripke–Joyal semantics'. The satisfaction clauses for $\vee$ and $\exists$, however, are closer in spirit to the semantics of Beth (see, e.g. Dummett (1977).) Kripke's semantics for first order intuitionistic logic appears in Kripke (1965). A good account of Beth–Kripke–Joyal semantics is to be found in Lambek and Scott (1986).

The concept of fullness given here is the straightforward result of an attempt to represent formally a geometric morphism to **Set**. Much more difficult is the problem of formally representing geometric morphisms to an arbitrary 'base topos' **E** within the internal language of **E**. As far as I know, this was first achieved (in a somewhat prolix way) by Zangwill (1980). A more practical solution, using partially defined terms, is presented in Chapman and Rowbottom (to appear).

Chapter 5. The basic results concerning sheaves in a topos are due to Lawvere and Tierney; an early account of their work appears in Freyd (1972). The construction of the sheafification functor within the internal language of a topos seems to have been first carried out in print by Veit (1981), although, again, the idea must have occurred to others.

For an account of how sheaves arise in set theory, see Chapter 9 of Johnstone (1977). For a systematic account of sheaf theory itself see

Tennison (1975). Gray (1979) is a fascinating history of the subject, with an exhaustive bibliography.

Chapter 6. The concept of an H-valued set and the fact that $\mathbf{Set}_H$ is a topos equivalent to $\mathbf{Sh}(H)$ is due to Higgs (1973) and independently to Fourman and Scott (1979). Theorem 6.4 was first proved as a consequence of the so-called 'relative Giraud theorem' (due, in its final form, to Diaconescu: cf. Chapter 4 of Johnstone (1977)); the proof presented here, using local set theories, is based on some unpublished work of Michael Brockway. Results (6.19), (6.22) and (6.23) are due to Bell (1982).

Chapter 7. The concept of a natural number object in a category is due to Lawvere, Theorem 7.7 to Freyd (1972), Theorem 7.10 to Johnstone (1979) and Example 7.11 to Tierney. The free topos and the free naturalized local set theory are discussed extensively in Lambek and Scott (1986).

For a detailed discussion of natural number objects in toposes, see Coste-Roy, Coste and Mahé (1980). For applications of properties of the real number object in a topos to independence proofs in analysis, see Fourman and Scott (1979) and Fourman and Hyland (1979).

Chapter 8. The chapter is based on Bell (1981, 1986); the last section incorporates ideas from Lawvere (1976).

Appendix. The notion of a geometric theory, and the existence of classifying toposes (for models in Grothendieck toposes) is due to Joyal and Reyes: see Makkai and Reyes (1977). Barr's theorem appears in Barr (1974). For a convenient summary of the basic properties of classifying toposes, see Blass and Ščedrov (1983). My account here is based on Cole (1979).

REFERENCES

This list contains only those works explicitly referred to in the text. For fuller bibliographies of topos theory, see Gray (1979), Johnstone (1977), or Lambek and Scott (1986).

Arbib, M. A. and Manes, E. G. (1975) *Arrows, structures and functors: the categorical imperative.* New York: Academic Press.

Barr, M. (1974) Toposes without points. *J. Pure and Applied Algebra* **5,** 265–80.

Barr, M. and Wells, C. (1985) *Toposes, triples and theories.* Berlin: Springer-Verlag.

Bell, J. L. (1981) Category theory and the foundations of mathematics. *Brit. J. Phil. Sci.* **32,** 349–58.

—— (1982) Some aspects of the category of subobjects of constant objects in a topos. *J. Pure and Applied Algebra* **24,** 245–59.

—— (1985) *Boolean-valued models and independence proofs in set theory* (2nd edn). Oxford: Clarendon Press.

—— (1986) From absolute to local mathematics. *Synthese* **69,** 409–26.

Bell, J. L. and Machover, M. (1977) *A course in mathematical logic.* Amsterdam: North-Holland.

Blass, A. and Ščedrov, A. (1983) Classifying topoi and finite forcing. *J. Pure and Applied Algebra* **28,** 111–40.

Boileau, A. (1975) *Types vs. topos.* Thesis, Université de Montreal.

Boileau, A. and Joyal, A. (1981) La logique des topos. *J. Symbolic Logic* **46,** 6–16.

Chapman, J. and Rowbottom, F. (to appear) A logical approach to topos theory. To appear.

Cole, J. C. (1979) Classifying topoi. Algebraische Modelle, Kategorien und Gruppoide. (*Studien zur Algebra und ihre Andwendungen* **7**). Akademie-Verlag Berlin, 5–20.

Coste, M. (1974) Logique d'ordre supérieur dans les topos élémentaires. *Séminaire de M. Benabou,* Université Paris-Nord.

Coste-Roy, M. F., Coste, M., and Mahé, L. (1980) Contributions to the study of the natural number object in elementary topoi. *J. Pure and Applied Algebra* **17,** 35–68.

Diaconescu, R. (1975) Axiom of choice and complementation. *Proc. Amer. Math. Soc.* **51,** 176–8.

Dummett, M. (1977) *Elements of intuitionism.* Oxford: Clarendon Press.

acLane, S. (1945) General theory of natural equivalences.
ath. Soc. **58,** 231–94.

74) *Connections between category theory and logic.* Thesis,
ty.

c of topoi. In J. Barwise (ed.) *Handbook of Mathematical*
m: North-Holland, pp. 1053–90.

Fourman, M. P. and Hyland, J. M. E. (1979) Sheaf models for analysis. In Fourman *et al.* (1979), pp. 280–301.

Fourman, M. P., Mulvey, C., and Scott, D. S. (eds.) (1979) *Applications of sheaves. Proc. L.M.S. Durham Symposium* 1977. Springer Lecture Notes in Math. 753.

Fourman, M. P. and Scott, D. S. (1979) Sheaves and logic. In Fourman *et al.* (1979), pp. 302–401.

Freyd, P. J. (1964) *Abelian categories: An introduction to theory of functors.* New York: Harper and Row.

—— (1972) Aspects of topoi. *Bull. Austral. Math. Soc.* **7,** 1–76 and 467–80.

Goldblatt, R. I. (1979) *Topoi: the categorial analysis of logic.* Amsterdam: North-Holland.

Gray, J. W. (1979) Fragments of the history of sheaf theory. In Fourman *et al.* (1979), pp. 1–79.

Grothendieck, A. and Verdier, J. L. (1972) *Théorie des topos* (SGA 4, exposés I–VI) (2nd edn). Springer Lecture Notes in Math. 269, 270.

Herrlich, H. and Strecker, G. E. (1973) *Category theory.* Boston: Allyn and Bacon.

Higgs, D. (1973) *A category approach to Boolean-valued set theory.* Lecture notes, University of Waterloo.

Johnstone, P. T. (1977) *Topos theory.* London: Academic Press.

—— (1979) Conditions related to De Morgan's law. In Fourman *et al.* (1979), pp. 479–91.

Kan, D. M. (1958) Adjoint functors. *Trans. Amer. Math. Soc.* **87,** 294–329.

Kock, A. and Wraith, G. C. (1971) *Elementary toposes.* Aarhus Univ. Lecture Notes Series No. 30.

Kleene, S. C. (1952) *Introduction to metamathematics.* Amsterdam: North-Holland and New York: Van Nostrand.

Kripke, S. (1965) Semantical analysis of intuitionistic logic I. In J. N. Crossley and M. A. E. Dummett (eds.) *Formal systems and recursive functions.* Amsterdam: North-Holland, pp. 92–130.

Lambek, J. (1981) The influence of Heraclitus on modern mathematics. In J. Agassi and R. S. Cohen (eds.) *Scientific philosophy today.* Dordrecht: Reidel.

Lambek, J. and Scott, P. J. (1986) *Introduction to higher order categorical logic.* Cambridge: Cambridge University Press.

Lawvere, F. W. (1963) Functorial semantics of algebraic theories. *Proc. Nat. Acad. Sci. USA* **50,** 869–73.

—— (1964) An elementary theory of the category of sets. *Proc. Nat. Acad. Sci. USA* **52,** 1506–11.

—— (1976) Variable quantities and variable structures in topoi. In *Algebra, topology and category theory: a collection of papers in honor of Samuel Eilenberg.* Academic Press, pp. 101–31.

MacLane, S. (1971) *Categories for the working mathematician.* Graduate Texts in Mathematics 5. Berlin: Springer-Verlag.

MacLane, S. and Birkhoff, G. (1967) *Algebra.* New York: Macmillan.

Makkai, M. and Reyes, G. E. (1977) *First-order categorical logic.* Springer Lecture Notes in Math. 611.

Mikkelsen, C. J. (1976) *Lattice-theoretic and logical aspects of elementary topoi.* Aarhus University Various Publications Series 25.

Mitchell, W. (1972) Boolean topoi and the theory of sets. *J. Pure and Applied Algebra* **2,** 261–74.

Osius, G. (1975) A note on Kripke–Joyal semantics for the internal language of topoi. In Lawvere *et al.* (eds.), *Model theory and topoi.* Springer Lecture Notes in Math. 445, pp. 349–54.

Tennison, B. R. (1975) *Sheaf theory.* L.M.S. Lecture Notes Series No. 20. Cambridge University Press.

Veit, B. (1981) A proof of the associated sheaf theorem by means of categorical logic. *J. Symb. Logic* **46,** 45–55.

Volger, H. (1975) Logical categories, semantical categories and topoi. In Lawvere *et al.* (eds.) *Model theory and topoi,* Springer Lecture Notes in Math. 445, pp. 87–100.

Yoneda, N. (1954) On the homology theory of modules. *J. Fac. Sci. Tokyo Sec. I.* **7,** 193–227.

Zangwill, J. (1977) *Local set theory and topoi.* M.Sc. thesis, Bristol University.

—— (1980) *Relative set theory.* Ph.D. thesis, Bristol University.

INDEX OF SYMBOLS

INDEX OF TERMS